Micromechanics and Nanosimulation
of Metals and Composites

Siegfried Schmauder · Leon Mishnaevsky Jr.

Micromechanics and Nanosimulation of Metals and Composites

Advanced Methods and Theoretical Concepts

 Springer

Professor Dr. Siegfried Schmauder
Institute for Materials Testing,
Materials Science and Strength
of Materials (IMWF)
University of Stuttgart
Pfaffenwaldring 32
D-70569 Stuttgart
Germany
siegfried.schmauder@imwf.uni-stuttgart.de

Dr.-Ing. habil. Leon Mishnaevsky Jr.
Senior Scientist
Risø.DTU National Laboratory
for Sustainable Energy
Technical University of Denmark
AFM-228
DK-4000 Roskilde
Denmark
leon.mishnaevsky@risoe.dk

ISBN: 978-3-540-78677-1   e-ISBN: 978-3-540-78678-8

Library of Congress Control Number: 2008936029

© Springer-Verlag Berlin Heidelberg 2009

With kind permission of Elsevier, Hanser, Sage, and Balkema.

Cover design: eStudio Calamar S.L.

Printed on acid-free paper

9 8 7 6 5 4 3 2 1

springer.com

Contents

Preface

The strength of metallic materials determines the usability and reliability of all the machines, tools and equipment around us. Yet, the question about which mechanisms control the strength and damage resistance of materials and how they can be optimised remains largely unanswered. How do real, heterogeneous materials deform and fail? Why can a small modification of the microstructure increase the strength and damage resistance of materials manifold? How can the strength of heterogeneous materials be predicted?

The purpose of this book is to present different experimental and computational analysis methods of micromechanics of damage and strength of materials and to demonstrate their applications to various micromechanical problems. This book summarizes at a glance some of the publications of the Computational Mechanics Group at the IMWF/MPA Stuttgart, dealing with atomistic, micro- and mesome-chanical modelling and experimental analysis of strength and damage of metallic materials.

In chapter 1, the micromechanisms of damage and fracture in different groups of materials are investigated experimentally, using direct observations and inverse analysis. The interaction of microstructural elements with the evolving damage is studied in these experiments. Chapter 2 presents different approaches to the micromechanical simulation of composite materials: embedded unit cells, multiphase finite elements and multiparticle unit cells. Examples of the application of these models to the analysis of deformation and damage in different materials are given. Chapter 3 deals with the methods of numerical modelling of damage evolution and crack growth in heterogeneous materials. Different methods of damage evolution modelling, in particular in materials with ductile (aluminium, cobalt) and brittle matrices, are applied to investigate the interrelations between microstructures and strength of these materials. Chapter 4 provides an insight into several methods of micromechanical computational modelling of materials with interpenetrating phases using graded materials. It defines the matricity model and demonstrates its application to the analysis of different materials. Multilayer models of graded materials, functionally graded finite elements, multiparticle unit cells with graded particle distribution and voxel based method of the 3D FE mesh generation are described in this chapter as well as models of graded materials used for milling applications.

Chapter 5 deals with methods of atomistics and dislocation modelling of the material behaviour and damage.

Siegfried Schmauder   Leon Mishnaevsky Jr.

Stuttgart, 13 August 2008   Roskilde, 13 August 2008

Acknowledgements

The investigations, described in this book, have been carried out in the framework of many research projects, funded by EU, ECSC (European Commission of Coal and Steel), DFG (Deutsche Forschungsgemeinschaft), Humboldt Foundation, and other funding bodies. In particular, the support by the German Bundesministerium für Bildung, Wissenschaft, Forschung und Technologie (BMBF) under grant No 1501029 and in connection with the COST-512 Program through contract 03K8004, by the European Commission via the Brite-Euram Project 8109 "Design of new tool materials with a structural gradient for milling application" and ECSC-Project 8834 "Influence of micromechanical mechanisms on strength and damage of tool steels under static and cyclic loading", from Deutsche Forschungsgemeinschaft through the Sonderforschungsbereich 381 "Charakterisierung des Schädigungsverlaufes in Faserverbundwerkstoffen mittels zerstörungsfreier Prüfung", through the projects 436RUS 17/68/04, Schm 746/16-1, Schm 746/25-1, Schm 746/12-1 and Schm 746/12-2 (Schwerpunktprogramm Gradientenwerkstoffe), RI 339/15-1, FI 686/1-1, Heisenberg fellowship "Werkstoffoptimierung auf dem Mesoniveau mittels numerischer Experimente unter Berücksichtigung von Effekten heterogener, komplexer und hierarchischer Phasenanordnungen" (MI 666-4/1) and other projects are gratefully acknowledged.

Some materials from previous publications of the authors, and other publications are reproduced here with kind permissions from Elsevier, Institute of Physics (IOP), Annual Reviews, EDP Sciences, Kluwer/Springer, Hanser and Sage publishers.

The authors want to express their gratitude to Dr. R. Balokhonov (Tomsk, ISPMS), Dr. P. Binkele (IMWF), Dr. M. Dong (Eberspächer), Dr. S. Hönle (Bosch), Dr. P. Kizler (MPA), Dr. C. Kohnle, Dr. C. Kohler (IMWF), Dr. N. Lippmann (Bosch), Dr. W. Lutz (Bosch), Dr. O. Minchev, Dr. J. Rohde, Dr. V. Romanova (Tomsk, ISPMS), Dr. E. Soppa (MPA),   and Dr. U. Weber (MPA/IMWF), as well as all their colleagues and former colleagues who contributed to our works.

The author L.M. gratefully acknowledges the valuable comments and suggestions by Lennart Mischnaewski.

The authors are grateful to Mrs. G. Amberg for her help during the preparation of the manuscript.

Siegfried Schmauder

Leon Mishnaevsky Jr.

Stuttgart, 13 August 2008

Roskilde, 13 August 2008

References

This book includes (fully or partially) the following publications of authors and their colleagues:

Chapter 1.

X. Ge, S. Schmauder, "Micromechanism of Fracture in Al/SiC Composites", J. Mat. Sci. 30, pp. 173-178 (1995).

L.L. Mishnaevsky Jr., N. Lippmann, S. Schmauder, P. Gumbsch, "In-situ Observation of Damage Evolution and Fracture in AlSi7Mg0.3 Cast Alloys", Eng. Fract. Mech. 63, pp. 395-411 (1999).

L. Mishnaevsky Jr., N. Lippmann, S. Schmauder, "Micromechanisms and Modelling of Crack Initiation and Growth in Tool Steels: Role of Primary Carbides", Zeitschrift f. Metallkunde 94, pp. 676-681 (2003).

Chapter 2.

S. Schmauder, "Computational Mechanics", Annual Rev. Mater. Res. 2002.32, pp. 437-465 (2002).

M. Dong, S. Schmauder, "Modeling of Metal Matrix Composites by a Self-Consistent Embedded Cell Model", Acta metall. mater. 44, pp. 2465-2478 (1996).

N. Lippmann, Th. Steinkopff, S. Schmauder, P. Gumbsch, "3D-Finite-Element-Modelling of Microstructures with the Method of Multiphase Elements", Computational Materials Science 9, pp. 28-35 (1997).

L. Mishnaevsky Jr., M. Dong, S. Hönle, S. Schmauder, "Computational Mesomechanics of Particle-Reinforced Composites", Computational Materials Science 16, pp. 133-143 (1999).

L. Mishnaevsky Jr., "Three-dimensional Numerical Testing of Microstructures of Particle Reinforced Composites", Acta Materialia 52/14, pp. 4177-4188 (2004).

V.A. Romanova, E. Soppa, S. Schmauder, R.R. Balokhonov, "Mesomechanical analysis of the elasto-plastic behavior of a 3D composite-structure under tension", Computational Mechanics 36, pp. 475-483 (2005).

Chapter 3.

S. Schmauder, "Crack Growth in Multiphase Materials", Encyclopedia of Materials: Science and Technology, Elsevier Science Ltd., pp. 1735-1741 (2001).

J. Wulf, S. Schmauder, H. Fischmeister, "Finite Element Modelling of Crack Propagation in Ductile Fracture", Computational Materials Science 1, pp. 297-301 (1993).

S. Aoki, Y. Moriya, K. Kishimoto, S. Schmauder, "Finite Element Fracture Analysis of WC-Co Alloys", Engineering Fracture Mechanics 55, pp. 275-287 (1996).

S. Hönle, S. Schmauder, "Micromechanical Simulation of Crack Growth in WC/Co Using Embedded Unit Cells", Computational Materials Science 13, pp. 56-60 (1998).

L. Mishnaevsky Jr., N. Lippmann, S. Schmauder, "Computational modeling of crack propagation in real microstructures of steels and virtual testing of artificially designed materials", International Journal of Fracture 120, pp. 581-600 (2003).

L. Mishnaevsky Jr., U. Weber, S. Schmauder, "Numerical analysis of the effect of microstructures of particle-reinforced metallic materials on the crack growth and fracture resistance", International Journal of Fracture 125, pp. 33-50 (2004).

C. Kohnle, O. Mintchev, S. Schmauder, "Elastic and Plastic Fracture Energies of Metal/Ceramic Joints", Computational Materials Science 25, pp. 272-277 (2002).

Chapter 4.

P. Leßle, M. Dong, S. Schmauder, "Self-Consistent Matricity Model to Simulate the Mechanical Behaviour of Interpenetrating Microstructures", Computational Materials Science 15, pp. 455-465 (1999).

S. Schmauder, U. Weber, "Modelling of Functionally Graded Materials by Numerical Homogenization", Arch. Appl. Mech. 71, pp. 182-192 (2001).

J. Rohde, S. Schmauder, G. Bao, "Mesoscopic Modelling of Gradient Zones in Hardmetals", Computational Materials Science 7, pp. 63-67 (1996).

L. Mishnaevsky Jr., "Functionally gradient metal matrix composites: numerical analysis of the microstructure-strength relationships", Composites Sci. & Technology 66/11-12, pp. 1873-1887 (2006).

L. Mishnaevsky Jr., "Automatic voxel based generation of 3D microstructural FE models and its application to the damage analysis of composites", Materials Science & Engineering A407/1-2, pp. 11-23 (2005).

S. Schmauder, A. Melander, P.E. McHugh, J. Rohde, S. Hönle, Or. Mintchev, A. Thuvander, H. Thoors, D. Quinn, P. Connolly, "New Tool Materials with a Structural Gradient for Milling Applications", J. Phys. IV France 9, pp. Pr9-147 - Pr9-156 (1999).

Chapter 5.

M. Ludwig, D. Farkas, D. Pedraza, S. Schmauder, "Embedded Atom Potential for Fe-Cu Interactions and Simulations of Precipitate-Matrix Interfaces", Modelling and Simulation in Materials Science and Engineering 6, pp. 19-28 (1998).

S.Y. Hu, M. Ludwig, P. Kizler, S. Schmauder "Atomistic Simulations of Deformation and Fracture of α-Fe", Modelling and Simulation in Materials Science and Engineering 6, pp. 567-586 (1998).

L. Farrissey, M. Ludwig, P.E. McHugh, S. Schmauder, "An Atomistic Study of Void Growth in Single Crystalline Copper", Computational Materials Science 18, pp. 102-117 (2000).

S. Nedelecu, P. Kizler, S. Schmauder, N. Moldovan, "Atomic Scale Modelling of Edge Dislocation Movement in the α-Fe-Cu System", Modelling and Simulation in Materials Science and Engineering 8, pp. 181-191 (2000).

Y. Furuya, H. Noguchi and S. Schmauder, "Molecular Dynamics Study on Low Temperature Brittleness in Tungsten Single Crystals", International Journal of Fracture 107, pp. 139-158 (2001).

S. Schmauder, P. Binkele, "Atomistic Computer Simulation of the Formation of Cu-Precipitates in Steels", Computational Materials Science 24, pp. 42-53 (2002).

C. Kohler, P. Kizler, S. Schmauder, "Atomistic simulation of the pinning of edge dislocations in Ni by Ni_3Al precipitates", Mat. Sci. and Engng. A400-401, pp. 481-484 (2005).

Chapter 1: Micromechanical Experiments

The purpose of this chapter is to analyse the micromechanisms of damage and fracture in heterogeneous materials, metals and composites, using direct observations of the damage evolution at the microlevel, combined with the macroscopic and/or computational analysis of the damage evolution.

In section 1.1, a SEM study of the micromechanism of fracture in SiC particle-reinforced 6061 aluminium composites is presented. The results lead to a better understanding of the micromechanism of particle breakage and interface debonding, and the special role of the particle effects in these composites.

In section 1.2, the mechanisms of damage initiation, evolution and crack growth in AlSi cast alloys are studied by in-situ tensile testing in a scanning electron microscope. It is shown that microcracks in these alloys are predominantly formed in the Si particles. Shear bands are seen to precede the breaking of the Si particles and the dislocation pile-up mechanism can thus be confirmed as the dominant damage initiating process in the matrix. Both micro- and macrocrack coalescence have been observed in the course of the experiments. The effect of the microstructure of the AlSi7Mg cast alloys on damage nucleation, crack formation and compliance reduction is analysed.

In section 1.3, micromechanisms of damage initiation and crack growth in high speed and cold work steels are investigated using scanning electron microscopy *in situ* experiments. The role of primary carbides in initiation and growth of cracks in tool steels is clarified. It is shown that initial microcracks in the steels are formed in primary carbides and then join together. A hierarchical finite element model of damage initiation, which included a macroscopic model of the deformation of the specimen under real experimental conditions and a mesomechanical model of damage in real microstructures of steels, was developed. Using the hierarchical model, the conditions of local failure in the steels have been obtained.

1.1 Micromechanisms of fracture in Al/SiC composites[1]

Engineering materials with a discontinuous second phase as a toughener [1] or reinforcement [2] have been widely studied in materials science and engineering. Investigations of the fracture characteristics of SiC particle-reinforced aluminium have shown that particle addition usually lowers the fracture toughness [3-5]. Reported fracture toughness' values for unreinforced aluminium alloys are in the range of 25-75 MPa m$^{1/2}$, while the composites have plane strain toughness values of 7-25 MPa m$^{1/2}$ [6, 7]. Many researchers have shown that the effect of microstructure on the fracture toughness is significantly affected by the details of the matrix microstructure, interface characteristics, and degree of clustering in the materials [8-9]. However, SEM fractography has revealed that the fracture surface consists of microvoids, corresponding to ductile fracture with dimples [10]. The sources of these dimples have been attributed to fracture of SiC particles [11], inclusions and precipitates or decohesion from the matrix as well as matrix failure [12, 13]. An attempt to explain these special failure characteristics of Al/SiC composite materials, which behave macroscopically brittle, but microscopically ductile, were the main purpose of this work. The fracture toughness tests on the composites were carefully designed with single-edge notched sheet (SENS) [14] specimens in the SEM. Both qualitative observations of void nucleation and quantitative measurements of crack profiles were made to assess the specific role of the particle-reinforcement mechanism in the composites. The microstructure analysis is proposed to understand and explain the particle effects during the crack initiation and propagation in these composites.

1.1.1 Experimental procedure

The composites used consisted of particle-reinforced aluminium alloy 6061 manufactured by extruding mixtures of aluminium powder and SiC particles. The volume fractions of particles in the composites were 0%, 10% and 20%. The mechanical properties of these composites are shown in Table 1.1. Distributions of measured SiC particle diameters are shown in Fig. 1.1a and b.

The SENS sample was designed according to the requirements of the SEM machine. The dimensions of the sample are shown in Fig. 1.2. The test was carried out in a Jeol JSM-35 scanning microscope. The machine automatically records the applied load versus displacement curves, and the monitor is used to examine the tip of the notch to understand the notch deformation, as well as nucleation, growth and coalescence of voids during loading. A record of the process is made by a video recorder.

[1] Reprinted from X. Ge, S. Schmauder, "Micromechanism of Fracture in Al/SiC Composites", J. Mat. Sci. 30, pp. 173-178 (1995) with kind permission from Springer

1.1.2 Results of Experiments and Analysis

Qualitative observations of void nucleation

General observations were made on the tip and root of the notch during the load-ing process. Voids nucleated in the middle of the notch root, as observed in the SEM, at K_0 / K_1 equal to 0.68, 0.784, and 0.85 for 0%, 10% and 20% SiC volume fraction composites, respectively, where K_1, is the stress intensity factor of the sample calculated according to Brown and Srawley [15] and K_0 is the fracture toughness. Measured data of K_0 and K_1 are shown in Table 1.2.

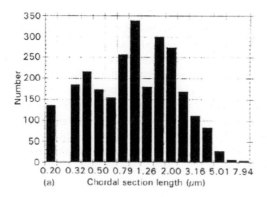

(a)

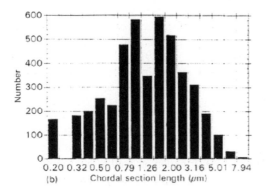

(b)

Fig. 1.1 Distributions of particle diameters. (a) 10% Al/SiC, (b) 20% Al/SiC (courtesy J. Wulf).

Table 1.1
Mechanical properties of Al/SiC composites used in the test[a]

Matrix	SiC particle (vol%)	Heat treatment[b]	Yield Strength σ_y (Mpa)	Ultimate Strength σ_u (MPa)	Elastic Modulus E_c (Gpa)
6061	0	T6	368,5	394	71,7
6061	10	T6	381,2	420	90,5
6061	20	T6	397	458	107,8

[a] Data in Table 1.1 are from Kobe-Steel Corporation.
[b] T6 heat treatment: solution treated at 803 K for 2h , water quenched, aged at 448 K for 8 h and air cooled.

During loading, the first void was observed in the centre of the notch root, Figs 1.3 and 1.4. Fig. 1.3 shows a stage of void growth at the notch root (arrows 1, 2, 3) as well as plastic deformation in the tip region of the notch in 0% SiC composites (arrow a). Fig. 1.4a shows void nucleation and growth in a 10% SiC composite sample. When the voids grow at the root of the notch, two possibilities exist for void growth to cause microcrack initiation in the adjacent free surface: one arises at the nearest point to the void in the free surface, characterizing the high stress concentration in the notch tip (point a); another, about 120 μm away, will form a microcrack (point b). As the loading increases, the voids at the notch root grow and coalesce towards the microcrack and combine directly with the microcrack. (Fig. 1.4b). Fig. 1.4c is the picture of a local amplification of point c in Fig. 1.4b, showing the crack propagation. Fig. 1.4d shows the propagation of the main crack. The crack in the Al/20% SiC sample propagates so rapidly that it is difficult to record more detail during loading.

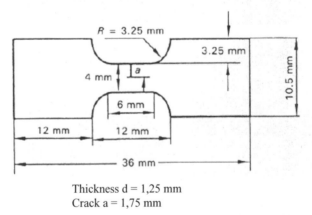

Thickness d = 1,25 mm
Crack a = 1,75 mm

Fig. 1.2 Dimensions of SENS specimen.

Fig. 1.3 Scanning electron micrograph of void nucleation, growth and coalescence in the notch root surface of a pure aluminium sample.

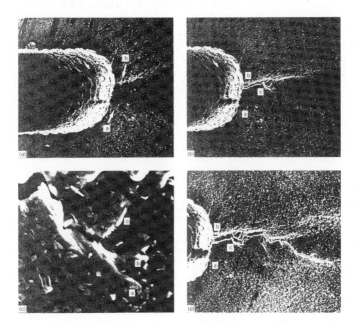

Fig 1.4 Scanning electron micrographs of void nucleation, growth and coalescence in Al/10% SiC. (a) Void nucleation, growth and coalescence, (b) void coalescence and crack initiation, (c) local magnification of point c in (b), (d) the main crack propagation.

Table 1.2
Fracture toughness of Al/SiC

	0% SiC	10% SiC	20% SiC
K_I (MPa m$^{1/2)}$)	38.7	25.76	22
K_0 (MPa m$^{1/2)}$)	26.4	20.2	18.9
K_0 / K_I	0.68	0.784	0.85

K_I and K_0 are calculated from [15]

Quantitative measurements of COD curves

CODs of the notch and *2u(x)* were measured for the specimens, where x is the distance behind the notch tip as indicated in Fig. 1.5. The results are directly measured from the scanning electron micrographs and are shown in Fig. 1.6.

$$u(x) = 8 \left[\frac{K_1}{(2\pi)^{1/2} E} \right]^{x^{1/2}} \tag{1.1}$$

The crack propagation profile is that associated with a plane stress crack with the correlated applied stress intensity factor, K_I [16] where K_I is shown in Table 1.2 and E is Young's modulus of the composites from Table 1.1. Equation 1.1 is plotted together with experimental data in Fig. 1.7. The experimental data are lower than that predicted by Equation 1.1. Fig. 1.8 is the stress intensity factor for three composites measured behind the crack tip during the R-curve determination.

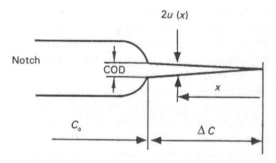

Fig. 1.5 Schematic drawing of the SENS specimen. Notch length C$_o$, crack extension ΔC, crack profile by COD, 2u (x) at a distance behind the crack tip.

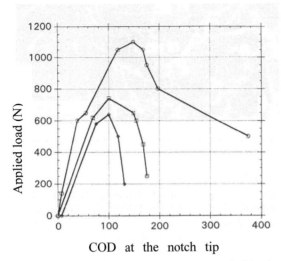

COD at the notch tip

Fig 1.6 Crack opening displacement versus applied load curves. (○) 0% SiC, (□) 10% SiC, (◊) 20% SiC.

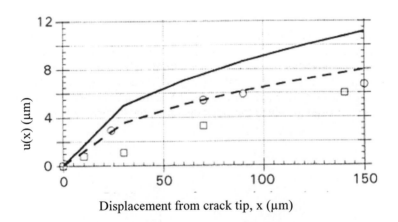

Displacement from crack tip, x (µm)

Fig. 1.7   Comparison between (——, - - - -) prediction from Equation 1.1 and (○,□) experimental data of the crack profile for (——,○) 10% SiC and (- - -, □) 20% SiC.

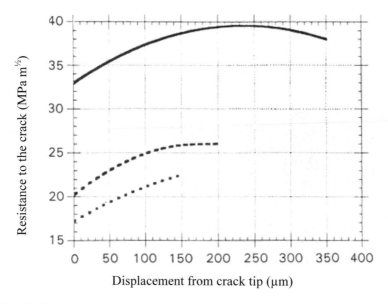

Fig. 1.8 Crack resistance curves for Al/SiC composites: (——) 0% SiC, (— — —) 10% SiC, (- - -) 20% SiC.

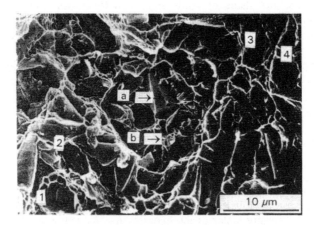

Fig. 1.9 Scanning electron micrograph showing the properties of the primary and secondary voids.

Primary and secondary voids

Void nucleation is significantly altered when the particles are present. From Fig, 1.9, there are at least two kinds of void observable in the fracture surface of Al/SiC composites. The first kind of void nucleates either at broken particles or at decohering interfaces (point a).

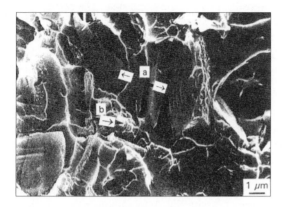

Fig. 1.10 Scanning electron micrograph of primary and secondary voids.

Both the size and shape of voids are found to be associated with the SiC particles [11], and this is called a primary void. Therefore, the surface of these voids characterizes cleavage (Fig. 1.10, point a). The other kind of void is nucleated in the matrix ligaments between the particles. The dimensions of these voids are about 0.3-1.2 μm for Al/20% SiC composites, as shown in 1.10. From Fig. 1.10, point b, it can be seen that these voids are constrained by interface bond forms around them and their sizes are very small compared with the primary void. These voids are termed secondary voids. They are dimples and characterize microscopic ductility. However, these void dimensions are affected significantly by particle size, volume fractions, and interface properties.

Discussion

In Table 1.2, K_0 is the toughness when a void nucleates as observed in SEM, and K_1 is the stress intensity factor of the material. K_0 / K_1 can be used to express the fracture behaviour of the composites. When a void nucleates and grows, the material can still sustain additional applied loading if $K_0 < K_1$. When K_0 / K_1 is smaller, i.e. the stages of void nucleation, growth and coalescence are longer, the composites show ductile behaviour. If K_0 / K_1 increases, and tends to 1 as the SiC particle volume fraction increases, the effect of void nucleation, growth and coalescence decreases, and the material becomes brittle. In Fig. 1.6 the notch opening displacement decreases rapidly as particle volume fractions increase. When the voids nucleate and grow at the notch root, the curve of load versus notch opening displacement does not change its shape markedly until the voids coalesce to microcrack, particularly in the curve of pure aluminium in Fig. 1.6. Equation 1.1 is used to describe the crack profile. Compared with curves in Fig. 1.7 and the data measured in the test, all data points are obviously lower than predicted by Equation 1.1, which states the real length of the crack in Al/SiC composites is longer than that given by Equation 1.1, and shows the brittle nature of the materials. Fig. 1.8 shows the R-curve of these composites in the plane stress state. The resistance to the crack propagation in pure aluminium is about twice that of Al/20% SiC. During practical tensile tests, the curve of applied load versus displacement characterizes ductile features for pure aluminium samples and brittle features for Al/20% SiC samples. When an applied load reaches a certain criterion value, the crack initiates at the tip of the notch and rapidly propagates in Al/20% SiC samples. This situation is very similar to crack growth in ceramic matrix composites [17].

During the tensile test, more detailed examinations of Al/10% SiC in the SEM were made of void nucleation, growth and coalescence, as well as the crack initiation. The void was first observed at the symmetry plane of the notch root surface (Fig. 1.4a) because there is a high stress constraint region. As the loading increases, the first void nucleates and grows and then the second and third voids are observed (Fig. 1.4b). Similar situations are found in Fig. 1.3 in the notch root of pure aluminium. Comparing Fig. 1.3 with Fig. 1.4a, although there is the same number of the voids observed between the centre of the notch root and the free surface in these two materials, the size of the voids is very different. The maximum size is about 20 μm for pure aluminium, 5 μm for Al/10% SiC. The ratio of both void sizes is 4. The deformation of the materials has been altered by the particles in the matrix, so the void size becomes small and the composite is brittle. But voids are not observed in the free surface of the notch tip region in Figs 1.3 and 1.4. When loading increases, the voids coalesce at the notch root, and at the same time microcracks initiate in the free surface of the notch tip. The shear failure near the notch root and free surface can be observed in Fig. 1.4b.

After voids coalesce and microcracking initiates, the main crack will be formed. The crack meanders microscopically whereas the failed surface is flat macroscopically, more so when the SiC volume fraction increases. Although apparently easy paths for crack propagation can develop early in the high straining process, the main crack does not necessarily follow these routes. The crack propagation is mainly affected by the microstructure of the composites. Fig. 1.4c indicates three possibilities for crack propagation: point a is a stress concentration region caused by a small group of cluster particles; point b shows interface debonding in the tip region of the crack; point c is a possible way to form the secondary crack connected with the main crack. The crack will follow the direction of the easiest propagation. Finally, the crack goes along the point c direction in Fig. 1.4d. There are two reasons to explain why the crack follows this route. On the top right in Fig. 1.4c, the direction of the crack has been influenced to turn left by interface debonding at the left tip region of the crack, which has inclined to the maximum principal strain direction, so it is reasonable for the crack to turn back to the original line; on the bottom right in Fig. 1.4c, a larger debonding is formed. The influence of this debonding on the crack path is greater than that of points a and b. The crack propagation in Al/SiC composites can be described as follows: first, the voids nucleate, grow and coalesce at the notch root, and a microcrack initiates at the tip of the notch in the free surface; then the voids coalesce and connect with the microcrack to form the main crack; third, the debonding or particle breakage in the tip region of the crack occur before the crack advances; these debonding or broken particles coalesce with the crack, and the crack propagates. Observed primary and secondary voids have been shown to explain exactly why the Al/SiC composites depict microscopically ductile features. The primary voids associate with particles, and the particle can be found to be located inside the primary void in Figs 1.9 and 1.10.

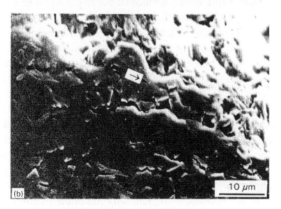

Fig. 1.11 Debonding and broken particles in Al/20% SiC: (a) in the notch tip region, (b) near the failed crack surface.

The sources of primary voids are the interface debonding or the cracked particles. Points 1 and 2 in Fig. 1.9 show that both the size and shape of the voids are associated with the SiC particles in it. Points 3 and 4 in Fig. 1.9 show the primary voids for particles that may be in the opposite fracture surface. From point a in Fig. 1.10, there are cracked particles, showing typical cleavage fracture. It is reported that the particles will crack at a relatively low strain level [18, 19]. The stress triaxial constraint around the particle makes the matrix harder than in the absence of particles. The linear elastic part of this constraint has been analysed theoretically and quantitatively [20]. The reason for debonding and breakage of particles at the tip depend on the spacing of particles and interface stress constraint. These regions are under a strong deformation constraint and plastic strain associated with void formation is smaller. For these reasons, the processes of nucleation, growth and coalescence of the secondary voids are then too small to exhibit tensile plastic features of the material. The low fracture toughness of these composites is mainly determined by primary voids and characterized as brittle events, while secondary voids have little influence on the ductile behaviour of the materials.

Because of the high stress concentration at the tip of the crack, interface debonding and particle cracking develop prior to the main crack arrival. Fig. 1.11 shows the region of the crack tip containing many broken particles and interface debonding, which will have a great effect on the crack paths. Fig. 1.11a shows the debonding or broken particles in this region of Al/20% SiC. Fig. 1.11b depicts the debonding and broken particles near the failed fracture surface. The arrows in Fig. 1.11 give the direction of crack propagation. This microstructural region ahead of the main crack experiences the rapid propagation conditions for macrocracks, and affects the crack path, although macroscopically the main crack follows the directions that the maximum principal stress would predict as shown in Fig. 1.4d. This is why the crack propagates so of the notch can be explained by the high constraint effect of triaxial stress [21]. HREM analysis of the interface in Al/SiC with T6 heat treatment shows the brittle Al_4C_3 precipitates [22]. So the primary void characterizes the brittle property of composites. Secondary voids occur in the spacing between particles during loading. These voids follow three stages of nucleation, growth and coalescence. However, the dimensions of these voids are very small (Fig. 1.9), and they can only be examined clearly by magnifying more than 1000 times in the SEM. Secondary voids, which show many small dimples in the fracture surface and behave in a ductile manner, rapidly in Al/20% SiC samples in the test. The many debonding and broken particles in this region of the crack tip is the main reason far the brittle fracture and, hence is responsible for the low ductility of the composite materials. However, it is not yet clear quantitatively at what applied strain level, particle cracking occurred and which of the two types of behaviour, interface debonding or particle cracking, is predominant.

Conclusions

1. K_0 / K_1 can be used to express material toughness K_0 / K_1 is equal to 0.68, 0.784 and 0.85 for 0%, 10% and 20% Al/SiC materials. If K_0 tends toward K_1 the material toughness becomes lower.

2. Two kinds of void have been defined according to the properties of the voids in the fracture surface. The primary voids govern the brittle property of composites, while the secondary voids, governing the ductile property, have little influence on the fracture toughness.
3. The voids initially nucleate, grow and coalesce at the notch root surface. The crack propagation observed in the free surface of the samples is associated with debonding and particle cracking in the tip region of the crack.
4. The direction of crack propagation depends on the microstructure in the tip of the crack, and macroscopically on the maximum principal strain direction.

References

[1]   Flinn D B, Rühle M and Evans A G (1989), Acta Metall. 37, p. 3001.
[2]   Ge X, Zhang D and Ju D (1989), Advances in Constitutive Laws for Engineering Materials Vol. 1, edited by 1. Fan and S. Murakami (International Academic Publishers, Chongging, China, 1989), p. 497.
[3]   Nair S V, Tien J K and Bates R C (1985), Int. Metal. Rev 30, p. 275.
[4]   Divecha A D, Fishman S G and Kamarkar S D (1981), J Metals 33, p. 12.
[5]   Divecha A D, Fischman S G (1981), Sampe Q, p. 40.
[6]   da Vidson D L (1991), Metall. Trans. 22A, p. 113.
[7]   Roebuch B, Lord J D (1990), Mater. Sci Technol. 6, p. 1199.
[8]   Birt M J, Johnson W S (1990), Fundamental Relationship between Microstructure and Mechanical Properties of Metal Matrix Composites. Liaw P K , Gungor M N(ed), MMS, California, USA, p. 71.
[9]   Vasudevan A K, Richmond O, Zok F and Embury J D (1989), Mater Sci Eng A 107, p. 63.
[10] Lewandowski J J, Liu C and Hunt H W Jr (1989), ibid A107, p. 241.
[11] Mummery P, Derby B (1990), Fundamental Relationship between Micro-structure and Mechanical Properties of Metal Matrix Composites. Liaw P K , Gungor M N (ed) MMS, California, USA, p. 161.
[12] Handwerker C A, Vandin M D, Kattner U R and Lee D J (1990), Metal-Ceramic Interface, In: M. Rühle M, A. G. Evans A G,.Ashby M F and Hirx J P (ed) Acta-Scripta Metall. Proc. Ser 4, (Pergamon Press, Oxford, p. 129.
[13] Lee J-D, Vandin M D, Handwerker C A and Kattner U R (1988), High Temperature/High Performance Composites. edited by Lemkey F D , Fishman S G, Evans A G and Strife J R (ed) MRS, Pittsburgh, USA, p. 357.
[14] Logsdon W A, Liaw P K (1986), Eng. Fract. Mech. 24, p. 737.
[15] Brown F Jr, Srawley J E (1966), ASTM STP40 (American Society for Testing and Materials) Philadelphia, PA..
[16] Rudel J, Kelly J F and Lawn B R (1990), J Am. Ceram. Soc, p. 3313.
[17] Ge X, Zhang D and Ju D (1989), Proc ICCM-7 4, p. 36.
[18] Lagage H, Lloyd D J (1989), Can. Metall.Q 28, p. 145.
[19] Wang Z, Zhang R J (1991), Metall. Trans. 22A, p. 1585.
[20] Ge X, Schmauder S (1993), Mater Sci. Eng. A 168, pp. 93-97.
[21] Poech M H, Fischmeister H F (1992), Acta Metall. Mater. 40, p. 487.
[22] van den Burg M, Hosson J Th M (1992), Acta Metall. Mater. 40 Suppl., p. 281.

1.2 In-situ observation of damage evolution and fracture in AlSi cast alloys[2]

AlSi7Mg cast alloys are now widely used as structural materials in industry and their production increased remarkably during the last years [1]. The practical application of these materials is based on their strength and fracture resistance. To further improve the strength of the material, it is necessary to have a good understanding of the damage evolution and fracture of these alloys. The purpose of the paper is to clarify the mechanisms of damage initiation, damage evolution, compliance reduction and fracture of two AlSi7Mg cast 0.3 alloys with almost identical composition but very different microstructure.

Several investigations of the failure of AlSi cast alloys have been reported in the literature. Gurland and Plateau [2] studied the ductile rupture of an AlSi13 alloy and obtained an expression for the particle cracking stress and a relation between fracture strain and the fraction of broken particles. Yeh and Liu [3] have studied the cracking of silicon particles in aged AlSi7Mg (A357) alloys. They analysed the mechanism responsible for the cracking of the Si particles and the effects of strain and stress on the fraction of broken Si particles. Yeh and Liu have shown that the dislocation pile-up mechanism is the most probable one among other theories of the failure of Si-particles. Höner and Groß [1] studied the fracture behaviour and tensile strength of AlSi alloys and its dependence on the microstructure, which is determined in turn by the conditions of melt processing and casting.

The outline of this paper is as follows: in section 2 we briefly discuss the mechanisms and models of crack initiation and crack growth in ductile materials. Section 3 gives a description of the experimental details. An analysis and discussion of our results is given thereafter.

1.2.1 Failure mechanisms of ductile materials

The course of failure in ductile materials is generally divided into three stages [1, 4-6]:

1. Nucleation of microcracks/microvoids at random positions in the loaded body independent of their location relative to the other microcracks. The growth rate of the microcrack density usually increases with increasing density of microcracks [7]. This nucleation stage is followed by:
2. The link-up and coalescence of microcracks and the formation of the `macro-cracks', which leads to:
3. The growth of the macrocracks until one of the cracks reaches a critical size and begins to grow auto catalytically. This finally results in the complete failure of the body. All these processes proceed not only successively, but

[2] Reprinted from L.L. Mishnaevsky Jr., N. Lippmann, S. Schmauder, P. Gumbsch, "In-situ Observation of Damage Evolution and Fracture in AlSi7Mg0.3 Cast Alloys", Eng. Fract. Mech. 63, pp. 395-411 (1999) with kind permission from Elsevier

also simultaneously. For instance, growth of a crack can proceed simultaneously with the increase in the microcrack density.

Microcrack nucleation

The mechanisms of microcrack initiation are discussed in detail by Knott [8] and Tetelman and McEvily [4]. Tetelman and McEvily have shown that large tensile stresses as well as shear stresses develop at the tip of a glide band if the band is blocked by a strong obstacle. They determined the conditions of crack nucleation, and showed that the tensile stresses are less important during the nucleation process than the shear stresses. Knott [8] described and analysed several micromechanisms of fracture. Among them are microcrack nucleations by squeezing together of dislocations at the head of the slip band (Stroh model) or by the interaction of a pile-up with a carbide particle (Smith model). The model developed by Yeh and Liu to describe the failure of Si-particles in their AlSi7Mg alloys is similar to the Smith model and explains the microcrack initiation as a result of particle failure caused by stress concentrations originating from dislocation pile-ups at these particles.

Void and microcrack coalescence

Thomason [9] describes void coalescence with a mathematical model for the coalescence of square holes arranged into a square array in a rigid-plastic matrix under tension. The voids are elongating in the direction of the tensile load and approach each other in the direction normal to the tensile load. When the voids get close, necking occurs between them and they coalesce rapidly. The deformation of the material between the voids therefore controls void coalescence and necking. Finkel [5] has studied experimentally the necking between microcracks. He showed that the failure of the necks between microcracks is caused by shear stresses. Seidenfuss [10] recognised three possible mechanisms of void coalescence: local plastic constriction of material between voids, failure of the layer between voids caused by shear bands and failure by formation of secondary smaller voids on very small inclusions in the material between available voids. He noted that the joining of voids proceeds in a stepwise manner. The first mechanism dominates when the density of voids is high; the third prevails at relatively low void densities. Seidenfuss concluded that the main mechanism of void coalescence is the formation of shear bands in which small secondary voids can sometimes be observed. Shear bands causing the failure of the layers between coalescing voids have also been observed by Roberts et al. [11]. With regard to the interaction of macrocracks, it is known that: "originally collinear mode I cracks seem to avoid each other" [12]. On the basis of the analysis of the stability of straight crack paths, Melin has shown that the tip to tip coalescence of cracks does not occur for two collinear cracks [12].

Formation and growth of macrocracks

Crack growth as a result of the superposition of stress fields from a large crack and from a pile- up of dislocations in the vicinity of the crack tip has been modeled by Yokobori [13]. In this model the dislocation pile-up acts as a stress concentrator similar to a microcrack. The stress required for macrocrack growth is much higher than the stress required for microcrack formation from a pile-up of dislocations. As a consequence, ductile fracture is usually determined by the formation of a microcrack in the vicinity of a large crack, and not by crack growth itself. Ebrahimi and Seo [14] have shown that crack propagation in ductile materials may also involve the cleavage of favourably oriented grains ahead of the main crack tip followed by ductile tearing of remaining ligaments. The formation of river patterns on the fracture surface has been attributed [4] to the joining of the main crack with secondary new cracks nucleated ahead of the main crack. Joining can occur by tearing (tongue formation) or secondary cleavage.

From the above considerations, it follows that although the main models of fracture are based on the assumption that the crack grows in a continuous material (or, as a version, in continuous material with a plastic zone in the vicinity of the crack tip) [8], the real physical mechanisms of fracture can differ significantly from this assumption.

1.2.2 Experimental procedure

The course of damage evolution in AlSi7Mg cast alloys was observed by in-situ tensile testing in a scanning electron microscope (SEM). Fig. 1.12 gives the geometry and the dimensions of the notched CT-specimens used in our experiments. The notches in the specimens were sawn with a saw blade with a defined radius of the saw teeth equal to the required radius of the notch. Such notching ensures high quality of the notch surface. The side surface of the CT- specimens was ground and subsequently polished with diamond paste down to the 3 mm grade. The specimens were taken from cast components.

The heat treatment of the components included a solution annealing in an air circulation kiln for 12 h at 540°C and the artificial ageing in an air circulation kiln for 12 h at 170°C. The specimens were loaded in an in-situ tensile stage (Fa. Raith, Dortmund, Germany) in the SEM. A constant crosshead speed of 0.3 mm/s was used. The specimens were loaded until a large crack was visible on the surface of the specimens. During loading, the highly deformed region in the notch root (about 100 x 100 μm) of the CT-specimens was monitored. The experiments were conducted on AlSi cast alloys of the type AlSi7Mg0.3 with both lamellar and globular microstructure of the Si-particles.

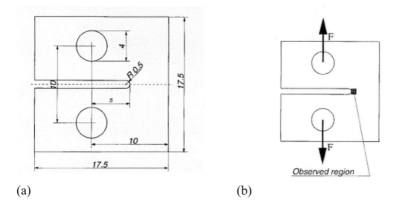

(a)   (b)

Fig. 1.12. Shape and size of the CT-specimen (a) and the small area in the notch root (b) which was observed during the tests. Dimensions are given in mm.

In the lamellar structure, the Si inclusions had an aspect ratio (length to width) of about 4...20 and the small dimension of the particles were between 2.5 and 5 µm. The Si particles can be transformed into a more rounded shape by the addition of Sb. The particles in the globular microstructure were approximately circular with a diameter of about 3-6 µm. The diameter of the Al grains was between 50 and 130 µm. The Si-particles in the alloys were located preferentially (but not exclusively) on the Al grain boundaries corresponding to the fact that the growth of Al grains during the formation of the alloy is inhibited by the available Si-particles. Fig. 1.13 shows micrographs of the two different (globular and lamellar) microstructures. Even in the alloy with globular microstructure the Si-particles form a network and each cell of the network corresponds to a grain of the Al-matrix.

Simultaneously with the loading of the specimens, the externally applied displacement of the crosshead and corresponding load were recorded. The displacements reached 0.5 mm for the specimens with the lamellar microstructure and 0.9 mm for those with the globular microstructure. In Fig. 1.14 typical load-displacement curves are shown for both microstructures. Both have the characteristic signatures of load-displacement curves for softening materials [15]. At first, the applied force increases almost linearly with increasing displacement and then reaches a peak. The following decease was most pronounced in the alloy with lamellar microstructure. The macrocracks formed later and grew when the curve was already descending considerably. These processes will be detailed in the following section.

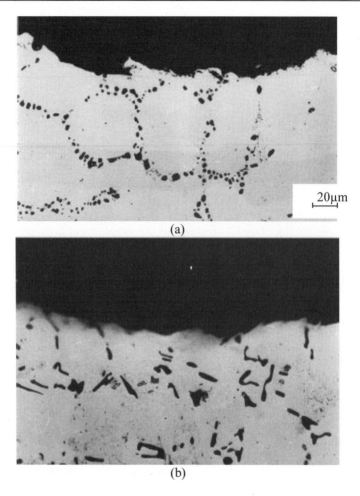

Fig. 1.13 Micrographs of the two microstructures: (a) globular and (b) lamellar (a scale is the same).

1.2.3 Experimental Observations

Microcrack initiation and evolution

At the initial stages of loading, the formation of shear bands was observed in the ground notch region near the surface. Then, several microcracks appeared at some distance from the notch surface. This stage is indicated with label A in Fig. 1.14 and corresponds to the displacement of 0.13 and 0.16 mm for the alloys with lamellar and globular structures, respectively. The applied force was approximately

500 N for both microstructures. The microcracks initiated exclusively in the Si-particles. The amount of broken Si-particles in relation to the total number of particles in the observed area of the alloy with lamellar structure was about 4%.

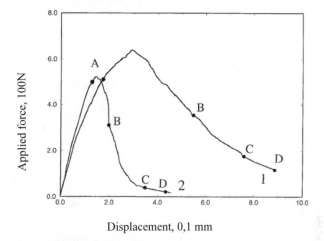

Displacement, 0,1 mm

Fig 1.14 Load-diplacement curve fot the specimen with globular (1) and lamellar (2) microstructure. The following SEM micrographs correspond to the points marked with the letters A±D. Figs 1.15 und 1.16 (point A on both curves), Figs 1.17 and 1.18 (point B) Figs 1.19 and 1.20 (points C and D)

A similar calculation for the globular structure was not possible since the cluster arrangement of the Si-particles on the boundaries of the Al-grains did not allow distinguishing individual broken particles. Thereafter, additional matrix shear bands formed and the density of microcracks increased. Figs. 1.15 and 1.16 show the microcracks formed at the initial stage of damage evolution.

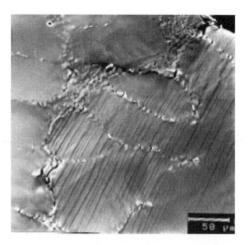

Fig. 1.15 Area in the vicinity of the notch root (globular microstructure; magnification x350; displacement 0.185 mm).

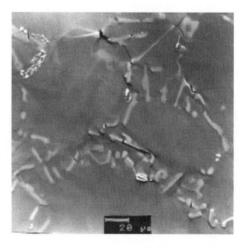

Fig. 1.16 Area in the vicinity of the notch root (lamellar microstructure; magnification x500; displacement 0.16 mm).

The micrographs represent small areas rather close to the notch surface, and correspond to the alloys with the globular and lamellar microstructure, respectively. The microcracks are more or less randomly distributed up to a distance of about 300 mm from the notch surface (0.6 notch radii). The microcracks thus are oriented mostly along the lines of maximum shear stress. Upon further increasing the applied load, the nucleation and accumulation of microcracks at some distance from the notch is accompanied by nucleation and growth of cracks from the notch surface.

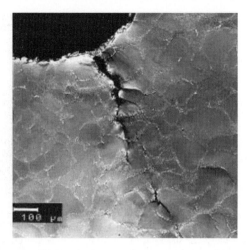

Fig. 1.17 Crack formation in the vicinity of notch root and crack initiation at some distance from the notch root (globular microstructure; magnification x100; displacement 0.55 mm).

In summary, during the initial stages of damage evolution in the AlSi7Mg cast alloys the destruction of the Si-particles was the prevailing mechanism of microcrack nucleation. Microcracks formed predominantly at 'random' sites throughout the stressed area and not at the root of the notch.

Crack growth and coalescence

At the next stage of damage evolution, relatively large cracks formed from initially small surface cracks. Simultaneously, the density of microcracks at some distance from the notch surface increased and several small cracks formed there. Figs. 1.17 and 1.18 show the large crack which starts at the notch surface and the smaller cracks and microcracks formed at some distance from the notch surface. The micrographs shown on Figs. 1.17 and 1.18 correspond to the points marked with the letter B on the force-displacement curve (Fig. 1.14). Comparing Figs. 1.17 and 1.18, it is seen that the direction of the cracks is the same for both microstructures. The angle between the crack direction and the axis of symmetry of the CT- specimen was about 308. Shear bands are apparent between the large cracks and the small cracks (cf. Fig. 1.18). The large crack then grew and joined with the small cracks in front. The path of the large crack corresponds to the direction from the crack tip to the available microcracks and shear bands. The small cracks and microcracks then joined and formed a second large crack located about 0.4-0.5 mm away from the notch surface (about 0.9 notch radii). The formation of the second large crack was observed at a load point displacement of about 0.24 mm and at an applied force of 100 N for the lamellar microstructure and at about 0.7 mm and 180 N for the globular microstructure.

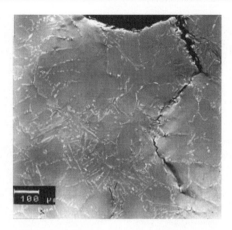

Fig. 1.18 Crack formation in the vicinity of notch ground (lamellar microstructure). The microcracks set the crack path (magnification x115; displacement 0.20 mm).

Upon further loading, the large crack from the notch surface and the second crack join. Figs. 1.19 and 1.20 show this coalescence for the globular and the lamellar microstructure, respectively. Figs. 1.19a and 1.20a depict the earlier stages when there are still two separate large cracks in the materials.

This stage corresponds to the points marked with the letter C in Fig. 1.14. Figs. 1.19b and 1.20b present the later stage when the cracks coalesced and formed one single large crack, labelled with the letter D in Fig. 1.14.

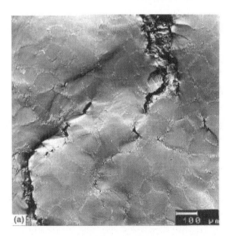

Fig. 1.19 Crack coalescence in the alloy with globular microstructure. Two cracks (a) before and (b) after coalescence (magnification x100; displacement (a) 0.76 and (b) 0.89 mm).

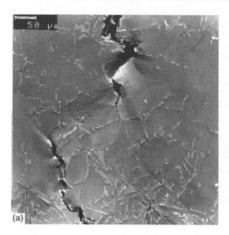
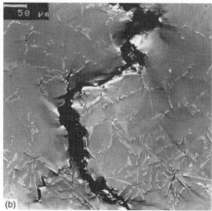

Fig. 1.20 Crack coalescence in the alloy with lamellar microstructure. Two cracks (a) before and (b) after coalescence (magnification x200; displacement (a) 0.35 and (b) 0.43 mm).

It is of interest to note that almost the same crack patterns formed in both materials, although the structures of the materials differ significantly. The crack paths followed the location of the microcracks in front of the tip only until a second large crack was available at close distance. Then the interaction between the stress fields of the cracks determined their paths.

1.2.4 Analysis of results

Microcrack nucleation

The prevailing mechanisms of microcrack initiation in the AlSi7Mg cast alloys investigated in this study can be clarified by comparison and correlation with the results of other authors. The first step is to identify the mechanisms and the criteria controlling the breaking of the Si- particles.

One may assume that the Si-particle failure is caused by a critical level of the maximal tensile stress. In this case, one would expect that the first microcracks initiate directly at the centre of the notch. This is clearly inconsistent with our observations that the microcracks are formed mostly in random sites at some distance from the notch surface. Therefore, it follows that local fluctuations of the stress field caused by dislocation pile-ups and Si-particles influence the microcrack nucleation much more than the overall stress distribution.

Following Yeh and Liu [3], the critical length of a dislocation pile-up which causes the breaking of the particles can be calculated. They derived the following formula for the externally applied stress se required for particle failure due to a one-plane pile-up of dislocations where E is the Young's modulus, G is the shear modulus, b is the Burgers vector, n is the Poisson ratio, M is the Schmid factor and L is the length of the pile-up.

$$\sigma_e = \sqrt{\frac{EGb}{2(1-v)\pi^2 M^2 L}} \qquad (1.2)$$

In our experiments, the first broken Si-particles were observed in the notch region at an externally applied load of about 500 N (see Figs. 1.15 and 1.16). At this load the Al matrix in the notch ground starts yielding, which for this specific matrix alloy has been shown to begin at 218 MPa [16]. Taking the angle at which particle cracking first occurs from the micrographs (Figs. 1.15, 1.16; y2308), one calculates a Schmid factor of

$$M = \sin\Theta\cos\Theta \cong 0.43 \qquad (1.3)$$

Substituting the values E = 100 GPa, n = 0.33, G = 40 GPa [3], b = 0.286 nm [17] and the local stress se = 218 MPa into Eq. (1.2), one obtains: L = 9.9 mm. This value is rather close to the pile- up length calculated by Yeh and Liu for the as-quenched conditions (L = 14.4 mm [3]). This length corresponds to dislocations in the pile-up, where τ-shear stress

$$n = \frac{(1-v)\pi\tau L}{Gb} \qquad (1.4)$$

If τ is again taken to be of the order of the yield stress (218 MPa), Eq. (1.4) gives n ≈ 250 dislocations in the pile-up. Our results differ here from the results of Yeh and Liu, who have used the value of the external stress (17.2 MPa). Since our in-situ observations show that the aluminium matrix already deformed plastically we claim that the local stress level must be larger than the externally applied stress and of the order of the yield stress of the matrix. This value for the length of the pile-up is very reasonable since it is about 5-10 times lower than the linear size of Al-grains.

Together with the direct observation that the microcracks are located mostly along the lines of maximum shear stress and always appear together with the first shear bands, our data strongly support the dislocation pile-up mechanism as the origin of particle failure. Our in-situ observations (Figs. 1.15 and 1.16) linked with the load-displacement diagram (Fig. 1.14) confirms that the microcracks are responsible for the compliance reduction of the material. It is apparent that the microcracks are formed just before the peak load is reached and macroscopic compliance reduction starts after the density of microcracks reached a reasonable value. The rate at which the compliance reduction proceeds, however, is determined by the microcrack coalescence.

Crack growth and coalescence

The formation of (small) cracks through the coalescence of microcracks as described above complies rather with the descriptions of microcrack coalescence given by Finkel [5], Seidenfuss [10] and Roberts et al. [11] than that by Thomason [9]. Microcrack coalescence was observed only after shear bands were formed between them, and not due to their expansion or growth.

It was also observed that the macrocracks are initiated on and propagate from the notch surface, although the microcracks are nucleated at some distance from the surface. Again, shear bands are responsible for the crack initiation there.

The final destruction of the material via the initiation and growth of the surface crack, the formation of the second crack from randomly distributed microcracks at some distance from the surface and the coalescence of the macrocracks bears many similarities with the mechanisms described in the literature [5, 14]. A few aspects, however, are very different: the microcracks in the material were not only formed in the vicinity of the crack tip as a result of the stress field from the crack, but also in "random" sites throughout the stressed volume consistent with the dislocation pile-up model for their nucleation. Both microstructures showed the same behaviour in this respect, which means that the following steps of damage evolution, the growth of the macrocracks and their coalescence, is determined by the global stress distribution in the loaded specimen (which is the same in both cases) rather than by the microstructure of the material.

In modelling the fracture processes in this alloy, one should therefore take into account not only microcracks which are formed in front of the growing crack and are absorbed by the crack, but also the distributed microcracks, which are formed simultaneously with the formation of the large crack and influence its path. The evolution of these distributed microcracks may lead to the formation of other large cracks, which interact with the existing ones.

Alloy microstructure, compliance reduction and fracture

Table 1.3 gives the consumed energies corresponding to the points marked on Fig. 1.14. The energy was calculated as the area under the curves of Fig. 1.14. Evidently, for the globular microstructure the energy needed to reach the descending branch of the force-displacement curve is almost four times larger than for the lamellar structure.

Comparing Figs. 1.17 and 1.18 with the data from Table 1.3, the relation between the fracture energies of both alloys can be estimated, since the cracks in Figs. 1.17 and 1.18 are approximately of equal size. The energies differ by almost a factor of 4: 0.23 and 0.058 J, for globular and lamellar alloys, respectively. Therefore the energy needed to form the crack and to create a unit crack area of the alloy with the globular structure is about 4 times larger than that of the alloy with the lamellar structure.

The question now arises, what determines the differences in fracture characteristics (cf. Fig. 1.14) between the two different alloys investigated here and whether there is potential for improvement.

Table 1.3
Consumed energy J corresponding to the points marked on the force-displacement curves (Fig. 1.14)

Points on curves of Fig. 1.14	Lamellar microstructure	Globular microstructure
Appearance of first, randomly distributed microcracks (points A)	0.03	0.046
Peak load	0.042	0.119
Formation of a large crack (points B)	0.058	0.236
Formation of two large cracks (points C)	0.083	.0292
Coalescence of large cracks (points D)	0.086	0.312

Up to this point many similarities between the two alloys have been discussed and no difference in failure mechanisms has been identified. Nevertheless, the descending branch in the load-displacement curves in Fig. 1.14 starts much earlier and compliance reduction is more pronounced in the alloy with the lamellar microstructure. The difference in reduction of stiffness of the material as a function of displacement is most pronounced at the initial stages of compliance reduction. In the lamellar structure, the load drops by a factor of ten when the displacement is increased from 0.15 to 0.3 mm. In contrast, the load only drops by a factor of 6 when the displacement is increased from 0.28 to 0.8 mm in the globular structure.

Having identified the necking along the shear bands formed between the microcracks as the main mechanism for their coalescence and the microcrack coalescence as the origin of the reduction of compliance, the different degree of the compliance reduction of the two alloys can be rationalised to some extent. The alloys differ mostly in the size of the Si-particles. The larger particles in the lamellar structure prohibit homogeneous slip transmittal and plastic deformation is concentrated on the few places at these particles where they are broken. This leads to less homogeneously distributed plastic deformation and microcrack distribution. A signature of this less even distribution is that the microcracks in the lamellar structure (Fig. 1.16) are opened much more clearly than those in the globular structure (Fig. 1.17) at similar strains. Consequently, the plastic strain in one individual slip band of the lamellar structure is significantly larger than in the globular structure and necking along these shear bands happens more readily.

The peak loads reached in the two alloys can be interpreted in a similar way. The peak load of the globular material is reached at a displacement which is approximately twice that of the lamellar material (0.281 and 0.144 mm, respectively). The peak load of the alloy with globular microstructure itself, however, is only 20% larger than that of the alloy with the lamellar structure (627 and 522 N, respectively). Again, if the coalescence of the microcracks is the decisive step for the compliance reduction and if coalescence is controlled by a critical plastic strain in the shear bands between the microcracks, the globular structure permits larger plastic strain since there are more paths for plastic deformation than in the lamellar structure and each of them has to carry comparatively less plastic strain. However, the corresponding stresses are only marginally higher since their level is mostly determined by the level at which yielding and microcrack initiation begin and this is similar for both alloys, irrespective of their microstructure.

Potential benefits for alloy development in this class of AlSi alloys are seen in two aspects, both are related to the microstructure refinement. Firstly, the reduction of the size of the Al- matrix grains should reduce the mean free path for dislocation pile-ups and might therefore prolong the initiation of the first microcracks to somewhat higher loads. Secondly, the size of the Si-particles and their shape should be pushed towards more rounded and more evenly distributed particles to avoid strain localisation in shear bands and to prolong the compliance reduction of the material to higher total strains.

Conclusions

The course of crack initiation and growth in two AlSi7Mg cast alloys which are distinguished by their globular and lamellar microstructure was investigated in-situ in a scanning electron microscope and can be described as follows:

- Nucleation of microcracks: microcracks are initiated predominantly by failure of Si-particles caused by dislocation pile-ups. This occurs at 'random' positions throughout the strained volume. The microcracks are oriented mainly along the lines of maximum shear stress.
- Formation of an initial crack: the initial crack starts to grow from the notch root; simultaneously, the microcrack density at some distance from the notch surface increases. This stage of damage evolution corresponds to the beginning of the descending branch on the force-displacement curve. The necking along the shear bands between the microcracks is identified as the mechanism controlling the coalescence of the microcracks.
- Crack growth: the direction of crack growth coincides initially with the shear band formed at the initial stage of deformation. Simultaneously with the growth of the first crack, the density of microcracks inside the specimen increases further. This is followed by the formation of the second large crack at some distance from the notch ground surface.
- Macrocrack coalescence: the large crack which starts from the notch surface joins with the second crack formed in the material due to the coalescence of microcracks far apart from the notch ground surface. The coalescence of these cracks leads to the formation of a large crack and finally to failure of the specimen.

The two macrocracks coalesce, although the microcrack distributions in front of each of them would direct them in different directions. The effect of the interaction of the cracks on their trajectories appears to be more powerful than the effect of the distributed microcracks and shear bands in front of the crack tips for both lamellar and globular alloys. Although the mechanisms of failure are the same in both alloys, they differ significantly in their overall mechanical response. Due to the more homogeneous distribution of plastic strain in the alloy with the globular microstructure, the coalescence of the microcracks is prolonged to larger total plastic strains and the overall response is significantly more ductile. Further alloy development should therefore aim at the reduction of matrix grain size as well as Si- particle size to achieve a more even distribution of the particles and of the plastic strain.

References

[1] Höner KE, Groß J (1992), Bruchverhalten und mechanische Eigenschaften von Aluminium-Silicum-Gußlegierungen in unterschiedlichen Behandlungszuständen. Gießereiforschung; 44(4), pp.146-60.

[2] Gurland J, Plateau J. (1963), Trans ASM; 56. pp. 442-54.

[3] Yeh JW, Liu WP. (1996), The cracking mechanism of silicon particles in A357 aluminium alloy. Met Mat Transactions A; 27A, pp. 3558-63.

[4] Tetelman AS, McEvily Jr AJ. (1967), Fracture of structural materials. New York: Wiley & Sons.

[5] Finkel VM. (1970), Physics of fracture. Moscow: Metallurgiya, (in Russian).

[6] Betekhtin VI, Vladimirov VI (1979), Kinetics of microfractures of crystallic bodies. In: Problems of strength and plasticity of solids. Leningrad: Nauka, pp. 142-54.

[7] Lemaitre J. (1992), A course on damage mechanics. Berlin: Springer.

[8] Knott JF. (1973), Fundamentals of fracture mechanics. London: Butterworths.

[9] Thomason PF. (1968), A theory for ductile fracture by internal necking of cavities. J Inst of Met; 96, pp. 360-5.

[10] Seidenfuss M. (1992), Untersuchungen zur Beschreibung des Versagenverhaltens mit Hilfe von Schädigungsmodellen am Bespiel des Werkstoffes 20 MnMoNi 55. Dissertation, MPA Stuttgart.

[11] Roberts W, Lehtinen B, Easterling KE (1976), An in situ SEM study of void development around inclusions in steel during plastic deformation. Acta Met Mat; 24, pp. 745-58.

[12] Hancock JW, Mackenzie AC. (1976), On the mechanisms of ductile fracture in high-strength steels subjected to multiaxial stress state. J Mech Phys Solids; 24, pp. 147-69.

[13] Melin S. (1983), Why do cracks avoid each other? Int J Fracture; 23, pp. 37-45.

[14] Yokobori T. (1968), An interdisciplinary approach to fracture and strength of solids. Groningen: Wolter-Noordhoff.

[15] Ebrahimi F, Seo HK (1996), Ductile crack initiation in steels. Acta Mater; 44(2), pp. 831-43.

[16] Lippmann N, Steinkopf Th, Schmauder S, Gumbsch P. (1997), 3D-finite element modelling of microstructures with the method of multiphase elements. Comp Mat Sci, pp. 28-35.

[17] Shtremel MA. (1982), Strength of alloys. 1. Defects of lattice. Moscow: Metallurgiya, (in Russian).

1.3 Micromechanisms of damage initiation and growth in tool steels[3]

The improvement of service properties of tool steels presents an important source of increasing the efficiency of metalworking industry. In order to develop a numerical model of damage or fracture in the steel, which should serve to predict the lifetime, or to improve the properties, one needs to know the mechanisms of damage and fracture in the steels [1-4]. The direct *in situ* observation of the fracture mechanisms of the steels under a microscope is quite difficult as compared with the case of more ductile materials, since the material fails abruptly. Then, not only qualitative parameters of fracture (like its mechanisms) but also quantitative ones (like critical damage parameters) are of interest.

The purpose of this work was to study the mechanisms and conditions of damage initiation and growth in the tool steels both qualitatively and quantitatively. The work includes the following steps:

- Scanning electron microscopy (SEM) *in situ* experiments on 3-point bending of specimens with inclined notches.
- Finite element (FE) simulation of the deformation of the specimens on macro- and mesolevel, taking into account the real microstructure of the steels observed in the SEM -experiments.
- Numerical analysis of the effect of the arrangement of primary carbides in the tool steels on the fracture behavior.

1.3.1 Micromechanisms of damage initiation in tool steels

The mechanisms of local failure and critical values for failure of the constituents of the steel have been determined. The constitutive law and elastic constants of the steel constituents are already known from literature and from our previous investigations [3, 4, 6]. The analysis of the mechanisms of damage initiation in the tool steels includes SEM *in situ* experiments and FE simulation of the deformation of the specimens on macro- and mesolevel. The SEM *in situ* observation of the damage initiation seeks to clarify the micromechanisms of damage initiation, whereas the hierarchical finite element model (macro- and mesomodel) is applied to determine the failure conditions for steel constituents using the real loading conditions and real microstructures of the steel.

[3] Reprinted from L. Mishnaevsky Jr., N. Lippmann, S. Schmauder, "Micromechanisms and Modelling of Crack Initiation and Growth in Tool Steels: Role of Primary Carbides", Z. Metallkunde 94, pp. 676-681 (2003) with kind permission of Carl Hanser Verlag

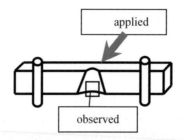

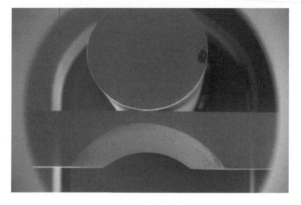

Fig. 1.21 3-point bending specimen: (a) scheme and (b) side view of loading device.

In order to clarify the mechanisms of damage initiation and growth in the steels, a series of SEM *in situ* experiments was carried out. 3-point bending specimens with an inclined notch, as described in [5], were used in these tests. These specimens allow observing the micro- and mesoprocesses of local deformation and failure of carbides and the matrix of steels during loading of macroscopic specimens in the SEM. The shape of the specimens is shown schematically in Fig. 1.21a. A photograph of the specimen under loading is given in Fig. 1.21b. The advantage of the specimen with the inclined notch is that the most probable location of first microcrack initiation in the specimen notch can be simply predicted (which is not the case in the conventional 3-point bending specimens). Therefore, one can observe this location with high magnification during loading and identify very exactly the load and the point in time at which the first microcracks form. Specimens made from the cold work steel.

X155CrVMo12-1 (in further text denoted as KA) and the high speed steel HS6-5-2 (denoted as HS) have been used. In the experiments, the specimens with different orientations of primary carbide layers were studied. Since the tool steels are produced in the form of round samples and because they were subject to hot reduction after austenitization and quenching, they are anisotropic: the carbide layers are oriented typically along the axis of the cylinder (this is the direction of hot reduction). Therefore, the following designation of the specimen orientation was used: L - the direction along the carbide layers, R - radial direction in the work-piece, C - the direction along the workpiece axis. In the experiments, specimens with orientations CL (the specimen is oriented along the carbide layers; the ob-served area is oriented along to the ingot axis), LC (the specimen is oriented along the round ingot axis) and CR (the specimen is oriented along the carbide layers; the observed area is oriented normally to the ingot axis) have been used. The specimens have been subjected to the heat treatment (hardening at 1070°C in vac-uum and tempering 2 h at 510 0c), and then polished with the use of the diamond pastes till the roughness R_z of the surface of the specimens does not exceed 3μm. The notch region of the specimens was etched with 3 and 10% HN03 until the carbides were clearly seen on the surface.

Table 1.4
Critical forces in the tests.

Type of the specimen	Force at which a first mi-crocrack was observed in the specimen (N)	Force at which the specimen failed, (N)
KALC	95, 52, 37.5	155, 85,160
KACR	50, 55, 37.5	95, 95, 70
HSCR	45, 50, 50	95, 80, 95
HSLC	50, 72.52, 127	200, 190, 195

The force-displacement curves were recorded during the tests. The loading was carried out in small steps, with a rate of loading of about 1 mm/s. The places in the specimen notch where microcrack initiation was expected have been observed through SEM during the tests. It was observed that the first microcracks formed only in the primary carbides, and not in the "matrix" of the steel. Also, no micro-crack along the carbide/matrix interface was observed in the tests. The forces at which the failure of primary carbides was observed in each specimen are given in Table 1.4. Fig. 1.22 shows SEM micrographs of typical primary carbide in the notch region of steels before and after failure. Since the picture were taken frontal and the specimens used were with the inclined notch, the magnifications of micro-graphs in x- and y-directions in Fig. 1.22 are different. In some carbides, multiple cracking was observed as well (see Fig. 1.23).

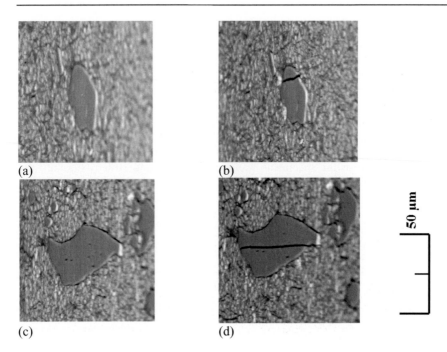

(a)   (b)

(c)   (d)

50 μm

Fig 1.22 Carbide grains before (a, c) and after failure (b, d). (Area size 40 μm ×100 μm).

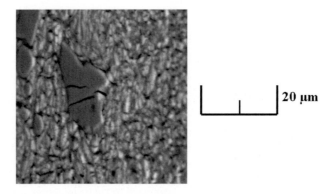

20 μm

Fig. 1.23 Multiple cracks in a primary carbide. (Area size 40 μm ×100 μm)

Generally, the course of failure of the specimens was as follows:
1. Formation of a microcrack at some carbide.
2. Formation of several microcracks at many carbides at different locations of the observed area (in so doing, the microcracks are formed rather at larger carbides at some distance from the boundary of the specimen, than in more strained macroscopically areas in the vicinity of the lower

boundary of the specimen; the local fluctuations of stresses caused by the carbides have evidently much more influence on microcracking than the macroscopic stress field).

3. After the failure of many carbides, the microcracks (or plastic zones in front of the microcracks) begin to grow into the matrix; just after this occurs, the specimens fail.

The failure of many carbides was observed just before the specimens failed. Fig. 1.24 shows a segment of the loaded area with many failed carbides.

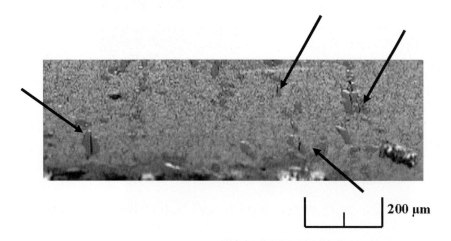

200 μm

Fig. 1.24 Microcracks (shown by arrows) in primary carbides.

1.3.2 Condition of failure of primary carbides in tool steels

To simulate the deformation of 3-point bending specimens with inclined notch, a three-dimensional (3D) FE model of the specimen was developed. The forces measured in the tests described above were applied in the simulations. The displacements from the boundary nodes of elements which are located in the vicinity of the symmetry plane and at the lower notch boundary (Fig. 1.21) are used as boundary conditions in the micromechanical simulation of carbide failure.

Then, the two-dimensional (2D) micromechanical simulations of carbide failure have been carried out for each microstructure and each load, measured in the experiments: a 2D model was created, which represents the cut-out at the notch region of the specimen. The real structure region of the micromodel contains 5000 elements of the plane strain type TRIP 6 and size 100 μm x 100 μm and is placed in the lower left corner of the macroscopic 3-point bending model, where the carbide was observed experimentally. As boundary conditions the displacements from the model of deformation of 3-point bending specimen were taken. Since the mesh density in the 2D case is higher, the calculated displacements have been linearly interpolated between the points which were available in the 3D simulation. The micromechanical simulation was performed with the use of the multiphase element method [8-11]. The micrograph of the carbide, obtained in SEM *in situ* experiments was digitized and then automatically imposed on the region of the real structure. The micrographs to be digitized were chosen in such a way that they were representative enough for the given materials. Due to the inclined notch surface, the micrographs in Fig. 1.22 have different scales in x- and y-directions. To take that into account, the micrographs were scaled with the use of the image analysis software XView accordingly to their scales in both directions. The properties of carbide and matrix are as follows [3 - 6] : (cold work steel) Young's modulus $E_c = 276$ GPa, $E_M = 232$ GPa, constitutive law of the matrix: $s_y = 1195 + 1390$ [l-exp (-εpl/0.0099)]; (high speed steels) $E_c = 286$ GPa, $E_M = 231$ GPa, constitutive law of the matrix: $s_y = 1500 + 471$ [l-exp(-εpl /0.0073)], Poisson's ratio - 0.19 (carbides) and 0.3 (matrix).

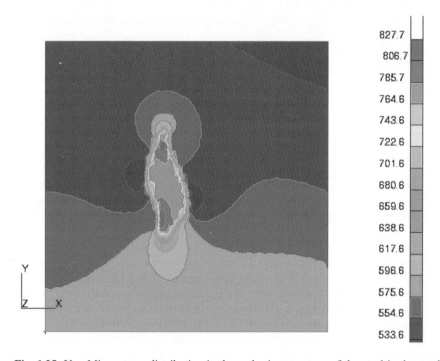

Fig. 1.25 Von Mises stress distribution in the real microstructure of the steel in the notch of the specimen (the boundary conditions in the 2D micromodel were taken from the 3D macromodel of the 3-point bending specimen).

Fig. 1.25 gives the distribution of von Mises stress in the real microstructure of the cold work steel at the loads at which the carbide failed. Supposing that failure of the carbides is determined by the action of maximal normal stresses, one obtains the failure stresses of carbides for different steels and orientations (see table 1.5).

Table 1.5
Calculated failure stresses of primary carbides in tool steels

Type of the steel	KALC	KACR	HSCR	HSLC
Failure stress of carbides (MPa)	1826	1840	1604	2520

The initial microcracks in the steels are formed within primary carbides (i. e., not along the carbide/matrix interface and not in the matrix). For our further simulation, this means that we can use the multi phase element method. The main input data for the simulation (i. e., the carbide failure condition) was determined with the use of the combined SEM *in situ* and FE model approach.

Conclusions

The mechanisms of damage initiation and growth in tool steels were investigated and the role of primary carbides in damage and fracture of the steels was clarified.

The experimentally observed course of damage evolution in the steels was as follows: One microcrack appears in carbide, and then several microcracks appear in other carbides at different locations of the observed area. In so doing, the microcracks are formed rather at larger carbides at some distance from the boundary of the specimen, than in the macroscopically more strained areas in the vicinity of lower boundary of the specimen. The local fluctuations of stresses caused by the carbides have evidently much more influence on the microcracking than the macroscopic stress field). Finally, after the failure of many carbides, the microcracks (as well as plastic zones in front of the microcracks) begin to grow into the matrix; just after this occurs, the specimens fail. The initial microcracks in the steels are formed in primary carbides (i. e., not along the carbide/matrix interface and not in the matrix).

References

[1] Mishnaevsky Jr. L, Schmauder S (1999), Steels and Materials for Power Plants. In: P. Neumann et al. (Eds.), Proc. EUROMAT-99, Vol. 7, Wiley-VCH Verlag, Weinheim, p. 269.
[2] Berns H, Broeckrnann C, Weichert D (1996), Key Eng. Mater. 118119, p. 163.
[3] Gross – Wege A., Weichert D, Broeckmann C (1996), Comp. Mater. Sci. 5, p. 126.

[4] Lehmann A (1995), Diploma Thesis, TU Bergakademie Freiberg

[5] Lippmann N, Lehmann A,. Steinkopff Th, Spies H.-J. (1996), Comp. Mater. Sci. 7, p. 123.

[6] Lippmann N (1995), Dissertation, Freiberg.

[7] Le Calvez Chr, Ponsot A., Lichtenegger G.., Mishnaevsky Jr L., Schmauder S., Iturriza I., Rodriguez Ibabe M. (2001), Final Report RTD EU Project, Influence of Micromechanical Mechanisms of Strength and Deformation of Tool Steels under Static and Cyclic Load, Creusot Loire, France.

[8] Mishnaevsky Jr L.,. Lippmann N, Schmauder S. (2001), Proc. 10th Int. Conf. Fracture, Honolulu, HA, CD-ROM.

[9] Mishnaevsky Jr. L.,. Schmauder S (2001), Appl. Mech. Rev. 54, p. 49.

[10] Mishnaevsky Jr L.., Dong M., Hoenle S., Schmauder S. (1999), Comp. Mater. Sci. 16,, p. 133.

[11] Mishnaevsky Jr. L., Weber U., Lippmann N., Schmauder S. (2001), in: M. Cross, J.W. Evans, C. Bailey (Eds.), Computational Modelling of Materials, Minerals, and Metals Processing, Proc. TMS Conference, p. 673.

[12] Mishnaevsky Jr. L., Lippmann N., Schmauder S. (2001), in: ASME Int. Mechanical Engineering Congress and Exposition, New York, CD, Vol. 2.

[13] Mishnaevsky Jr. L. (1998), Damage and Fracture of Heterogeneous Materials, Balkema, Rotterdam.

[14] Berns H., Melander A.,. Weichert D,. Asnafi N, Broeckmann C., Gross- Weege A. (1998), Comp. Mater. Sci. II, p. 166.

Chapter 2: Micromechanical Simulation of Composites

The microstructure (fixed by features such as the grain size, the presence of particles, layers/coatings, impurities, and internal interfaces) frequently determines the properties (such as strength, toughness, and fracture energy) of advanced materials. Because the microstructure is determined during production (by parameters such as temperature, time, and pressure) and the properties are determined in application (e.g., implementation and life-time), the microstructure/property-relationship (Figure 2.1) has been the focus of many investigations in the recent past. This relationship especially requires knowledge of how to link the different length scales in modeling and characterizing these materials.

The linkage of modeling on the nanoscale (nm-length scale) with that on the macroscale (length scale of real specimens) is a current challenge in materials science. Despite cheaper and faster computer resources, it is still difficult to inflate atomistic models of crystals to the size of specimens and components. Special problems of crack propagation have been successfully treated by dedicated coupled atomistic continuum methods [1, 2]. Nevertheless, a general method to connect theoretical calculations on the nanoscale with the macroscale is not available. Instead, at present it seems more promising to restrict oneself to selected materials problems and to connect different length scales by qualified physical laws that describe corresponding materials behavior [3, 4]. The terminology of relevant length scales, together with referring physical phenomena and methods, is illustrated in Figure 2.2 [5]. These methods require knowledge on the atomistic level.

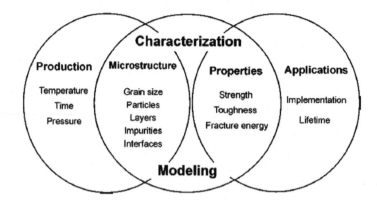

Fig 2.1 Microstructure/property-relationship.

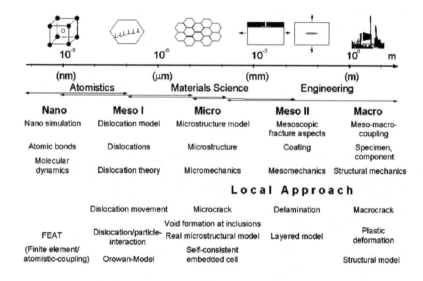

Fig 2.2 Numerical methods referring to different length scales.

Particulate-reinforced composites have become increasingly attractive in recent years for their high-strength and creep-resistant properties. These materials can be divided into three basic groups: (*a*) metal matrix composites (MMCs) such as Al reinforced with SiC-particles or with B-fibers, (*b*) brittle matrix composites (BMCs) such as Al_2O_3 toughened with Al, and (*c*) interpenetrating microstructural composites (IMCs) such as WC toughened with Co and others. Figure 2.3 depicts examples of fiber and particulate-reinforced composites. It should be noted that when continuum mechanical methods are applied, the relevant microstructural features such as particle sizes or phase areas have to be large enough (in the micrometer regime or above) in order to be modeled appropriately.

In section 2.1, the limit flow stresses for transverse loading of metal matrix composites reinforced with continuous fibers and for uniaxial loading of spherical particle reinforced metal matrix composites are investigated with the use of embedded cell models. A fiber of circular cross section or a spherical particle is surrounded by a metal matrix, which is again embedded in the composite material with the mechanical behavior to be determined iteratively in a self-consistent manner. Stress-strain curves have been calculated for a number of metal matrix composites with the embedded cell method and compared with literature data of a particle reinforced Ag-58vol.%Ni composite and for a transversely loaded uniaxially fiber reinforced Al-46vol.%B composite. Good agreement has been obtained between experiment and calculation and the embedded cell model is thus found to represent well metal matrix composites with randomly arranged inclusions.

Systematic studies of the mechanical behavior of fiber and particle reinforced composites with plane strain and axisymmetric embedded cell models are carried out to determine the influence of fiber or particle volume fraction and matrix strain-hardening ability on composite strengthening levels. Results for random inclusion arrangements obtained with self-consistent embedded cell models are compared with strengthening levels for regular inclusion arrangements from conventional unit cell models. It is found that with increasing inclusion volume fractions there exist pronounced differences in composite strengthening between all models.

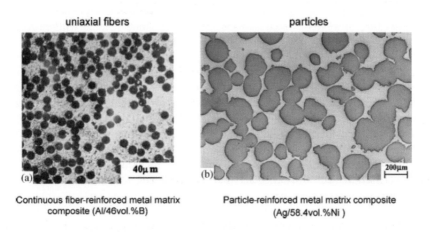

uniaxial fibers	particles

Continuous fiber-reinforced metal matrix composite (Al/46vol.%B)

Particle-reinforced metal matrix composite (Ag/58.4vol.%Ni)

Fig 2.3 Typical microstructures to be modelled in computational mechanics. (*a*) Al/B$_f$ fiber-reinforced MMC, (*b*) Ag/58.4vol.%Ni particle-reinforced MMC.

Finally, closed-form expressions are derived to predict composite strengthening levels for regular and random fiber or particle arrangements as a function of matrix hardening and particle volume fraction. The impact of the results on effectively designing technically relevant metal matrix composites reinforced by randomly arranged strong inclusions is emphasized.

In section 2.2, the 3D-multiphase finite element approach is presented. This approach permits finite-element-modelling of the plastic deformation of realistic 3D-microstructures. In contrast to conventional 'single phase elements' where the phase boundaries are simulated by element edges, the 'multiphase element' can be assigned to different materials when a phase boundary runs across it. The 3D-multiphase element is firstly applied to a simple test geometry. The efficiency of the 3D-multiphase element method is demonstrated by the analysis of a more complex 3D-microstructure. Finally, for a comparison of 2D- and 3D-simulations the stress distribution obtained in the 3D-calculation is compared with the results of a 2D-simulation of a representative intersection of the microstructure.

In section 3.3, different methods of automatic generation of 3D microstructural models of materials are discussed. A program for the automatic generation and the design of FE meshes for idealized 3D multiparticle unit cells (with spherical inclusions) is presented. Numerical testing of Al/SiC composites with random, regular, clustered and gradient arrangements of spherical particles is carried out. The fraction of failed particles and the tensile stress–strain curves were determined numerically for each of the microstructures. It is found that the strain hardening coefficient increases with varying the particle arrangement in the following order: gradient < random < clustered < regular microstructure. The variations of the particle sizes causes strong decrease in the strain hardening rate of the composite, and leads to a quicker and earlier damage growth in the composites.

Further, another approach to generate 3D FE models of composites based on the procedure of a step-by-step packing (SSP) of a finite volume with structural elements is discussed. This has been used to design the composite structure consisting of an Al(6061)-matrix with Al_2O_3-inclusions. A three-dimensional mechanical problem of the structure behaviour under tension has been solved numerically, using both an implicit finite-element method and an explicit finite-difference code. Special attention is given to the comparison of quasi-static and dynamic calculations. Evolution of plastic deformation in the matrix during tensile loading has been investigated. Qualitative and quantitative analysis of different components of stress and strain tensors is provided on the basis of mesomechanical concepts. Based on 3D-analyses, the conclusions regarding the approximations when considering deformation behaviour on meso and macro scale levels have been performed.

Finally in this chapter, yet another method for the reconstruction and generation of 3D microstructures of composites based on the voxel array data is presented. The geometry-based and voxel array based methods of reconstruction and generation of finite element models of 3D microstructures of composite materials are discussed and compared. With the use of the developed program, the deformation and damage evolution in composites with random and graded microstructures were numerically simulated. The tensile stress-strain curves, fraction of failed elements, and stress, strain and damage distributions at different stages of loading were determined for different random microstructures of the composites. It is shown that the stiffness, peak and yield stresses of a graded composite decrease with increasing the sharpness of the transition zone between the region of high volume content of the hard phase and the reinforcement free region. The critical applied strain, at which the intensive damage growth begins, is decreasing with increasing the volume content of the hard phase of the composite.

References

[1]   Kohlhoff S., Schmauder S. (1989), In Atomistic Simulation of Materials: Beyond Pair Potentials, ed. V Vitek, DJ Srolovitz, New York: Plenum, pp. 411–18.
[2]   Kohlhoff S., Gumbsch P., Fischmeister HF. (1991), Philos. Mag. A 64, pp 851–78.
[3]   Zohdi T., Oden J., Rodin G. (1996), Comp. Methods Appl. Mech. Eng. 138, pp. 273.
[4]   Thirteenth US National Congress of Applied Mechanics (US-NCAM), June 21– 26. (1998). University of Florida, Abstract Book, ISBN 0–9652609.
[5]   Kizler P., Uhlmann D., Schmauder S. (2000), Nucl. Eng. Design 196, pp. 175–83.

2.1 Embedded unit cells[1]

Metal matrix composites (MMCs) are defined as ductile matrix materials reinforced by brittle fibers or particles. Under external loading conditions the overall response of MMCs is elastic-plastic. MMCs are frequently reinforced by continuous fibers which are aligned in order to make use of the high axial fiber strength. However, the mechanical behavior of these composites under transverse loading is well behind their axial performance [1-7]. On the other hand, it is well known, both from experiment and calculations [8] that details of transverse strengthening in uniaxially fiber reinforced MMCs are a strong function of fiber arrangement. To derive the mechanical behavior of MMCs, a micromechanical approach is usually employed using cell models, representing regular inclusion arrangements. As regular fiber spacings are difficult to achieve in practice, most of the present fiber reinforced MMCs contain aligned but randomly arranged continuous fibers. Thus, the accurate modeling of the mechanical behavior of actual MMCs is very complicated in practice even if the fibers are aligned.

Initially, the transverse mechanical behavior of a unidirectionally continuous fiber-reinforced composite (A1-B) with fibers of circular cross section was studied by Adams [9] adopting finite element cell models under plane strain conditions: a simple geometrical cell composed of matrix and inclusion material is repeated by appropriate boundary conditions to represent a composite with a periodic microstructure. Good agreement was achieved between calculated and experimental stress-strain curves for a rectangular fiber arrangement. The influence of different regular fiber arrangements on the strength of transversely loaded boron fiber reinforced Al was analyzed in Refs [2, 5, 7]. It was found that the same square arrangement of fibers represents two extremes of strengthening: high strength levels are achieved if the composite is loaded in a 0° direction to nearest neighbors while the 45° loading direction is found to be very weak for the same fiber arrangement. .4 regular hexagonal fiber arrangement lies somewhere between these limits [2, 5, 7, 10, 11]. The transverse mechanical behavior of a realistic fiber reinforced composite containing about 30 randomly arranged fibers was found to be best described but significantly underestimated by the hexagonal fiber model [5]. One reason for the superiority of the hexagonal over the square arrangement in describing random fiber arrangements is due to the fact that the elastic stress invariants of the square arrangement agree only to first order with the invariants of the transversely isotropic material while the hexagonal arrangement agrees up to the second order [l0]. Dietrich [6] found a transversely isotropic square fiber reinforced Ag-Ni composite material using fibers of different diameters. A systematic study in which the fiber volume fraction and the fiber arrangement effects have been investigated was founded into a simple model in Ref. [11].

---

[1] Reprinted from M. Dong, S. Schmauder, "Modeling of Metal Matrix Composites by a Self-Consistent Embedded Cell Model", Acta Metall. Mater. 44, pp. 2465-2478 (1996) with kind permission from Elsevier.

The influence of fiber shape and clustering was numerically examined in some detail by Llorca et al. [12], Dietrich [6] and Sautter [13]. It was observed that facetted fiber cross sections lead to higher strengths compared to circular cross sections except for such fibers which possess predominantly facets oriented at 45° with respect to the loading axis in close agreement with findings in particle reinforced MMCs [14]. Thus, hindering of shear band formation within the matrix was found to be responsible for strengthening with respect to fiber arrangement and fiber shape [12]. In Refs [8, 11, 15, 16] local distributions of stresses and strains within the microstructure have also been identified to be strongly influenced by the arrangement of fibers. However, no agreement was found between the mechanical behavior of composites based on cell models with differently arranged fibers and experiments with randomly arranged fibers loaded in the transverse direction.

The overall mechanical behavior of a particle reinforced composite was studied with axisymmetric finite element cell models by Bao et al. [17] to represent a uniform particle distribution within an elastic-plastic matrix. Tvergaard [18] introduced a modified cylindrical unit cell containing one half of a single fiber to model the axial performance of a periodic square arrangement of staggered short fibers. Horn [19] and Weissenbek [20] used three dimensional finite elements to model different regular arrangements of short fibers and spherical as well as cylindrical particles with relatively small volume fractions ($f < 0.2$). It was generally found that the arrangement of fibers strongly influences the different overall behavior of the composites. When short fibers are arranged in a side-by-side manner, they constrain the plastic flow in the matrix and the computed stress-strain response of the composite in the fiber direction is stiffer than observed in experiments. If the fibers in the model are overlapping, strong plastic shearing can develop in the ligament between neighboring fibers and the predicted load carrying capacity of the composite is closer to the experimental measurements. In Ref. [21] the stress-strain curves based on FE-numerical solutions of axisymmetric unit cell models of MMCs are given in a closed form as a function of the most important control parameters, namely, volume fraction, aspect ratio and shape (cylindrical or spherical) of the reinforcement as well as the matrix hardening parameter.

One reason for the discrepancy between experiments and calculations based on simple cell models in the case of particle, whisker and fiber reinforced metals is believed to be the un-natural constraint governing the matrix material between inclusion and simulation cell border [5. 15, 17-19, 22-24] resulting in an unrealistic strength increase. The influence of thermal residual stresses in fiber reinforced MMCs under transverse tension was studied in Ref. [7] and found to lead to significant strengthening elevations in contrast to findings in particulate reinforced MMCs where strength reductions were calculated [25].

A limited study on the overall limit flow stress for composites with randomly oriented disk- or needleshaped particles arranged in a packet-like morphology is reported by Bao et al. [17]. In Refs [26, 27] a modified Oldroyd model has been proposed to investigate analytically-numerically the overall behavior of MMCs with randomly arranged brittle particles. Duva [28] introduced an analytical model to represent a random distribution of non-interacting rigid spherical particles perfectly bonded in a power law matrix. The Duva model is a self-consistent model and should be valid particularly in the dilute regime of volume fractions, $f < 0.2$.

In this work, cell models are applied to simulate, for a number of relevant parameters, the transverse behavior of MMCs containing fibers in a regular square or hexagonal arrangement as well as the mechanical behavior of MMCs containing particles in a regular arrangement. MMCs with randomly arranged inclusions are modeled by a recently introduced self-consistent procedure with embedded cell models. This method of surrounding a simulation cell by additional "equivalent composite material" was introduced in Ref. [6] for structures which are periodical in the loading direction and was recently extended to non-periodic two-dimensional [29-31] and three-dimensional composites [13, 27]. The method is known to remove the above described unrealistic constraints of cell models. An initial comparison of two- and three-dimensional embedded cell models in the case of perfectly-plastic matrix material depicts elevated strength levels for the three-dimensional case [27], similar to composites with regularly arranged fibers [11].

The purpose of the present paper is to investigate the mechanical behavior of MMCs reinforced with regular or random arranged continuous fibers under transverse loading, as well as the mechanical behavior of MMCs reinforced with regular or random arranged particles under uniaxial loading, and to systematically study composite strengthening as a function of inclusion volume fraction and matrix hardening ability. The finite element method (FEM) is employed within the framework of continuum mechanics to carry out the calculations.

2.1.1 Model Formulation

A continuum mechanics approach is used to model the composite behavior. The inclusion behaves elastically in all cases considered here and its stiffness is much higher than that of the matrix, so that the inclusion can be regarded as being rigid. In addition, the continuous fibers of circular cross section and spherical particles are assumed to be well bonded to the matrix so that debonding or sliding at the inclusion-matrix interface is not permitted. The uniaxial matrix stress-strain behavior is described by a Ramberg-Osgood type of power law

$$\sigma = E\varepsilon \qquad \varepsilon \leq \varepsilon$$

$$\sigma = \sigma_0 \left[\frac{\varepsilon}{\varepsilon_0} \right]^n \qquad \varepsilon > \varepsilon_0 \qquad (2.1)$$

where σ and ε are the uniaxial stress and strain of the matrix, respectively, σ_0. is the tensile flow stress, the matrix yield strain is given as , $\varepsilon_0 = \sigma_0/E$, E is Young's modulus, and $N = 1/n$ is the strain hardening exponent. Thus, $N = 0$ corresponds to a non-hardening matrix.

J_2 flow theory of plasticity with isotropic hardening is employed with a von Mises yield criterion to characterize the rate-independent matrix material. The von Mises equivalent stress and strain are given as:

$$\sigma_v = \sqrt{3J_2} = \sqrt{\frac{3}{2} s_{ij} s_{ij}}$$

$$\varepsilon_v = \frac{1}{1+v} \sqrt{\frac{3}{2} e_{ij} e_{ij}}$$

$$(2.2)$$

where $s_{ij} = \sigma_{ij} - \sigma_{kk}/3$, $e_{ij} = \varepsilon_{ij} - \varepsilon_{kk}/3$ and v is Poisson's ratio. In the analytical approach, the metal matrix is considered incompressible, so Poisson's ratio of the matrix will become 0.5 after reaching the yield stress. However, in reality Poisson's ratio of the composite remains below 0.5 as the matrix starts yielding. It changes from the elastic value v to the limit value 0.5 with increasing yielding zone in the matrix.

The Ramberg-Osgood type of matrix power law hardening is assumed to be valid for the matrix described in terms of von Mises equivalent stress and strain

$$\sigma_v = \sigma_{v0} \left[\frac{\varepsilon_v}{\varepsilon_{v0}} \right]^N$$

$$(2.3)$$

with the following relations between stress and strain under uniaxial loading and von Mises equivalent stress and strain.

(a) In the case of a two-dimensional (2D) plane strain condition for continuous fiber reinforced composites ($\sqrt{3}/2 \approx 0.866$, $2/\sqrt{3} \approx 1.1547$)

$$\sigma = \frac{\sqrt{3}}{2} \sigma_v \quad \text{with} \quad \sigma_0 = \frac{\sqrt{3}}{2} \sigma_{vo}$$

$$\varepsilon = \frac{2}{\sqrt{3}} \varepsilon_v \quad \text{with} \quad \varepsilon_0 = \frac{2}{\sqrt{3}} \varepsilon_{v0}$$

$$(2.4)$$

(b) In the case of three-dimensional (3D) axisymmetrical condition for particle reinforced composites

$$\sigma = \sigma_v \quad \text{with} \quad \sigma_0 = \sigma_{v0}$$

$$\varepsilon = \varepsilon_v \quad \text{with} \quad \varepsilon_0 = \varepsilon_{v0}$$

$$(2.5)$$

The global mechanical response of the composite under external loading is characterized by the overall stress $\overline{\sigma}$ as a function of the overall strain $\overline{\varepsilon}$. Moreover, to describe the results in a consistent way, the reference axial yield stress o. and yield strain a, of the matrix, as defined in equation (2.4) for the 2D case and in equation (2.5) for the 3D case, will be taken to normalize the overall stress and strain of the composite, respectively.

Following Bao *et al.* [17] the composite containing hard inclusions will necessarily harden with the same strain hardening exponent, *N*, as the matrix for the case of hard inclusions, when strains are in the regime of fully developed plastic flow. At sufficiently large strains the composite behavior is then described by

$$\overline{\sigma} = \overline{\sigma}_N \left[\frac{\overline{\varepsilon}}{\varepsilon_0} \right]^N \tag{2.6}$$

where $\overline{\sigma}_N$ is called the asymptotic reference stress of the composite which can be determined by normalizing the composite stress by the stress in the matrix at the same overall strain ε [equation (2.2)], as indicated in Fig. 2.4:

Composite with a Strain–hardening Matrix (N>0.0)

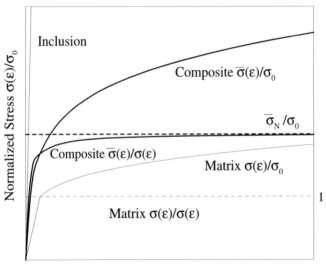

Fig 2.4 Composite strengthening.

$$\sigma_N = \sigma_0 \left[\frac{\overline{\sigma(\varepsilon)}}{\overline{\sigma(\varepsilon)}} \right] \qquad \text{for} \quad \overline{\varepsilon} \gg \varepsilon_0 \tag{2.7}$$

For a matrix of strain hardening capability N, the limit value $\overline{\sigma}_N / \sigma_0$ is defined as the composite strengthening level, which is an important value to describe the mechanical behavior of composites and which depends only on fiber and particle arrangement, inclusion volume fraction and matrix strain-hardening exponent.

Unit cell models

Before introducing the self-consistent embedded cell model, two regular aligned continuous fiber arrangements, namely square and hexagonal arrangements, and two regular particle arrangements, namely primitive cubic and hexagonal arrangements, are at first considered, as shown in Figs 2.20(a)-(d).

It is well known that the repeating unit cells 1 (or 3) and 2 in Fig. 2.20(a) can be extracted from the regular array of uniform continuous fibers to model exactly the composite with square fiber arrangement under 0° and 45° transverse loading, respectively, whereas the repeating unit cells 6 and 7 (or 8) can be taken from Fig. 2.20(b) to model exactly a hexagonal fiber arrangement under 0° and 30° transverse loading, respectively, if the appropriate boundary conditions are introduced making use of the symmetry conditions. Moreover, the unit cells 4, 9 and 10 can also represent geometrically regular arrangements.

 For further simplification the modified unit cells 5. 11 and 12 (circular and elliptically shaped cell models) may be derived from the cells 3. 4. 8, 9 and 10. An ellipsoidal unit cell has been used previously in Ref. [32] and shown to possess very complicated boundary conditions. As can be seen later these unit cells can be employed, however, in the embedding method to model the mechanical behavior of composites with random fiber arrangement.

The results of the unit cell models (as illustrated in Fig. 2.10(b) as an example for three different regular fiber arrangements with non-hardening matrix) show that the composite strengthening levels are quite different, especially at high volume fractions of fibers, with the composite strengthening for square arrangements under 0° loading being highest and for square arrangements under 45° loading being lowest. Further results regarding these three different regular fiber arrangements using unit cell models are given in Ref. [11]. A comparison of the stress-strain curves for the composite Al 46 vol.% B [in Fig. 2.8(a)] shows that the stress-strain curve from random fiber packing given in Ref. [5] lies between the curves from square unit cell modeling under 0°. loading and hexagonal unit cell modeling. The curves from square unit cell modeling under 45°. loading are even lower. As mentioned above, the strength of composites with randomly arranged inclusions cannot be described by modeling regular inclusion arrangements.

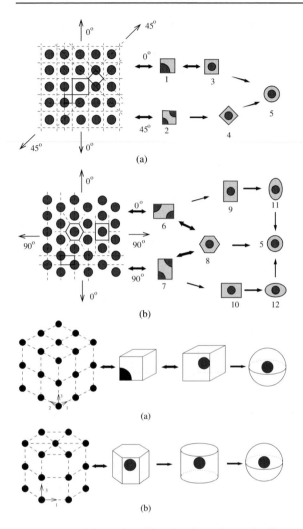

Fig 2.5 Modeling of unidirectional continuously fiber reinforced composites with: (a) square and (b) hexagonal fiber arrangements under transverse loading conditions and particle reinforced composites with: (c) primitive cubic and (d) hexagonal particle arrangements.

Rather, a new model must be employed, which describes approximately the geometry and the mechanical behavior of real composites with randomly arranged inclusions, if models with many inclusions have to be avoided for modeling and computational reasons. For particle reinforced composites, the primitive cubic particle arrangements can be modeled exactly by a conventional cell model with appropriate symmetry and boundary conditions [Fig. 2.20(c)], however. the hexagonal arrangement has to be simulated by approximate cell models [Fig. 2.20(d)]

[17]. The primitive cubic and axisymmetric unit cells have been used in this paper to model two representatives of regular particle arrangements. Further simplifications for 3D modeling are spherical unit cells shown in Figs 2.20(c) and (d) which will be employed, however, for embedded cell modeling.

Embedded cell models

In the present work, 2D and 3D self-consistent embedded cell models will be applied to model the mechanical behavior of composites with random continuous fiber and particle arrangements. Figure 3(a) describes schematically a typical plane strain (2D) embedded cell model with a volume fraction of $f = (d/D)^2$ or axisymmetric (3D) embedded cell model with a volume fraction of $f = (d/D)^3$, where instead of using fixed or symmetry boundary conditions around the fiber-matrix or particle-matrix cell, the inclusion-matrix cell is rather embedded in an equivalent composite material with the mechanical behavior to be determined iteratively in a self-consistent manner.

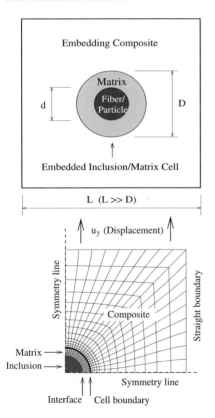

Fig 2.6 (a) Embedded cell model and (b) finite element mesh for an embedded cell model.

If the dimension of the embedding composite is sufficiently large compared with that of the embedded cell, e.g. $L/D = 5$ as used in this paper, the external geometry boundary conditions introduced around the embedding composite are almost without influence on the composite behavior of the inner embedded cell. Indeed, there exists no difference in the calculated results whether the vertical surfaces are kept unconstrained or remain straight during uniaxial loading.

To investigate the influence of the geometrical shape of the matrix phase on the overall behavior of the composite, different shapes of cross section of the embedded cell were chosen with a circular shaped continuous fiber surrounded by a circular, square, elliptical or rectangular shaped metal matrix (Fig. 2.9). A typical FE mesh and corresponding symmetry and boundary conditions are given in Fig. 2.6(b), where a circular fiber or a spherical particle is surrounded by a circular (for 2D) or spherical (for 3D) shaped metal matrix, which is again embedded in the composite material with the mechanical behavior to be determined.

Iterative modelling procedure

Under axial displacement loading at the external boundary of the embedding composite (Fig. 2.6) the overall response of the inner embedded cell can be obtained by averaging the stresses and strains in the embedded cell or alternatively the reaction forces and displacements at the boundary between the embedded cell and the surrounding volume.

The embedding method is a self-consistent procedure, which requires several iterations as shown in Fig. 2.22.

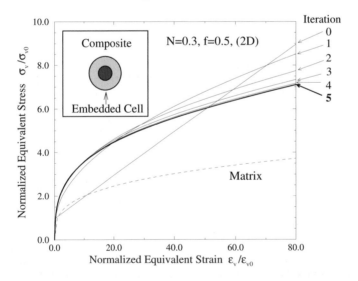

Fig 2.7 Iterative modeling procedure: stress-strain curves for different iteration steps.

An initially assumed stress-strain curve (iteration 0 in Fig. 2.22) is first assigned to the embedding composite, in order to perform the first iteration step. An improved stress-strain curve of the composite (iteration 1) will be obtained by analyzing the average mechanical response of the embedded cell. This procedure is repeated until the calculated stress-strain curve from the embedded cell is almost identical to that from the previous iteration. The convergence of the iteration occurs typically at the fifth iteration step, as illustrated in Fig. 2.22

It has been found from systematic studies that convergence of the iteration to the final stress-strain curve of the composite is independent of the initial mechanical behavior of the embedding composite (iteration 0). From an arbitrary initial stress-strain curve of the embedding composite the required composite response can be reached after 4-5 iterations for all cases.

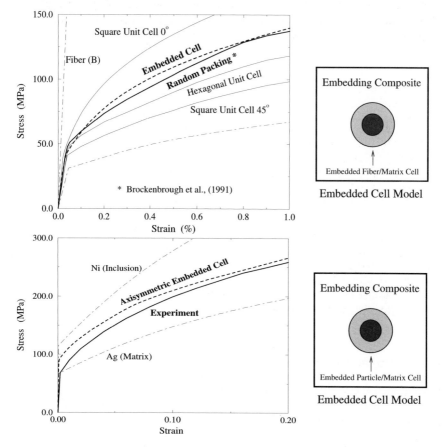

Fig. 2.8 Comparison of the mechanical behavior (a) of an Al-46vol.%B fiber reinforced composite ($N = 1/3$, $f = 0.46$) under transverse loading from different models; and (b) of a Ag-58 vol.%Ni particulate composite from embedded ceil model and experiment.

The LARSTRAN finite element program [33] was employed using 8-noded plane strain elements (for 2D) as well as axisymmetric biquadrilateral elements (for 3D) generated with the help of the pre- and postprocessing program PATRAN [34]. A DEC-Alpha work station 3000/300L was used to carry out the calculations, which typically took 30-50 min to obtain a stress-strain curve with 100-150 loading steps in one iteration loop.

Comparison with experiments

An example of the composite Al 46vol.%B with random fiber packing taken from Ref. [5] has been selected to verify the embedded cell model. This MMC is a 6061-O aluminum alloy reinforced with unidirectional cylindrical boron fibers of 46% volume fraction. The room temperature elastic properties of the fibers are Young's modulus, $E^{(B)} = 410$ GPa, and Poisson's ratio, $v^{(B)} = 0.2$. The experimentally determined mechanical properties of the 6061-O aluminum matrix are: Young's modulus, $E^{(Al)} = 69$ GPa, Poisson's ratio, $v^{(Al)} = 0.33$, 0.2% offset tensile yield strength, $\sigma_0 = 43$ MPa, and strain-hardening exponent, $N = 1/n = 1/3$.

Figure 5(a) shows a comparison of the stress-strain curves of the composite Al-46 vol.% B under transverse loading from simulations of a real microstructure together with results from different cell models. The stress-strain curve from the embedded cell model employed in this paper shows good agreement with that from the calculated random fiber packing in the elastic and plastic regime, which lies between the curves from square unit cell modeling under 0° loading and hexagonal unit cell modeling.

Furthermore, the stress-strain curve from another experiment [27] on the composite Ag-58vol.% Ni with random particle arrangement (Young's modulus, $E^{(Ni)} = 199.5$ GPa, $E^{(Ag)} = 82.7$ GPa, Poisson's ratio, $v^{(Ni)} = 0.312$, $v^{(Ag)} = 0.367$, and yield strength, $\sigma^{(Ni)} = 193$ MPa, $\sigma^{(Ag)} = 64$ MPa) has been compared in Fig. 2.8(b) with that from the self-consistent embedded cell model. Good agreement in the regime of plastic response is obtained, although the Ni-particles in the experiment were not perfectly spherical.

These results indicate that the embedded cell model can be used to successfully simulate composites with random inclusion arrangements and to predict the elastic-plastic composite behavior.

Geometrical shape of embedded cell

As mentioned above, different shapes of cross section of the embedded cell model with a circular shaped fiber, as shown in Fig. 2.9(a), are also taken into account to investigate the influence of the geometrical shape of the embedded cells on the overall behavior of the composite. The stress-strain curves of all embedded cell models with different geometrical shapes are plotted in Fig. 2.9(b). With an exception of the square 45° embedded cell model the stress-strain curves are very close for all embedded cell shapes, namely, square-0°, circular, rectangular-0°, rectangular-90°, elliptic-0° and elliptic-90°.

From the calculated results of the embedded cell models localized flows have been found around the hard fiber with preferred yielding at 45°. Because of the special geometry of the square 45° embedded cell model with the cell boundary parallel to the preferred yielding at 45°, the overall stresses of the composite with such a geometrical cell shape are therefore reduced, such that a relative lower stress-strain curve has been obtained from the modeling.

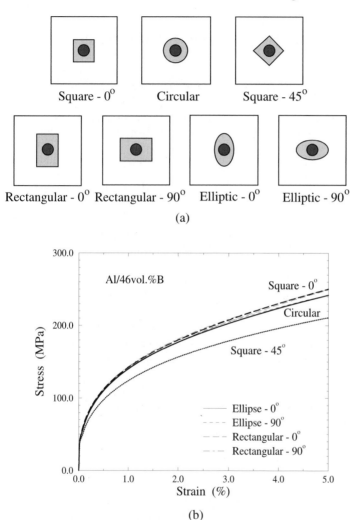

(a)

(b)

Fig 2.9 Embedded cell models: influence of (a) different matrix shapes on (b) stress-strain curves for an Al-46vol.%B ($N = 1/3, f = 0.46$) composite.

The almost identical responses of all other embedded cell models indicate that, besides the special shape of matrix with 45° cell boundaries, the predicted mechanical behavior of fiber reinforced composites under transverse loading is independent of the modeling shape of the embedded composite cell. That allows us to employ any embedded cell shape to model the mechanical behavior of the composite with random fiber arrangement. In the following systematic studies the circular shaped embedded cell model with circular cross section of rigid fiber will thus be taken to predict the general transverse mechanical behavior of the composite with random fiber arrangement. In the same way, a spherical shaped embedded cell model containing a spherical particle will be used to predict the axial mechanical behavior of composites with random particle arrangement.

2.1.2 Systematic studies with self-consistent embedded cell models

Composite strengthening is dependent on fiber and particle arrangement, inclusion volume fraction and the matrix strain-hardening exponent. The effect of fiber and particle arrangement is best demonstrated by considering different cell models. The effect of the inclusion volume fraction has been taken into account by applying different ratios of the circular (2D) and spherical (3D) matrix and the inclusion in the embedded cell model.

The effect of the matrix strain hardening exponent has been investigated by changing the parameter, N, of the material hardening law for the matrix in equation (2.1). Some results from the systematic studies with embedded cell models are given in Figs 2.10-2.13.

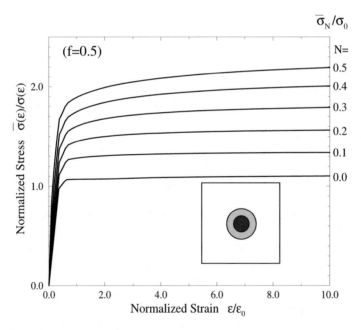

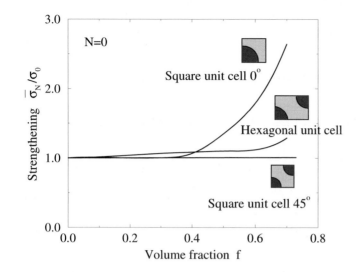

Fig 2.10 (a) Normalized stress-strain curves from embedded cell models (2D) for different N-values (f = 0.5) and (b) composite strengthening from different cell models (2D, N = 0).

The influence of matrix strain-hardening is shown in Figs 2.10(a) and 2.11(a). The predicted transverse (2D) and axial (3D) overall stress-strain curves (normalized by yield stress and yield strain of the matrix, respectively) for the case of strain-hardening exponents between $N = 0.0$ and 0.5 are depicted in Fig. 2.10(a) (2D) for a fiber volume fraction off $f = 0.5$ and in Fig. 2.11(a) (3D) for a particle volume fraction of $f = 0.4$. At sufficiently large strains (e.g. at E = lot,) the normalized overall stresses approach constant values, i.e. composite strengthening levels, as illustrated on the right-hand side of Figs 2.10(b) and 2.11(b) for $\bar{\varepsilon} = 10\varepsilon_0$. The strength of the composite is seen to increase with N and similar trends of nearly linear increase with N are found for all particle volume fractions, f [Figs 2.11(a) and 2.13].

The dependence of composite strengthening on inclusion volume fraction obtained by the self-consistent embedded cell modelling is shown in Fig. 2.10(b) for a non-hardening matrix with continuous fiber reinforcement and in Fig. 2.11(b) for a strain-hardening matrix ($N = 0.2$) with particle reinforcement. For comparison, the corresponding values of composite strengthening for regular fiber and particle arrangements taken from unit cell modelling and from two approximate models, namely Duva's model [28] and the modified Oldroyd model [26,27] are also drawn as a function of fiber and particle volume fraction in these figures, respectively. A detailed discussion of this comparison is given in Ref. [27].

Duva's model is a self-consistent model and valid particularly in the dilute regime, $f < 0.2$. Composite strengthening by the Duva model is given as Composite strengthening levels of the modified Oldroyd model are taken from Ref. [27].

$$\frac{\overline{\sigma}_N}{\sigma_0} = (1 - f)^{-(2.11N + 0.39)} \tag{2.8}$$

A summary of the dependence of composite strengthening on inclusion volume fraction and matrix strain-hardening for randomly arranged continuous fibers and particles is depicted in Figs 2.12 and 2.13(a) for matrix strain-hardening exponents $N = 0.0$-0.5, and compared with predictions from 2D and 3D unit cell models. In addition, a comparison with predictions from Duva's model and the modified Oldroyd model for the case of particle reinforced composites is shown in Fig. 2.13(b).

For continuous fiber reinforced composites with a non-hardening matrix the hexagonal arrangement provides slightly higher transverse composite strengthening over the square arrangement at volume fractions $f < 0.5$, whereas the random arrangement from the embedded cell model supplies an intermediate value between the two regular arrangements except at volume fraction $0.38 < f < 0.5$ (Fig. 2.12). At volume fractions $0.38 < f < 0.5$, the random arrangement from the embedded cell model possesses the lowest composite strengthening level. On the contrary, the composite strengthening of the square arrangement is higher than that of the hexagonal arrangement for volume fractions $f > 0.4$. For the composite with a small exponent of strain-hardening matrix, $N < 0.2$, similar relations among the three arrangements were found, with the cross points of any two curves being closer to $f = 0$ with increasing N. For composites with a higher strain-hardening exponent of the matrix, $N > 0.2$, the strengthening values are almost the same for the three arrangements considered at volume fractions of $f < 0.2$. At volume fractions of $f > 0.2$, the square arrangement provides again the highest strengthening and the hexagonal arrangement the lowest strengthening. A comparison of composite strengthening in Fig. 2.12 for three continuous uniaxial fiber arrangements can be summarized as follows: in the case of low strain-hardening of the matrix, composites with hexagonal fiber arrangement behave stronger than those with random fiber arrangement, which behave, however, at low volume fractions stronger than those with random arrangement. With increasing strain-hardening ability of the matrix, the composites of all three arrangements behave similarly at low fiber volume fractions. At higher inclusion volume fractions the difference in composite strengthening is becoming larger, with the square arrangement possessing the highest, the random arrangement the intermediate and the hexagonal arrangement the lowest level.

The stress and strain distributions in a typical embedded cell model (2D) are shown in Fig. 2.14, In the embedded cell the localized flow stresses with preferred yielding in the 45° direction are seen apparently around the hard fiber and extending into the embedding composite in the vicinity of the embedded cell.

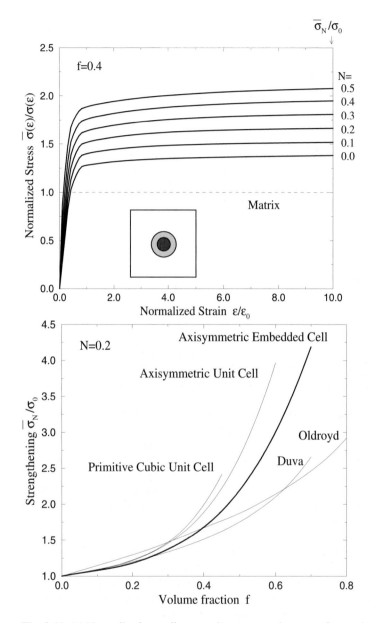

Fig. 2.11 (a) Normalized overall composite stress-strain curves from axisymmetric embedded cell models (3D) for different N-values ($f = 0.4$) and (b) composite strengthening for matrix strain-hardening $N = 0.2$ from different cell models (3D).

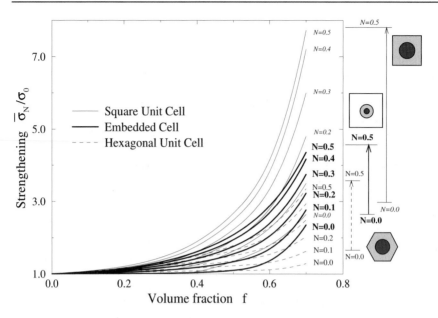

Fig 2.12 Comparison of composite strengthening by continuously aligned fibers from different 2D cell models: embedded cell model, square and hexagonal unit cell model.

It characterizes the local response of the composite under transverse loading. However, the stresses and strains are homogeneously distributed in the embedding composite far from the embedded cell. This represents the overall behavior of the corresponding composite. In the case of random fiber arrangement, the situation around every single fiber is, of course, more complicated. However, the real situation can be considered in average as a fiber-matrix cell surrounded by corresponding equivalent composite material, which represents the average overall composite behavior.

From comparisons for particle reinforced composites in Figs 2.28(a) and (b) it can be seen that the composite strengthening levels from the primitive cubic, the axisymmetric unit cell models and the self-consistent axisymmetric embedded cell models are very close at low particle volume fractions, f, and low matrix strain-hardening exponents, N. With increasing particle volume fraction, f and matrix strain-hardening exponent, N, the strengthening level of the composites increases for all the models considered. However, the primitive cubic unit cell provides the highest composite strength, while the self-consistent axisymmetric embedded cell predicts the softest composite response among these three models. This effect can be explained by considering the constraint of neighbouring particles through the necessary strict boundary conditions of the models with regularly arranged particles. As the particle volume fraction increases, the distance between the particles decreases until they touch at a volume fraction of $f = 0.5236$ for the primitive cubic cell model (primitive cubic particle arrangement), and a volume fraction of $f =$

2/3 for the axisymmetric cell model (hexagonal particle arrangement). In the self-consistent axisymmetric embedded cell model the particles remain surrounded by the matrix up to an extreme volume fraction of $f = 1$, as seen in Fig. 2.6. The equivalent composite surrounding the embedded inclusion-matrix cell provides thus less constraint on the inclusion-matrix cell compared to conventional unit cell models of the same volume fraction. In most practical relevant composites, the inclusions are randomly arranged and on average, no such restricting constraints exist as in the regular primitive cubic and hexagonal arrangements. For this reason, the self-consistent axisymmetric embedded cell model is believed to be the most realistic approximation to the geometry of real composites containing randomly arranged spherical particles.

At low particle volume fractions ($f < 0.2$) and low matrix strain-hardening ability ($N < 0.2$) the composite strengthening levels from the present selfconsistent axisymmetric embedded cell model and the Duva model are very similar. With increasing particle volume fraction, f, the differences in composite strengthening between these two models are significant for low matrix strain-hardening ability, however, for $N \sim 0.5$ the two strength predictions are comparable. The modified Oldroyd model provides a very narrow distribution of composite strengthening with respect to N of matrix strain-hardening and an almost linear strength dependence on particle volume fraction, f. Good agreement between the self-consistent axisymmetric embedded cell model and the modified Oldroyd model exists only at low volume fractions and high strain-hardening rates ($N \sim 0.5$) of the matrix.

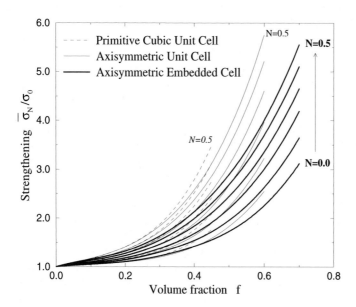

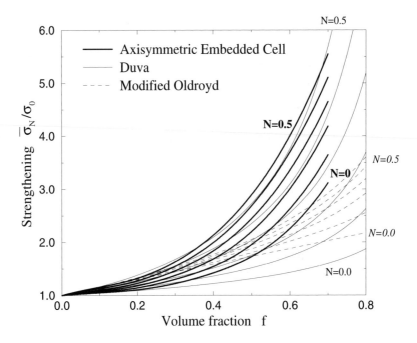

Fig 2.13 Comparison of composite strengthening by spherical particles from different 3D cell models: (a) embedded cell model, primitive cubic and axisymmetric unit cell model and (b) embedded ceil model, Duva model and Oldroyd model.

From above comparisons between the models it is clear that at low particle volume fractions ($f < 20\%$) the strengthening of metal matrix composites reinforced with randomly or regularly arranged particles can be predicted with either the axisymmetric unit cell model, the embedded cell model or the simple Duva model. However, for higher particle volume fractions the self-consistent embedded cell model should be applied to obtain the overall mechanical response of technically relevant composites reinforced by randomly arranged spherical particles

Strengthening Model

The strength of MMCs reinforced by hard inclusions under external mechanical loading has been shown to increase with inclusion volume fraction and strain-hardening ability of the matrix for all inclusion arrangements investigated. From the presented numerical predictions a strengthening model for aligned continuous fiber reinforced MMCs with random, square ($0°$) and hexagonal arrangements, as well as for spherical particle reinforced MMCs with random, primitive cubic and hexagonal arrangements, can be derived as a function of the inclusion volume fraction f, and the strain-hardening exponent, N, of the matrix [31]:

$$\overline{\sigma}_N = \sigma_0\left[(1 - \frac{f}{c_1(2-N)})^{-(c_2 N + c_3)} - c_4(f + \frac{N}{5})\right] \qquad (2.9)$$

where σ_0 is the matrix yield stress, and c_1, c_2, c_3 and c_4 are constants summarized in Table 2.1

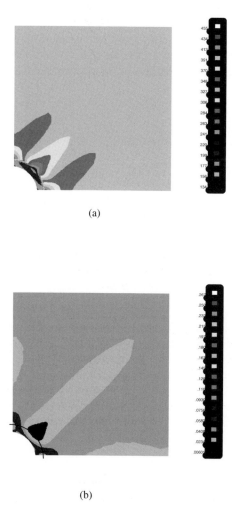

(a)

(b)

Fig 2.14 (a) Equivalent stress and (b) effective plastic strain in an embedded cell model with circular rigid fiber ($f = 0.5$) and metal matrix ($E = 100$ GPa, $\varepsilon_0 = 0.1$ %. $\sigma_0 = 100$ MPa, $N = 0.2$) at 3.8% total strain.

Equation (2.10) represents best fits to the calculated composite strengthening values C,y for matrix strain-hardening exponents N in the limit of $0.0 < N < 0.5$ for square 0°, hexagonal and random fiber arrangements, respectively, and fiber volume fractions, f in the range of $0.0 < f < 0.7$. A comparison of this strengthening model [equation (2.10)] for random fiber arrangements with the values calculated by using self-consistent embedded cell models shows close agreement with an average error of 1.25% and a maximum error of 6.95% [31].

Table 2.1:
Constants for strengthening models

	c_1	c_2	c_3	c_4
2D				
Self-consistent embedded cell model	0.361	1.59	0.29	0.1
(Random fiber arrangement)	0.405	2.35	0.65	0.22
Square unit cell model (0°)	0.305	1.3	0.05	0.0
Hexagonal unit cell model				
3D				
Self-consistent axisymmetric embedded cell model	0.45	2.19	0.84	0.53
(Random particle arrangement)	0.34	2.3	0.65	0.5
Primitive cubic unit cell model	0.38	2.5	0.7	0.66
Axisymmetric unit cell model				

Equation (2.10) is also available for matrix strain-hardening exponents N in the limits of $0.0 < N < 0.5$ for self-consistent axisymmetric embedded cell models (particle volume fractions f in the range of $0.05 < f < 0.65$ with an average error of 1.59% and a maximum error of 6.68% for the extreme case $f = 0.05$, $N = 0.5$), axisymmetric unit cell models (particle volume fractions f in the range of $0.05 < f < 0.55$ with an average error of 1.22% and a maximum error of 6.18% for the extreme case $f = 0.55$, $N = 0.5$) and for primitive cubic unit cell models (particle volume fractions f in the range of $0.05 < f < 0.45$ with an average error of 1.43% and a maximum error of 6.38% for the extreme case $f = 0.05$, $N = 0.5$).

Conclusions

The transverse elastic-plastic response of metal matrix composites reinforced with unidirectional continuous fibers and the overall elastic-plastic response of metal matrix composites reinforced with spherical particles have been shown to depend on the arrangement of reinforcing inclusions as well as on inclusion volume fraction f, and matrix strain-hardening exponent, N. Self-consistent plane strain and axisymmetric embedded cell models have been employed to predict the overall mechanical behavior of metal matrix composites reinforced with randomly arranged continuous fibers and spherical particles perfectly bonded in a power law matrix. The embedded cell method is a self-consistent scheme, which typically requires 4-5 iterations to obtain consistency of the mechanical behavior in the embedded cell and the surrounding equivalent composite. Experimental findings on an aluminum matrix reinforced with aligned but randomly arranged boron fibers (Al-46 vol.% B) as well as a silver matrix reinforced with randomly arranged nickel inclusions (Al-58 vol.% Ni) and the overall response of the same composites predicted by embedded cell models are found to be in close agreement. The strength of composites with aligned but randomly arranged fibers cannot be properly described by conventional fiber-matrix unit cell models, which simulate the strength of composites with regular fiber arrangements.

Systematic studies were carried out for predicting composite limit flow stresses for a wide range of parameters, f and N. The results for random 3D particle arrangements were then compared with regular 3D particle arrangements by using axisymmetric unit cell models as well as primitive cubic unit cell models. The numerical results were also compared with those from the Duva model and from the modified Oldroyd model. The strength of composites at low particle volume fractions has been shown to be in very close agreement except for the modified Oldroyd model. With increasing particle volume fractions, f, and strain hardening of the matrix, N, the strength of composites with randomly arranged particles cannot be properly described by conventional particle-matrix unit cell models, as those are only able to predict the strength of composites with regular particle arrangements.

Finally, a strengthening model for randomly or regularly arranged continuous fiber reinforced composites under transverse loading and particle reinforced composites under axial loading is derived, providing a simple guidance for designing the mechanical properties of technically relevant metal matrix composites: for any required strength level, equation (2.10) will provide the possible combinations of particle volume fraction, f, and matrix hardening ability, N. Thus, for the near future, strong impact of the present work on the development of new particle reinforced metal matrix composites is expected.

References

[1] Harrington W. C. Jr, (1993), Mechanical Properties of Metallurgical Composites (edited by S. Ochiai), p. 759. Marcel Dekker Inc., NY.
[2] Brockenbrough J. R. and Suresh S. (1990), Scripta metall. mater. 24, p. 325.
[3] Rammerstorfer F. G., Fischer F. D. and. Böhm H. J (1990). Discretization Methods in Structural Mechanics (edited by G. Kuhn et al.), IUTAM/IACM Symposium; Vienna, Austria, 1989, p. 393. Springer, Berlin, Heidelberg.
[4] Evans A. G. (1991), Mater. Sci. Engng, A 143, p. 63.
[5] Brockenbrough J. R., Suresh S. and Wienccke H. A. (1991), Acta metall. mater. 39, p. 735.
[6] Dietrich C. (1993), VDI-Fortschrittsberichte, Reihe 18, Nr. 128. VDI-Verlag, Düsseldorf.
[7] Nakamura T. and Suresh S. (1993), Acta metall. mater. 41, p. 1665.

[8]   Dietrich C., Poech M. H, Schmauder S. and. Fischmeister H. F. (1993), Verbundwerk-stoffe und Werkstoffverbunde (edited by G. Leonhardt et al.), p. 611. DGM-lnformationsgesellschaft mbH. Oberursel.

[9]   Adams D. (1970), J. Compos. Mater. 4, p. 310.

[10] Jansson S. (1992), Int. J. Solids Struct. 29, p. 2181.

[11] Zahl D. B., Schmauder S. and McMeeking R. M. (1994), Acta metall. mater. 42, p. 2983.

[12] Llorca J., Needleman A. and Suresh S. (1991), Acta metall. mater. 39, p. 2317.

[13] Sautter M. (1995), PhD. Dissertation, University of Stuttgart.

[14] Povirk G. L., Stout M. G., Bourke M., Goldstone J. A., Lawson A. C., Lovato M.,. Macewen S. R., Nutt S. R and Needleman A. (1992), Acta metall. mater. 40, p. 2391.

[15] Bohm H., Rammerstorfer F. G. and Weissenbek E.(1993), Comput. Mater. Sci. 1, p. 177.

[16] Böhm H. J., Rammerstorfer F. G., Fischer F. D. and Siegmund T. (1994), ASME J. Engng Mater. Technol. 116, p. 268.

[17] Bao G., Hutchinson J. W. and McMeeking R. M. (1991), Acta metall. mater. 39, p. 1871.

[18] Tvergaard V. (1990), Acta metall. mater. 38, p. 185.

[19] Horn C. L. (1992), J. Mech. Phys. Solids 40, p. 991.

[20] Weissenbek E. (1993), PhD. Dissertation, Technical University of Vienna.

[21] Li Z., Schmauder S., Wanner A. and Dong M. (1995), Scripta metall. mater. 33, p. 1289.

[22] Christman T.. Needleman A. and Suresh Y. (1989), Acta metall. mater. 37, p. 3029.

[23] Suquet P. M. (1993), MECAMA T 9.1, lnt. Seminar on Micromechanical Materials. p. 361, Editions Eyrolles. Paris.

[24] Thebaud F. (1993), PhD. Dissertation, Université de Paris Sud.

[25] Zahl D. B. and McMeeking R. M. (1991), Acta metall. Mater 39, p. 1171.

[26] Poech M. H. (1992), Scripta metall. mater. 27, p. 1027.

[27] Farrissey L., Schmauder S., Dong M., Soppa E., McHugh P. and Poech M. H. (1999), Investigation of the Strenghening of Particulate Reinforced Composites using Different Analytical and Finite Element Models. Comput. Mater. Sci. 15, pp 1-10.

[28] Duva J. M. (1984), Trans. ASME Series H, J. Engng Muter.Technol. 106, p. 317.

[29] Sautter M., Dietrich C., Poech M. H., Schmauder S. and Fischmeister H. F. (1993) Comput. Mater. Sci. I, 225.

[30] Zahl D. B. and Schmauder S. (1994), Comput. Mater. Sci. 3, p. 293.

[31] Dong M. and Schmauder S., Comput. Mater. Sci. (to be published, 1996).

[32] Jirvstrlt N. (1993), Comput. Mater. Sci. 1, p. 203.

[33] LASSO Engineering Association, Markomannenstr. 11, D-70771 Leinfelden-Echterdingen, Germany.

[34] PDA Engineering, 2975 Redhill Avenue, Costa Mesa. California 92626, U.S.A.

2.2 Multiphase finite elements

2.2.1 3D multiphase finite element method[2]

Introduction

Inhomogeneous stress distributions which occur as a result of the interaction of the individual microstructural constituents in multiphase materials are frequently simulated with the finite-element (FE-) method [1]. Such microstructural analyses contribute to the understanding of the stress and strain distributions in each constituent and can be used to depict critical loads or critical configurations. So far, FE modelling of realistic microstructures is essentially limited to 2D-models [2-5]. However, 2D-simulations are not able to represent the morphological complexity of a realistic 3D-microstructure. Inhomogeneities located underneath the surface of a modelled microstructure as well as the inherent 3D-nature of the microstructure may introduce significant changes in comparison to 2D-simulations. Therefore, fully 3D-analyses are currently requested in order to evaluate the predictions based on 2D-simulations. FE-modelling of two-phase materials based on single-phase elements is limited to the simplest 3Dstructures such as single inclusions or regular microstructures [6-10]. Conventional 3D-modelling of realistic microstructures is practically impossible since it would require fantastically complex FE meshes.

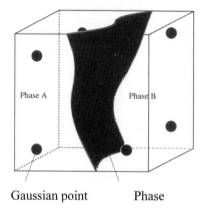

Gaussian point   Phase

Fig. 2.15 8-node hexahedral multiphase element with bilinear interpolation and full integration.

The first 3D-simulations of more realistic microstructures have been reported by Hollister and Kikuchi [11]. They applied digital image based modelling for the

[2] Reprinted from N. Lippmann, Th. Steinkopff, S. Schmauder, P. Gumbsch, "3D-Finite-Element-Modelling of Microstructures with the Method of Multiphase Elements", Computational Materials Science 9, pp. 28-35 (1997), with kind permission from Elsevier

micromechanical study of bone tissues. A 3D-model is developed by combining 2D-images. Each pixel in a 2D-image is recognized as a voxel which is identified as a finite element in the analysis.

Recently, the method of multiphase elements (MME) where the Gaussian points of one finite element can be assigned to different material properties has been successfully applied for the 2D-simulation of realistic microstructures [2-5]. Multiphase elements were introduced since the reconstruction of phase boundaries with irregular shape in singlephase FE-models is very complicated. The main advantage of the multiphase elements over conventional singlephase elements are the flexibility in the construction of the FE-model as well as the numerical efficiency. Based on our experiences with this method, the new 3D-multiphase element has been developed in the framework of the non-linear FE-code LARSTRAN [12]. The 3D-multiphase element is defined for the following element types:

- 8-node hexahedral elements with trilinear interpolation and full integration (eight Gaussian points, Fig. 2.15.
- 27-node and 20-node hexahedral elements with triquadratic interpolation Gaussian points).

Method

Consider the main ideas of the MPFE (multiphase finite element) method. Commonly, each element of the FE mesh is attributed to one phase; the same material properties are assigned to all integration points of an element and the phase boundaries are supposed to coincide with the edges of finite elements. The idea of the method of multiphase elements is that the different phase properties are assigned to individual integration points in the element. Therefore, the FE mesh in this case is independent of the phase arrangement of the material, and one can use relatively simple FE meshes in order to simulate the deformation in a complex microstructure.

The possibility of using initial meshes of arbitrary simple structures for simulation of the mechanical behaviour of materials with complex microstructure is the main advantage of the method of multi phase elements.

In the case of 3D simulations of the mechanical behaviour of heterogeneous materials with arbitrarily arranged phase boundaries, it is almost impossible to construct the FE mesh in such a way that the edges of finite elements correspond to the phase boundaries. Therefore, the possibilities, which the multiphase element method offers are especially important in the 3D case.

In the 3D MPFE code, finite elements of low order (with linear or square interpolation functions) are used. The form and distribution of the phases are introduced automatically in the FE model from digitised micrographs of the microstructure.

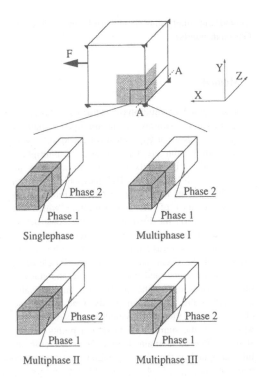

Fig 2.16 FE-models for the verification of the multiphase elements.

Several material layers are metallographically removed equidistantly from the surface of the specimen by careful polishing. Micrographs from each polished surface are made. The thickness of the removed layers must be known for each layer as exactly as possible. Then, the micrographs are digitised with the use of image analysis software, and the 3D real structure of the material is reconstructed from the digitised micrographs of the sections.

Each pixel in 2D digitised micrograph is assigned to some volume of material, and therefore determines the correspondence of Gaussian points to the phases in the material. The mechanical properties of the phases are automatically assigned to the integration points.

One multiphase element can be virtually divided into n subdomains, where n is the number of Gaussian points in this element. The subdomain k of a Gaussian point is fully assigned to the phase in which the integration point r_k is located. A closed integration (e.g. to provide the stiffness) in the multiphase element is not possible. Therefore, the Gaussian quadrature, which is the weighted summation of the function values at selected integration point's x_k, is used as integration scheme:

$$\int f(x)dx = \sum_k w_k \cdot f_{vk}(x_k) \tag{2.10}$$

The weights w_k correspond to the volume of the subdomains and f_{vk} indicates that the function for the phase v of the subdomain k is used.

In the following, the reliability of the 3D-muhiphase elements is studied by 3D-FE-modelling of a simple model and subsequent comparison with the results of singlephase FE-modelling. In this case we investigate the influence of a strong stress jump on the simulations. Furthermore, the development of a 3D-model of a realistic two-phase microstructure is presented. Finally, the results of this simulation are compared with the results of a 2D-simulation of a representative intersection of the microstructure.

Examples

Stress distribution due to a cubic inclusion in a matrix

The method of multiphase elements would usually be not applied to the modelling of such simple geometries as the cubic inclusion in a matrix which is presented in this first example. However, since this geometry can also be modelled by single-phase FE's it is suitable to demonstrate both the reliability as well as limitations of the multiphase elements by a direct comparison of both methods. A small number of elements in the models permit an intensive evaluation of results.

A cubic inclusion (phase 1) with an edge length 1 is embedded into an Al-matrix (phase 2) cell of the size 21 x 21 x 21 (Fig. 2.16). In this linear-elastic calculation the Young's modulus of Silicon (168.2 GPa) [13] is assumed for the inclusion and 70 GPa for the matrix (Al). The Poisson ratios of the matrix and inclusion are chosen to be 0.3 and 0.25, respectively. Boundary conditions are imposed which simulate the stress distribution in the model under uniaxial tension and plane strain conditions (symmetric boundary conditions). The 3D-models (see Fig. 2.16) are created with 20-node hexahedral elements as follows:

(1) A singlephase model with connected nodes at the phase boundary which has 64 conventional elements.

(2) The multiphase I model consisting of 27 multiphase elements. The central nodes of the multiphase elements containing the phase boundary are located on the phase boundary.

(3) The multiphase II model corresponds to the singlephase model in the number of elements and discretisation. The phase boundary is located at the element edge.

(4) Model multiphase III comprises 36 multiphase elements of various sizes. The phase boundary is located between the nodes and Gaussian points of the elements containing the phase boundary.

The row of elements along the z-axis in Fig. 2.16 is chosen to evaluate the maximum stress-component σ_x, for which a pronounced stress jump is expected at the phase boundary. The edge length of the elements is equivalent in x- and y-direction in all models to eliminate additional effects of stress gradients. Both, the results at the Gaussian points as well as the element averaged results are evaluated.

Results at the Gaussian points

Fig. 2.17 shows the calculated maximum normal stress σ_x along the z-direction (cf. Fig. 2.16, path A-A) for the Gaussian points with the lowest x- and y-distance to the z-axis. Generally, the same behaviour in the inclusion and matrix is observed in all simulations.

The stresses in the inclusion (phase 1) increase with decreasing distance to the phase boundary. The largest difference (less than 10%) occurs between single-phase and multiphase III model. The stress jump at the phase boundary is almost identical in the singlephase and the multiphase II simulation. Depending on the nearest distance of a Gaussian point to the phase boundary a more or less steep stress gradient is obtained in multiphase III and multiphase I simulations. The stresses in the matrix increase with the distance from the phase boundary.

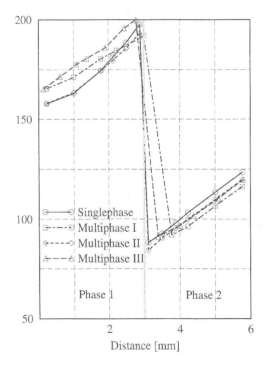

Fig 2.17 Results at the Gaussian points.

Element averaged results

Element averaged results (Fig. 2.18) which are used in post-processors such as PATRAN[3] to produce fringe plots are calculated on the basis of the averaged results at the nodes.

Almost equivalent values of stresses are obtained for the singlephase and multiphase II simulations whereas in the multiphase I and III simulations the stress jump is not sufficiently reproduced. Since different material properties occur in these multiphase elements the stress maxima or minima are 'smeared' in the phase boundary region. This behaviour can not be avoided but reduced by using smaller elements where either the matrix or the inclusion dominates in the calculation of the averaged results.

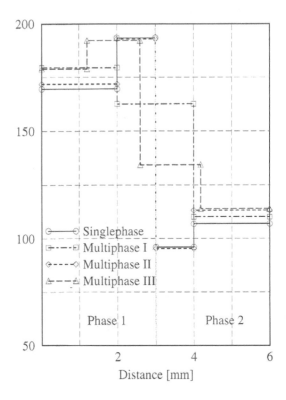

Fig 2.18 Element averaged results.

Comparison of a 2D- and 3D-simulation of a realistic microstructure

As an example of a more realistic application of multiphase elements, the microstructural behaviour of an AlSi-cast alloy is simulated in a 3D-FE-model and

---

[3] PDA Engineering, 2975 Redhill Avenue, Costa Mesa, California 92626, USA

compared with a 2D-simulation. In order to describe appropriate macroscopic boundary conditions, embedded cell simulations [4] are performed. A 3D-model of size 240 x 240 x 30 μm is generated as follows (Fig. 2.19):

- The central region of the model of size 120 x 120 x 30 μm with 9600 g-node hexahedral elements (c.f. Fig. 2.19(1)) is used for the simulation of the two-phase microstructure.
- The microstructure in the model is developed on the basis of 4 intersections of the microstructure at each 10 μm starting from the surface of the specimen.

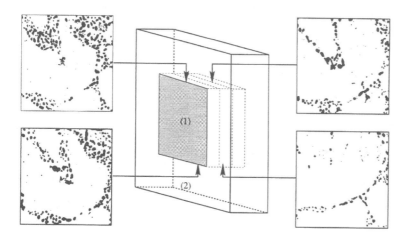

Fig 2.19 Generation of the 3D-multiphase model (schematically)

(A)   (B)

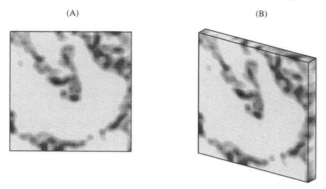

Fig 2.20 Designation of the phases to the elements: (A) 2D-FE model, (B) 3D-FE model

- The cell representing the microstructure is surrounded by 7296 single phase g-node hexahedral elements (c.f. Fig. 2.19(2)).

- All nodes at the free surfaces are fixed in z-direction (perpendicular to the model surface) to obtain appropriate plane strain boundary conditions.

The actual FEZ-mesh used in the calculations is given in Ref. [14]. For comparison to the 3D-simulation, a 2D-multiphase model of size 240 x 240 μm is created:

- With 1600 4-node quadrilateral plane strain elements in the central region. The edge length of these elements is equal to the edge length in the 3D-model.
- The two-phase microstructure in this model is chosen according to one of the intersections which are considered in the 3D-model (z = 10 μm). The cell which represents the microstructure is surrounded by 1216 4-node quadrilateral elements.

Table 2.2

Elastic-plastic material properties of the constituents

Property	Al-Matrix	Al/Si7Mg
Yield stress σ_0	218 MPa	250 Mpa
Yield strain ε_0	0.325 %	0.357%
Constitutive relation	$\sigma = \sigma_0 (\varepsilon / \varepsilon_0)^{0.14}$	$\sigma = \sigma_0 (\varepsilon / \varepsilon_0)^{0.08}$

The linear-elastic material properties in both models are imposed according to Section 3.1. Additionally, the constitutive relation for the simulation of the elastic-plastic properties of the Al-matrix and the AlSi7Mg-alloy (surrounding) are given in table 2.2. The constitutive relation for the AlSi7Mg-cast alloy was determined from experimental stress-strain diagrams. The mechanical properties of the matrix were determined separately from a macroscopic specimen with the chemical composition of the matrix.

The preprocessing for the 3D-model is almost the same as for the simple model with the inclusion, which demonstrates the efficiency of the method of multiphase elements in the preprocessing stage. This efficiency of model construction is needed for the 3D-analyses of realistic microstructures.

The designation of phase properties to the elements is depicted in Fig. 2.20 for both models. The color-coding is proportional to the number of Gaussian points in the phase: the darker the color the more Gaussian points of the element are assigned to silicon. The phase distribution in the 2D- and 3D-model is equivalent, since the same discretisation is used in both models.

The resulting stress distributions (σ_x Fig. 2.21) for both models show stress concentrations in the Si-eutectic as a result of the higher stiffness of Silicon. The maximum stresses σ_x and the extension of the stress maxima are somewhat higher in the 3D-model. However, generally the stress distributions are almost identical for both models. This may be understood as a result of the simulated microstructure, where large matrix cells exceeding the model dimension in z-direction are surrounded by Si-eutectic.

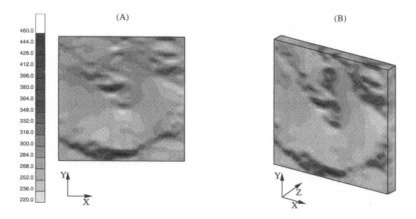

Fig 2.21 Stress distribution σ_x, loading axis corresponds to x-axis: (A) 2D-FE model, (B) 3D-FE model.

Since the model is limited to an intersection of such a cell, the material properties do not change significantly in the third dimension.

The results presented here are the first applications of the 3D-multiphase elements. In the future, we plan a direct comparison of 3Dsimulations with experiments on the basis of displacement fields. The displacements determined experimentally in the scanning electron microscope during the in-situ deformation of an AlSi-specimen are directly compared to the displacements calculated in a 3D-microstructural simulation of these experiments.

Conclusions

A 3D-multiphase element was introduced which enables the 3D-analysis of realistic microscopic as well as macroscopic structures. A simple inclusion model was used to demonstrate the reliability of 3D-multiphase elements by direct comparison with a 3D-singlephase model. The results obtained for the Gaussian points show very good agreement between singlephase and multiphase simulations.

Compared to the digital image based modelling (singlephase method, voxels) [11] the multiphase method distinguishes the different phases with better resolution, since the smallest unit to represent the second phase is the Gaussian point. Thus, even the effect of particles which are smaller than the element size can be taken into account in calculating stress distributions. In the case of equivalent mesh density the volume fraction of phases is more accurately reproduced in the multiphase method. Advantages of the multiphase elements in comparison to the conventional FE-modelling are:

- Higher flexibility of the models because one regular mesh can be used for the simulation of any microstructure.

- Increased resolution of material gradients since different constitutive relations can be used in one element.
- Easier assignment of failure criteria when crack initiation and crack propagation is modelled because those can be based on individual Gaussian points.
- Increased numerical efficiency: if one would decompose one 8-node multiphase hexahedral element with eight Gaussian points into eight 8-node singlephase elements, the CPU-time for the calculation would be about 16-times higher in the singlephase simulation.

The element averaged results obtained in the multiphase simulation differ from the singlephase results: as a consequence of the different material properties in the elements stress concentrations are reduced and stress jumps are 'smeared'. For certain applications this is actually intended; among them are gradient materials. The 3D-multiphase elements seem to be a tool for the stress-strain analysis of complex and irregular 3D-microstructures. However, occurrences at phase boundaries such as debonding can not be simulated with this method.

Finally, the creation of a 3D-model of a realistic microstructure was presented. This model shows the efficiency of the method of multiphase elements with respect to preprocessing and a realistic stress analysis of complex heterogeneous materials. In comparison to the 2D-simulation the maximum normal stresses in the 3D-simulation are somewhat higher for the simulated microstructure. We assume that in the case of microstructures with isolated particles the results of 2D- and 3D-simulations will differ more significantly.

References

[1] Fischmeister H.F., Karlsson B. (1977), Z. Metallk. 68, p. 311.
[2] Steinkopff Th., Sautter M. (1995), Comput. Mater. Sci. 4, p.10.
[3] Wulf J., Steinkopff Th., Fischmeister H.F., Acta Metall. 44 (1995) p.1765.
[4] Sautter M. (1995), Modellierung des Verformungsverhaltens mehrphasiger Werkstoffe mit der Methode der Finiten Elemente, Fortschrittberichte VDI, vol. 5, No. 398, VDI-Verlag, Düsseldorf.
[5] Lippmann N., Schmauder S., Gumbsch P.(1996), in: 4th International Conference Locahsed Damage '96, Fukuoka, Japan, June 3-5.
[6] Le Hazif R., de Montpreville C.T.(1983), 3D models for the calculation of the mechanical behavior of two-phase alloys, in: Bilde-Sorensen J.B., Hansen N., Horsewell A., Leffers T., Lilhoh H. (Eds.), Deformation of Multi-Phase and Particle-Containing Materials, Riso National Laboratoty, Roskilde, Denmark.
[7] Horn C.L., McMeeking R.M. (1991), Int. J. Plastic. 7, p. 255.
[8] Bao G., Hutchinson J.W., McMeeking R.M (1991), Acta Metall. 39, p.1871.
[9] Brockenbrough J.R., Hunt Jr W.H.., Richmond O. (1992), Ser. Metall. 27, p. 385.
[10] Brockenbrough J.R., Wienecke H.A., Romanko A.D., Alcoa Report No. 519303, pp. l-9.
[11] Hollister S.J., Kikuchi N. (1994), Biotechnol. Bioeng. 43, p. 586.
[12] LARSTRAN, LASSO lngenieurgesellschaft, Leinfelden-Echterdingen, Germany.
[13] Johansson S. (1988) , Micromechanical properties of silicon, Doctoral dissertation, Uppsala University.

[14] Lippmann N., Steinkopff Th., Schmauder S., Gumbsch P. (1996), 3D-FE-modelling of microstructures with the method of multiphase elements, in: Proceedings of General Workshop COST 512, Modelling in Materials Science and Processing '96, Davos, Switzerland, 29 Sept.-2 Oct.

2.2.2 Multiphase finite element method and damage analysis[4]

In this work, the possibilities of the methods of multiphase finite elements [1-5] and element elimination technique (EET) for studying the damage initiation and evolution, and fracture in two-phase materials for both 2D and 3D cases are discussed.

Element elimination technique and generalised damage parameter

Conditions of the element elimination: Generalised damage parameter

The element elimination technique (EET), described in Chapter 3 in more details, is based on the quasi-removal of finite elements, which satisfy some failure condition (which is to be defined for each material to be considered). In such a way the formation, growth and coalescence of voids or microcracks, and the crack growth are simulated. As criteria of local failure, both global (external loads or displacement) and local (i.e., defined for a given element; for instance, plastic strain, von Mises stress, hydrostatic stress, etc.) values as well as any combination of these values can be used.

A criterion of element elimination which is appropriate for each material should be chosen by comparison of numerical and experimental results as well as by experimentally studying micromechanisms of damage initiation. For instance, Wulf used the condition of critical plastic strain as criterion for element elimination [3]. Lippmann et al. [9] and Hönle et al. [10] have used a two-criteria model of element elimination for the simulation of Al/Si cast alloys: elements which were assigned to hard (Si) particles were eliminated on the basis of a normal stress criterion, and the critical value for failure of the ductile matrix phase was simulated using a damage-parameter, which is based on the work of Rice and Tracey [11] and Hancock and Mackenzie ([12], see also [13]). The damage parameter can be written as ([14], see also [3]) with failure initiation at a critical damage parameter value of D_c. Here ε_{pl} is the effective plastic strain, $\varepsilon_{pl,c}$ the critical plastic strain, η the stress triaxiality, $\eta = \sigma_H/\sigma_v$, σ_H the hydrostatic stress, and σ_v is the von Mises equivalent stress. The damage parameter fulfils all demands on locality, triaxiality of the stress-strain field, as well as taking into account the complete failure history.

$$D = \int_0^{\varepsilon_{pl,c}} e^{\frac{3\eta}{2}} d\tilde{\varepsilon}_{pl} \qquad (2.11)$$

[4] Reprinted from L. Mishnaevsky Jr., M. Dong, S. Hönle, S. Schmauder, "Computational mesomechanics of particle-reinforced com-posites", Computational Materials Science 16, pp. 133-143 (1999), with kind permission from Elsevier

The damage parameter (1) can be generalised in order to take into account the so-called "failure curves" of materials in the following way. Failure curves (see Fig. 2.22) present a relation between equivalent plastic strain and stress triaxiality at the crack initiation point inside the specimen, obtained on the basis of combined experimental and numerical investigations on tensile specimen ([15], see also [10]). The failure curve separates the equivalent plastic strain-stress triaxiality space into two parts. Below the curve the material is saved, no failure will occur. Points on the curve indicate failure initiation for a given stress-strain field. If one expresses the failure curves as shown in Fig. 2.22 as

$$\varepsilon_{pl.c} = A \cdot e^{-B\eta_c} \tag{2.12}$$

where A and B are two material-dependent parameters, and assuming that usually $\varepsilon_{pl} \leq \varepsilon_{pl.c}$ and $\eta \leq \eta_c$ holds during loading, one can derive the modified damage parameter [10]:

$$D = \frac{1}{A} \int_0^{\varepsilon_{pl.c}} e^{B\eta} d\tilde{\varepsilon}_{pl} \tag{2.13}$$

with the point of failure initiation at a critical damage parameter value of $D_c = 1$.

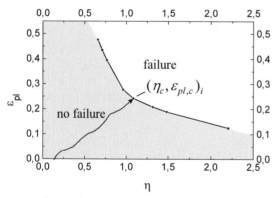

Fig 2.22 Scheme of a failure curve

Determination of the condition of element elimination: an example

To demonstrate how the condition of element elimination can be determined from experimental data, let us consider a simple macroscopical simulation of crack growth in Al/SiC (20 vol%) composites under 3-point-bending. An FE mesh similar to the mesh given in [3] was taken in the simulation. However, a criterion of element elimination different from that used in [3] was applied: instead of the critical plastic strain, the Rice and Tracey's damage parameter was chosen as criterion for element elimination. The calculations were carried out for a material with averaged elastic properties of the specimen *(E* =99.4 GPa and v = 0.323).

The difference between the critical plastic strain and Rice and Tracey's damage criterion for element elimination is determined by the fact that the Rice and Tracey's damage criterion is very sensitive to the degree of triaxiality in the deforming material, while this is not the case for the critical plastic strain criterion. The finite elements which are located on the loading surface were assumed to be elastic and were not subject to damage in order to simplify the simulation of the specimen/holder contact conditions. Based on the available data about the critical values of damage [3,7,9] we used three different values of critical damage parameters: D_c = 0.5, 0.15 and 0.2. As a result, the force-displacement curves for the specimen were obtained. The curves for the different values of the critical damage parameter are shown in Fig. 2.23. Although the peak points of the force-displacement curve were determined correctly and correspond to the experimental data, the appearance of the descending branch of the curve differs significantly from the experimentally observed results [3].

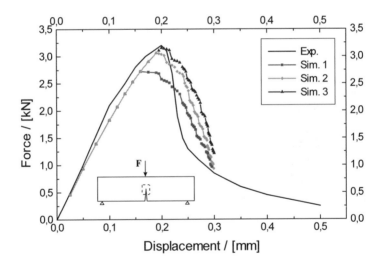

Fig 2.23 Force-displacement curves for 3-point bending. Sim.1 $(D_c = 0.2)$, Sim. 2 $(D_c = 0.15)$, Sim.3 $(D_c = 0.10)$.

The values of the calculated peak load for the different critical damage parameters as well as the experimentally obtained peak load are given in table 2.3. It is found that the critical damage parameter D_c = 0.2 results in simulating the correct peak load. Therefore, our results confirm the data from [3,10], that the critical value of Rice and Tracey's damage parameter for Al/20 vol% SiC should be taken as $D_c = 0.2$. The described procedure shows how the criterion of element elimination can be determined by comparing the experimental and numerical force-displacement curve.

Mesomechanical modelling of damage and failure in real structures

In this section, the model of damage initiation and growth, and fracture in a material with real structure is presented. In this model, both the multiphase element method and the EET are used. The simulation is carried out for WC/Co hard metals, which present particle-reinforced composites with coarse microstructure; these materials are characterised by a typically high content of hard inclusions and their relatively large size as compared with the thickness of areas of Co binder. A model microstructure of WC/Co material was taken and used in the simulations. The WC/Co specimen possesses a cobalt volume fraction of 16% and an average carbide size of 1.5 µm.

The microstructure is meshed using multiphase elements (MPFEs) and is embedded in an environment with the elastic material behaviour of the composite and a pre-crack just in front of the real structure (Fig. 2.24).

The material properties of the carbide are elastic and the elastic-plastic behaviour of the cobalt is represented by a modified Voce-type flow law with an additional Hall-Petch term [10], according to where $\sigma_y = 270$ MPa - the yielding stress of cobalt, ($\sigma_s = 970$ MPa, $\varepsilon^* = 0.06$, $k_y = 7$ N mm$^{-3/2}$ (material constants) and an average binder layer thickness of $L = 0.5$ µm were taken from [16].

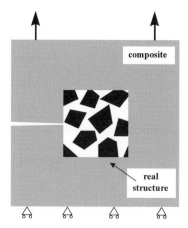

Fig 2.24 Micromechanical model for failure simulation in realistic structures.

Critical plastic strains and stress triaxialities, which were derived using crystal plasticity theory [17] at a critical void volume fraction of 15% [18] are shown in Fig. 2.22.

$$\sigma = \sigma_y + (\sigma_s - \sigma_y) \cdot \left[1 - \exp(-\frac{\varepsilon}{\varepsilon^*}) \right] + k_y L^{-1/2} \qquad (2.14)$$

These results lead to the modified damage parameter,

$$D = \frac{1}{1.04} \int_0^{\varepsilon_{pl.c}} e^{1.12\eta} d\tilde{\varepsilon}_{pl} \tag{2.15}$$

In the following, results of a crack propagation simulation based on Eq. (2.15) are described. As a first attempt, a hard meta I with a high cobalt content and no contact between carbide particles was modelled and investigated under external tensile loading (see Fig. 2.25(a)). This study focused on the failure behaviour of the ductile cobalt phase. Brittle fracture in the carbide phase was thus suppressed. At the considered level, (mesolevel) the structure of material was practically random, and therefore, no special consideration of the texture dependence was required in this case. The results achieved on the level of crystal plasticity were calculated for a wide range of microscopic arrangements and crystallographic slip system arrangements, and thus describe an averaged behaviour of the material based on crystal plasticity theory.

Figs. 2.25(b)-(e) show crack initiation and crack propagation in this structure. The crack enters the real structure by initiating a void (Fig. 2.25(b)), which starts to grow under increasing load. Further increase of the applied load leads to void initiation in front of the crack tip (Figs. 2.25(c) and (d))and coalescence with the main crack. Crack propagation is found to be a consequence of nucleation, growth and coalescence of the voids. This numerical study is in agreement with experimental findings on WC/ Co hard metals [19]. The force-displacement curve for this crack propagation depicts the experimental macroscopic failure behaviour of WC/Co hard metals as a quasi-brittle failure. The applied load increases nearly linearly, while it drops immediately when the critical load is reached.

Thus, local damage behaviour of the ductile cobalt phase is introduced in microscopic crack propagation simulations by making use of failure curves.

Table 2.3
Peak loads on the force-displacement curves at different critical values of Rice and Tracey's damage parameter

Critical damage parameter, D_c	0.1	0.15	0.2	Experiment [3]
Peak loads (kN)	2.76	2.95	3.1	3.1
Displacement (mm)	0.22	0.24	0.26	0.2

Meso-macromechanical modelling of damage and failure of two-phase composites

Microdamage in Al/Si cast alloys: unit cell simulation

In Al/Si cast alloy specimens, the damage process is determined by the strength of Si-particles and the Al matrix as well as the shape, size, arrangement and volume fraction of the Si-particles.

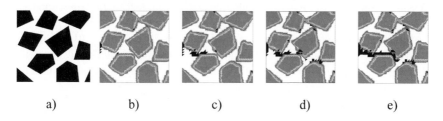

a)   b)   c)   d)   e)

Fig 2.25 (a) Idealised real structure (WC black, Co white), (b)-(e) void nucleation, growth and coalescence.

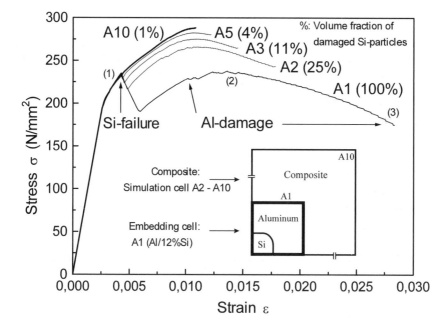

Fig 2.26 Local stress-strain relations of the Al/12%Si cast alloy concerning both particle cracking and matrix failure for different volume fractions of locally damaged material.

The Si-particles behave elastically and the Al-matrix elasto-plastically. It is known that only large Si-particles fracture during external loading [8,9]. Large particles are present at a low volume fraction of about 1-5% in the alloy, while the overall volume fraction of Si amounts to 12%.

In the mesomechanical model, a unit cell containing a silicon particle and the aluminum matrix is embedded in an equivalent homogeneous material with the same mechanical behaviour as that of the embedded cell (Fig. 2.26). Under the tensile displacement loading of the upper extern al boundary of the embedding composite in the cell shown in Fig. 2.26, the overall response of the inner embedded cell is obtained by averaging the stresses and strains in the embedded cell [20].

Microdamage of the two-phase material in the embedded cell [20] is simulated by a hybrid local approach for brittle cracking of silicon particles and for ductile failure of the aluminum matrix (Figs. 2.26 and 2.27). The normal stress criterion (σ^n_{max} = 320 Mpa) obtained from comparison of simulation and experiment in [21]) and node release technique are applied to simulate Si-particle cracking whereas the damage parameter D and the EET are applied for simulating ductile void growth in the Al-matrix. The damage parameter involves the loading state as well as the loading history and its critical value can be derived from the corresponding experiment and simulation (in this case, D_c = 0.7 [22]).

At a total strain of ε_{tot} = 0.43% the Si-particle is cracking orthogonally with respect to the tensile loading direction according to the normal stress criterion and then matrix damage occurs subsequently under increasing loading. The matrix damage propagates in the same direction as the Si particle crack. Local stress-strain relations of the Al/12% Si cast alloy concerning both particle failure and matrix damage are thus obtained for different volume fractions of locally damaged material (Fig. 2.26), which are further taken into account in the macromechanical model.

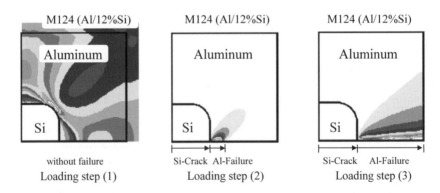

Fig 2.27 Local damage parameter distribution in the embedded cell for three different loading steps according to Fig. 2.26. White areas: $D = 0$, very bright areas: $D = 0.6 \dots 0.7$, dark areas: $D = 0.4 \dots 0.5$, medium dark till areas: $D = 0.1 \dots 0.3$.

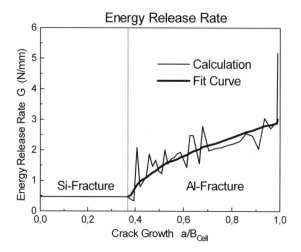

Fig 2.28 Energy release rate of matrix damage after silicon particle cracking

The local damage-parameter distribution in the embedded cell for the three different loading steps (Fig. 2.27) shows the development of damage-parameter concentration, which results in void growth and crack propagation in the matrix. The released energy is an important magnitude for comparison with the non-destructive evaluation of damage by acoustic emission.

From local energy considerations, the energy release rate by further damage propagation can be calculated as a function of microcrack length (Fig. 2.28). After fracture of the silicon particle, the energy release rate of the matrix damage increases with sub-critical microcrack growth until rupture of the material.

Macromechanical model of failure in Al/Si cast alloys

On the basis of the above-described mesomechanical model, macromechanical models of tensile or compressive specimens are set up to investigate crack growth. The mechanical behaviour of the Al/12%Si cast alloy with damage evolution was assigned to each finite element taking into account damage initiation and propagation calculated from the mesomechanical models presented above. In this way, the macroscopical development of microdamages until rupture of the specimen can be simulated. As an example, an initial Si-failure is introduced at a corner of a homogeneous specimen. The propagation of Si failure is calculated according to the critical normal stress obtained from micromodelling.

The simulations demonstrate two stages of damage evolution: At first the Si-particles fail along shear bands which nucleate from the initial failure side in the specimen at the macroscopic plastic flow stage, as shown in the stress distribution in Fig. 2.29. After Si-failure, the specimen can still carry increased loading and stretching.

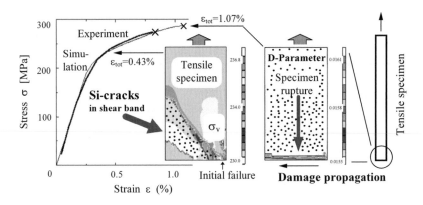

Fig 2.29 Macroscopical development of microdamages at different loading stages (areas of cracked Si-particles are represented by black dots).

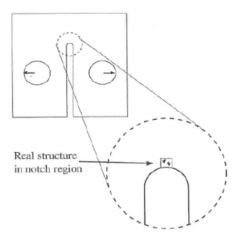

Fig 2.30 Scheme of loading of CT specimen and location of the area with real microstructure.

Under further loading, the damage parameter concentration increases in the cross section orthogonally to the loading direction and the crack starts to propagate perpendicular to the loading direction until failure of the specimen in agreement with the observations of fracture in such materials [20] (Fig. 2.29)

3D FE-simulation of deformation in real structures

In this section, the results obtained by Mishnaevsky Jr. et al. [6] in the framework of the COST -Project "Microstructural investigation of failure mechanisms in Al/Si cast alloys by 3D FE modelling" are reported and compared with above-described 2D models of mechanical behaviour of composites. To simulate the deformation of specimens from Al/Si cast alloys with different microstructures, 3D MPFE has been applied.

The small volume in which the real structure was reconstructed is located in the notch region of loaded compact tension (CT) specimen. The scheme of loading CT specimen is shown in Fig. 2.30. For a small volume (100 μm x 100 μm) in the notch ground region the microstructure of material was reconstructed with the use of digitised micrographs of polished sections as described above. The properties of components (Al-matrix and Si-particles) have been assigned to finite elements in correspondence with distribution of "black" and "white" areas on digitised micrographs.

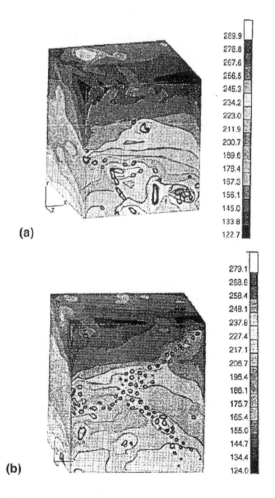

Fig 2.31 Stress distribution (σ_y) in the regions with real microstructure: (a) lamellar; (b) globular microstructure. Only particles at the surface are shown (for details see text).

The rest of the specimen was assigned the averaged properties of Al/Si cast alloy. In the area of real microstructure, there were 4000 finite elements, in the full specimen 16000 FE. The simulation was carried out for two alloys: one with lamellar and one with globular microstructures. The load was increased in five loading steps, each was 100 N. The mechanical properties and stress strain curve of Al-matrix and Si-particles have been determined experimentally as described in [6].

Some results of the simulation are shown in Fig. 2.31: the von Mises stress distribution in the region with real structure for both types of the alloy microstructure. The von Mises stress distribution is presented in order to study the effect of Si-particles on the stress field in the material. From Fig. 2.31 one can see that the Si-particles cause high local stress concentrations, especially on the notch surface. One may suppose that the high local stress concentrations lead to damage initiation in the vicinity of Si-particles. The experimental investigations of damage evolution in Al/Si cast alloys [6, 21, 23] confirm this assumption: microcracks have been formed in the vicinity of Si-particles, often near the notch surface, and then grow and form large cracks.

The developed approach permits 3D FE-simulations of deformation and local effects in heterogeneous materials, using the reconstructed real structures and the method of multiphase elements. However, the accuracy of 3D simulation depends strongly on the degree of fineness of the FE mesh and on the amount and quality of layers (sections) from which the 3D real structure is reconstructed. For both factors, there are unavoidable limitations (computational and experimental, respectively), which limit the possibility of this model.

Conclusions

In this paper, the results of application of advanced numerical methods (MPFE and EET) for the simulation of the mechanical behaviour and failure of particle-reinforced composites were presented.

Several aspects of FE simulation of particle-reinforced composites were considered:

- method of determination of conditions of local failure;
- modelling of deformation and fracture in coarse particle-reinforced composites (WC/Co hard metals, in this case);
- multilevel modelling of deformation and fracture of composites with fine particles; transition from a mesomechanical model of behaviour of composites with relatively fine particles (Al/Si) to the macroscopical model of material;
- possibility of generalisation of 2D methods of FE simulation of materials with real structures to the 3D case.

The methods are applied to model the mechanical behaviour of different types of materials: quasi-homogeneous (macromodels of Al/Si and Al/SiC), coarse particle-reinforced composites (WC/Co hard metals) and fine particle-reinforced composites with different types of microstructures (lamellar, globular) (Al/Si).

A mesomechanical model of damage and fracture of hard metals made it possible to determine the mechanisms of fracture. It is shown numerically that crack propagation can be simulated as a result of nucleation, growth and coalescence of voids. The force-displacement curve for this crack propagation depicts the experimental macroscopic failure behaviour of WC/Co hard metals as a quasi-brittle failure: the applied load increases nearly linearly, while it drops immediately when a critical load is reached. The local damage behaviour of ductile cobalt phase in hard metals was taken into account in the simulations by using the failure curves.

The possibility of generalisation of the FE model of the mechanical behaviour of material for the 3D case has been explored. Although there are some essential limitations in the used experimental and numerical techniques, related with the difficulties of obtaining many micrographs of sections of a material volume with strictly equal distances between them, and the limited amount of elements in the mesh, which must be however very fine, this first trial of 3D simulation of deformation of real structures has demonstrated possibilities of this approach.

Although the purely one-scale level models are appropriate for solving some partial problems (like the quasi-homogeneous model for the determination of the condition of element elimination in the material with low-filler content, and the mesomodel for the simulation of coarse composites like hard metals), the multilevel approach seems to be more promising, and can allow to describe the behaviour of material taking into account real physical mechanisms of material behaviour.

Thus, the meso-macromechanical model allows simulating the damage evolution and failure in an Al/Si alloy. The mechanism of failure of the alloy has been clarified. It has been shown numerically, that the failure of a large Si-particle causes subsequent damage of the Al-matrix on the microlevel. It is found that Si-particle cracking takes place at much higher loading levels in compression than in tension and no further Al-matrix damage occurs. After cracking of the Si-particle, the material can be further loaded. Sub-critical crack growth induces plastic deformation in the matrix. Si-particles fail in macroscopic shear bands which are initiated at crack tips before ductile rupture of the specimen in the main crack plane perpendicular to the loading direction.

References

[1] Steinkopff Th.,. Sautter M, Wulf J.(1995), Mehrphasige Finite Elemente in der Verformungs- und Versagenanalyse grob mehrphasiger Werkstoffe, Arch. Appl. Mech. 65, pp. 495-506.
[2] Steinkopff Th., Sa.utter M (1995), Simulating the elasto-plastic behavior of multi phase materials by advanced finite element techniques, Comput. Mater. Sci. 4, p. 1022.
[3] Wulf J (1995), Neue Finite-Elemente-Methode zur Simulation des Duktilbruchs in Al/SiC MPI für Metallforschung, Stuttgart, p. 205.
[4] Sautter M. (1995), Modellierung des Verformungsverhaltens mehrphasiger Werkstoffe mit der Methode der Finiten Elemente, Fortschr.-Ber.VDI, Reihe 5, Düsseldorf, p. 210.
[5] Lippmann N., Steinkopff Th., Schmauder S., Gumbsch P. (1997), 3D-finite-element-modelling of microstructures with the method of multiphase elements, Comput. Mater. Sci. 9, pp. 28-35.

[6] Mishnaevsky Jr L., Lippmann N., Gumbsch P., Schmauder S., Spranger H.-J., Arzt E. (1998), Mikrostrukturelle Untersuchung der Versagensmechanismen in Al/Si Gußlegierungen mit 3D-FE-Analysen, BMBF Project report, MPA Stuttgart, p. 102.

[7] LARSTRAN, LASSO Ingenieurgesellschaft, Leinfelden Echterdingen, Germany.

[8] Mishnaevsky Jr L., Minchev O., Schmauder S.(1998), FE-simulation of crack growth using damage parameter and the cohesive zone concept, in: M.W. Brown, E.R. de los Rios, K.J. Miller (Eds.), ECF 12 - Fracture from Defects, Proceeding of the 12th European Conference on Fracture Sheffield, vol. 2, EMAS, pp. 1053-1059.

[9] Lippmann N., Schmauder S., Gumbsch P. (1996), Numerical and experimental study of the failure mechanisms in Al/Si-cast alloys, in: H. Nisitani et al. (Eds.), Localized Damage IV, Computational Mechanics Publications, Southampton, UK, pp. 333-340.

[10] Hönle S., Dong M., Mishnaevsky Jr L., Schmauder S. (1998), FE-simulation of damage evolution and crack growth in two-phase materials on macro- and microlevel on the basis of element elimination technique and multiphase finite elements, in: A. Bertram et al. (Eds.), Proceedings of the Second European Conference on Mechanics of Materials, Magdeburg, pp. 189-196.

[11] Rice J.R., Tracey D.M. (1969), On the ductile enlargement of voids in triaxial stress fields, In. J. Mech. Phys. Solids 17, pp. 201-217.

[12] Hancock J.W., Mackenzie A.C (1976), On the mechanisms of ductile failure in high-strength steels subjected to multiaxial stress-states, J. Mech. Phys. Solids 24: 147-167.

[13] Fischer F.D., Kolednik O., Shan GX, Rammerstorfer F.G. (1995), A note on calibration of ductile failure damage indicators, Int. J. Fracture 73, pp. 345-357.

[14] Gunawardena S.R., Jansson S., Leckie F.A. (1993), Modeling of anisotropic behavior of weakly bonded fiber reinforced MMCs, Acta Meth. Mater. 41, pp. 3147-3156.

[15] Arndt J., Grimpe F., Dahl W. (1996), Influence of strain history on ductile failure of steel, in: J. Petit et al. (Eds.), Mechanisms and Mechanics of Damage and Failure of Engine, Materials and Structure, Proceedings of the International Conference, ECF-II, val. 2, EMAS, Landon, pp. 799-804.

[16] Poech M.-H., Fischmeister H.F., Kaute D., Spiegler R.(1993), FE modelling of the deformation behaviour of WC/Co alloys, Comput. Mater. Sci. I, p. 213.

[17] Connolly P.J., McHugh P.E. (1997), Fracture modelling in metal ceramic composites using the Gursan model and crystal plasticity theory, Presentation at Seventh International Workshop on Computational Modelling of Materials, Vienna.

[18] Bracks W., Hao S., Steglich D. (1996), Micromechanical modelling of the damage and toughness behaviour of nodular cast iran, J. de Physique IV C6, pp. 43-52.

[19] Sigl L.(1986) Das Zähigkeitsverhalten von WC-Co-Legierungen, VDI Fortschritt-Berichte, val. 5, 110. 104, VDI Verlag, Düsseldorf.

[20] Dong M., Schmauder S.!996) Modeling of metal matrix composites by a self-consistent embedded cell model, Acta Mater. 44: 2465-2478.

[21] Lippmann N., Schmauder S., Gumbsch P. (1996) Numerical and experimental study of early stages of the failure of Al/Si cast alloy, J. de Physique IV C6: 123-131.

[22] Schmauder S., Dong M. (1997), Micro-, meso- and macromechanical modeling of damage in an Al/Si cast alloy, in: A.S. Khan, J. Petit et al. (Eds.), Proceeding of the Sixth International Symposium on Plasticity and its Current Applications, Neat Press, Fulton, Maryland, pp. 275-276.

[23] Mishnaevsky Jr., L. Lippmann, N. Schmauder S., Gumbsch P. (1999), In situ observation of damage evolution and fracture in Al/Si7MgO.3 cast alloys, Eng. Fracture Mech. 63 (4), pp. 395-411.

2.3 Automatic generation of 3D microstructure-based finite element models

2.3.1 Idealized microstructures of particle reinforced composites: multiparticle unit cells with spherical inclusions[5]

Numerical simulations of deformation, damage and fracture of composites present an important tool for the prediction of materials behavior, and the optimization of mechanical properties of materials. In most cases, such numerical simulations are carried out two-dimensionally [1]. However, it has been shown in several works that results of 2D approximations are generally not correct for 3D case [1–5,9]. So, Jung et al. [5] compared results (stress and strain distributions) of the 2D and 3D FE simulations, and found that a 2D approximation gives results which are sufficiently different from the 3D solution. Therefore, the necessity of using three-dimensional simulation methods to analyze three-dimensional problems becomes apparent.

Several methods and concepts of FE simulation of deformation and damage of multiphase materials taking into account their microstructure have been developed in last years. table 2.4 gives a short overview of these methods. One can see from the table, that the mesh design for complex 3D microstructures and incorporating the microstructure data into FE models are one of main challenges of the 3D numerical testing and design of materials, which is met by using different techniques (multiphase finite elements, hierarchical models, Voronoi cell finite elements, etc.).

The purpose of this work was to develop a simple and efficient method of automated microstructure generation and mesh design, and to carry out systematic numerical testing of microstructures of composites. To simplify and automate the design of meshes for the 3D virtual testing of microstructures, a program "Meso3D" was developed. The program works with the commercial software MSC/PATRAN and produces artificial microstructures (i.e., different arrangements of round and ellipsoidal inclusions in a matrix) on the basis of given parameters and probability distributions of particle coordinates and sizes, and generates databases for the computational (finite element) testing of the materials with the required artificial microstructures. The designed microstructures are meshed with tetrahedral elements using the free meshing technique [27].

The microstructures with the random particles arrangement were generated using the uniform random number generator. Each coordinate was produced independently, with another random number seed. After the coordinates of a first particle were defined, the coordinates of each new particle were determined both by using the random number generator, and from the condition that the distance

---

[5] Reprinted from L. Mishnaevsky Jr., Three-dimensional Numerical Testing of Microstructures of Particle Reinforced Composites, Acta Materialia 52/14, pp. 4177-4188 (2004) with kind permission from Elsevier

between the new particle and all available particles is no less than 0.1 of the given particle radius. If the condition was not met, the seed of the random number generator was changed, and the coordinates of the new particle were determined anew. In order to avoid the boundary effects, the distance between a particle and borders of the box was required to be no less than 0.05 particle radius. The coordinates of the centers of particles for the regular (and any other pre-defined) particles arrangements are read from a text input file.

Table 2.4
Some 3D approaches to the micromechanical modelling of materials

Approach	Examples and peculiarities of the approaches
Unit cell with a single inclusion	Classical approach, based on the assumption that the local behavior of the material follows a periodic pattern
Unit cells with many inclusions	The multiparticle unit cell model (see [2–4, 6 ,7]) allows a simple generalization of conclusions obtained with the unit cells on the whole specimen. The simulations can require rather high computational resources
Hierarchical models	Example: a ''super-element'' approach developed by Geni and Kikuchi [8]. In the framework of this approach, a composite is modeled with a box-shaped super element, consisting of many different cylindrical unit cells with different particle volume fraction and shapes of particles
Voronoi cell finite element models (VCFEM)	FE mesh is created by Dirichlet tessellation. Each polygon, formed by such tessellation (''Voronoi cell'') contains one inclusion at most and is used as a finite element. In the 3D version, the microstructures of composites are ''reconstructed by assembling digitally acquired micrographs obtained by serial sectioning''. 3D equivalent microstructures (with real particles replaced by ellipsoids) are tesselated into a mesh of Voronoi cells, and the deformation and damage in these equivalent structures are simulated [9–11]
3D multiphase finite element method (MPFE) for real structure simulations	A 3D real microstructure of a material is included into a FE model by assigning the phase properties to the integration points according to the digitized images of microstructures [13,14]. The microstructure of the material is reconstructed using the image

analysis of micrographs of many sections (cuts) of materials

Design of materials on the basis of virtual testing and the statistical genetic algorithm

In the framework of the approach, developed by Zohdi et al. [15–17], large-scale micromechanical simulations are carried out by decomposing the global domain into a set of computationally smaller, decoupled problems, which deal with subdomains of the global domain. Microstructures of the materials are included into the numerical models of the subdomains on the basis of the Gauss point method. The optimal shape of inclusions was determined numerically using the statistical genetic algorithm

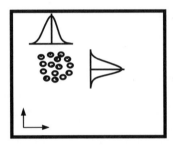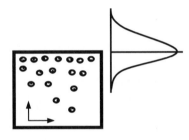

Fig. 2.32 Schema of the design of clustered (a) and gradient (b) microstructures.

In order to generate the localized particle arrangements, like clustered, layered and gradient particle arrangements, the coordinates of the particle centers were calculated as random values distributed by the Gauss law. The mean values of the corresponding normal distribution of the coordinates of particle centers were assumed to be the coordinates of a center of a cluster (for the clustered structure), or the Y- or Z-coordinate of the border of the box (for the gradient microstructure). Fig. 2.32 shows schematically examples of such a design of the microstructures. The standard deviations of the distribution can be varied, from highly clustered or highly gradient arrangements (very small deviation) to the fast uniformly random particle arrangements (a deviation comparable with the box size).

Finite element simulations and results

FE model, damage mechanisms and properties of phases

The problem was solved in the framework of the embedded cell approach [1, 26]. The main idea of the embedded cell approach is that microfields in inhomogeneous materials are determined, using a model consisting of a core ("local heterogeneous region"), embedded in an outer region that serves mainly for introducing loads into the core and has the averaged properties of the composite [13, 26]. According to Böhm [26], this strategy "avoids some of the drawbacks of the periodic field approach, especially the requirement that the geometry and all microfields must be strictly periodic". Then, the embedding allows avoiding the boundary effects on the particle failure. In our case, the FE meshes of the composites with different microstructures (a given amount of SiC particles in a box 10x10x10 mm, filled with elastoplastic Al matrix), generated with the use of the program "Meso3D" and commercial code MSC/PATRAN, were placed in a bigger box 14x14x14 mm. The embedding zone behaved as a composite with averaged properties, i.e., as an elastic–plastic material (Al/10%SiC).

The SiC particles behaved as elastic isotropic damageable solids, characterized by Young modulus E_P = 485 GPa, Poisson's ratio 0.165 and the local damage criterion, discussed below. The Al matrix was modeled as isotropic elasto-plastic solid, with Young modulus E_M = 73 GPa, and Poisson's ratio 0.345. The experimental stress–strain curve for the Al matrix was taken from [28]. The elements in the embedding were assigned the averaged mechanical properties of the Al/ SiC composite, with Young modulus E_{Av} = 75.7 GPa (for the volume content 10%) and E_{Av} = 88.4 GPa (for 15%), and Poisson's ratio 0.323 taken from [12, 13, 24]. The elasto-plastic stress–strain curve for the composite (embedding) was taken from [12] as well. The experimental stress–strain curve for the matrix as well as the constitutive law for the composite (embedding), taken from [12] were approximated by the deformation theory flow relation (Ludwik hardening law):, σ_y = σ + $h\varepsilon_{pl}$ where σ_y – the actual flow stress, σ_{yn} – the initial yield stress, and ε_{pl} the accumulated equivalent plastic strain, h and n – hardening coefficient and the hardening exponent. The parameters of the curve for the matrix were as follows: σ_{yn} = 205 MPa, h = 457 MPa, n = 0.20. For the composite (embedding), the parameters were: σ_{yn} = 216 MPa, h = 525.4 MPa, n = 0.25.

The nodes at the upper surface of the box were connected, and the displacement was applied to only one node. The model was subject to the uniaxial tensile displacement loading, 2.0 mm.

We considered cells with 5, 10 and 15 particles, the volume content of the inclusion phase was 2.5%, 5%, 10% and 15%. Totally, the models contained about 30,000 elements. Each particle contained about 400 finite elements. The radii of particles were calculated from the prescribed volume content and particle amount in the box, and were as follows: 1.1676 mm (volume content/ VC = 10%, N =

15), 0.9267 mm (VC = 5%, N = 15), 1.3365 mm (VC = 10%, N = 10) and 1.0608 (VC = 5%, N = 10).

The uniaxial tensile response of each microstructure was computed by the finite element method. The simulations were done with ABAQUS/Standard.

The damage was modelled as a local weakening of a finite element in which the damage criterion (maximum principal stress) exceeded a critical value. An ABAQUS Subroutine USRFLD, which allows simulating the local damage growth as a weakening of finite elements was developed. After an element failed, the Young modulus of this element was set to a very low value (50 Pa, i.e., about 0.00001% of the initial value). Only the elements in which the maximum principal stress exceeded a critical value (and not all the elements in the particle) were weakened in the framework of this algorithm [18–20].

Baptiste [21], Hu et al. [22] and Derrien et al. [23] have demonstrated that the main mechanism of damage initiation for the AlSiCp systems is the particle failure. According to Derrien and co-workers [23, 24], the failure of specimen occurs by linking of the microcracks initiated in the matrix from the broken particles. In the matrix near the broken particle, the cavities and microvoids nucleate, grow and coalesce, and that leads to the failure of the matrix ligaments between particles. A similar result was obtained by Kobayashi et al. [25]: „since the interfacial bonding strength is sufficiently high in the case of an Al/SiC system, the predominant factor for toughening is the fracture strength of reinforcements''. Therefore, only damage in SiC particles was considered at this stage of the work. According to [21–23], the SiC particles in AlSiC composites become damaged and ultimately fail, when the critical maximum principal stress in the particle material exceeds 1500 MPa. This value was used in our simulations as a criterion of damage of SiC particles as well. As output parameters of the numerical testing of the microstructures, the effective response of the materials and the amount of failed particles NF versus the far-field strain curves were considered. The displacement loads, at which the first particle fails and at which most particles fail, as well as the rate of failure of particles (determined as a slope of the NF–u curve in its linear part) were considered as parameters of the damage growth rate in the materials, which depend on their microstructures. To characterize the "rate of particles failure" v_F quantitatively, we calculated it as the total amount of failed particles in the cell divided by the displacement difference between the point where the first particle failed ($u_{1st\_particle\_fail}$) and the last particle failed ($u_{all\_particles\_fail}$).

The generation of a microstructure and pre-processing for each cell took about 15 min (2–3 min of this time were interactive) on the Compaq personal computer and 2 GB of RAM. The analysis of a model took about 10 h on the SERVus cluster of the University of Stuttgart.

Deformation and damage behavior of composites with the random arrangement of particles

The deformation and damage evolution of the Al/SiC composites with random SiC particle arrangements were simulated with the use of the above described model. The purpose of this part of the investigation was to verify whether the "random" particle arrangements have peculiarities as compared with regular or localized particle arrangements, and whether these peculiarities are stable, reproducible and typical for the random arrangements.

Since the random particle arrangements were generated from a pre-defined random number seed parameter (idum), (which should ensures reproducibility of the simulations), variations of this parameter lead to the generation of new microstructures. Five realizations of the random microstructures with 15 spherical particles and volume content of ceramic phase 10% (produced with different random numbers seeds) were generated and tested.

Fig. 2.34(a) shows the tensile stress–strain curves for the five random arrangements (15 particles, volume content 10%). For comparison, we included also the curves for the regular and gradient particle arrangements. Fig. 2.34(b) gives the amount of failed particles plotted versus the far-field applied strain. Fig. 2.33 shows the equivalent plastic strain in a cell with randomly arranged 10 particles (VC = 5% and 10%) both on the boundaries of the box and on the matrix/particle interfaces.

One can see from Fig. 2.34, that the effective responses of the materials with random microstructures (even in different realizations) lay very close one to another and differ from that for the regular or localized microstructures. However, some variations of both flow stress and damage behavior of different random microstructures are observed as well, especially, after the far-field strain exceeds 0.1. The difference between the stresses for different realizations of the same random structure falls in the range of 2% even at the rather high far-field strain ($\varepsilon = 0.2$). For comparison, the difference between the regular and gradient particle arrangement is about 16% at the far-field strain 0.2, and 9% at the far-field strain 0.1 (see Fig. 2.39). Therefore, although the stress–strain curves diverge a little bit when the strain is higher than 0.1, the differences between realizations of the random microstructure are still much smaller, than the difference of the mechanical response between the different types of the microstructures.

In the following parts, at least three to five realizations of random microstructures will be simulated and averaged, when a random microstructure is compared with other microstructures. The rate of particle failure is lower for all the considered random particle arrangements than for the regular and clustered microstructures: the fraction of failed particles increases from 40% to 80%, when the far-field strain increases 2.8 times in the case of the random particle arrangement, and increases from 20% to 86% when the far-field strain increases 1.5 times for the regular particle arrangement.

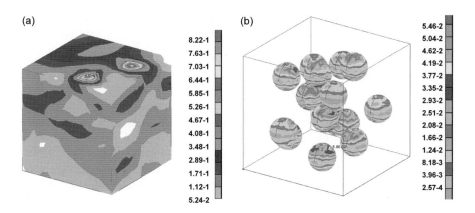

(a)

8.22-1
7.63-1
7.03-1
6.44-1
5.85-1
5.26-1
4.67-1
4.08-1
3.48-1
2.89-1
1.71-1
1.12-1
5.24-2

(b)

5.46-2
5.04-2
4.62-2
4.19-2
3.77-2
3.35-2
2.93-2
2.51-2
2.08-2
1.66-2
1.24-2
8.18-3
3.96-3
2.57-4

Fig. 2.33 Equivalent plastic strain in a cell with randomly arranged 10 particles: whole unit cell (a) and on the matrix/particle interfaces (b). VC = 10%. Total strain = 0.25.

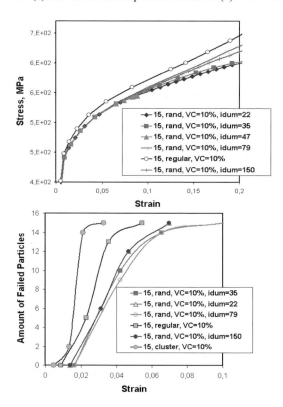

Fig. 2.34 Stress–strain curves (a) and the amount of failed particles plotted versus the far field strain (b) for the five random arrangements (15 particles, VC = 10%) and for the regular particle arrangement.

One can see that the flow stress for the regular microstructure of the composite is higher than that for the random microstructures (in the following paragraphs, we could see that the regular microstructure ensures highest flow stress among all considered microstructures). It is of interest that in the simulations by Segurado et al. [7] the flow stress for the regular BCC particle arrangement was significantly lower than for the random particle arrangement. The difference between our result and the results by Segurado et al. [7] can be caused by the fact that the model from [7] does not take into account the stress redistribution in the composite due to the reinforcement fracture. The varied distance between particles together with the effect of the interaction of cracks in neighboring particles in the case of the random particle arrangement can lead to the formation of weakened regions with high density of failed particles. The deformation of the weakened regions determines the low stiffness of the whole cell. This is not the case if the particles are placed equidistantly, as in the regular microstructures, and the local weakening in a particle is averaged over the entire specimen. Therefore, the constant large distance between particles, typical for the regular microstructures, can prevent the formation of weakened areas in our case, but not in the model by Segurado et al. [7].

Effect of the amount of particles and the volume content on the deformation and damage in the composite

At this stage of the work, the effects of volume content of hard phase and the amount of particles on the effective response and damage behavior of the composite were considered.

Fig. 2.35 shows the tensile stress–strain curves for the random particle arrangements with varied amount of particle (volume content of particles 5%, amount of particles 5, 10 and 15, Fig. 2.35(a) and the same for the volume content 10%, Fig. 2.35(b)). Fig. 2.36 gives the tensile stress–strain curves for the regular particle arrangement with varied volume content of particles (10 particles, volume content varied from 2.5% to 15%, Fig. 2.36(a), and 15 particles, volume content was varied from 5% to 15%, Fig. 2.36(b)). Fig. 2.37 shows the amount of failed particles in the box plotted versus the far-field applied strain (10 particles, varied volume content, Fig. 2.37(a) and 15 particles, varied volume content, Fig. 2.37(b)).

One can see from Fig. 2.36 that the flow stress of composite increases with increasing the volume content of particles. An increase in the volume content by 5% leads to the increase of the flow stress by 4%, both for the cells with 10 and 15 particles.

The amount of particles (at the same volume content) influences the effective response of the material only at rather high stage of particle failure, when many particles failed already. In this case, the difference between the response of the composite with 5, 10 and 15 particles increases with increasing load, and the flow stress is higher for a composite with a higher amount of particles.

The curves of the amount of failed particles plotted versus the applied strain have an (almost) linear part (up to 10–12 particles of 15 fail) and an"asymptoti" part (when the amount of failed particles slowly approaches the total amount of particles). Since the last part of the curve ("asymptotic") hardly reflects the real damage growth process, one may define a "critical applied strain" as a far-field applied strain at which the linear part of the curve goes into the „asymptotic" part of the curve. In most cases, this takes place when 80% of particles (12 particles of 15, or 8 particles of 10) fails.

The critical applied strain depends on the volume content of particles as well. One can see from Figs. 2.37(a) and (b) that the higher the volume content of particles, the lower is the critical applied strain. An increase of the volume content of particles by 5% leads to the decrease of the critical applied strain by 4; . . . ; 5%, both for the cells with 10 and 15 particles.

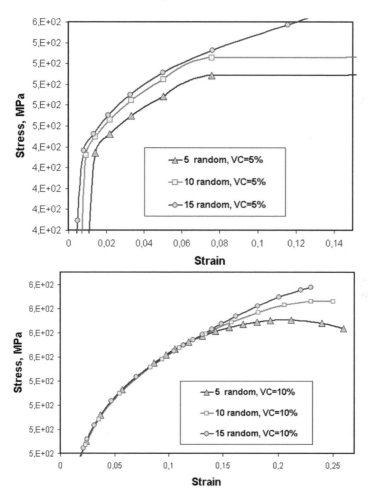

Fig 2.35 Stress–strain curves for the random particle arrangement, VC = 5% (a) and 10%, (b), amount of particles 5, 10 and 1

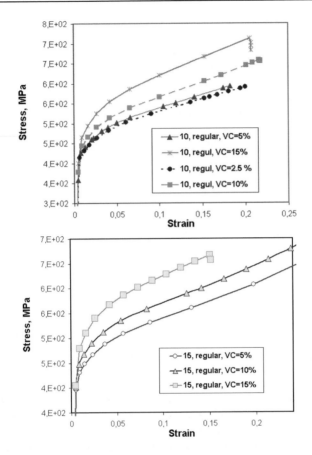

Fig. 2.36 Stress–strain curves for the regular particle arrangement: (a) 10 particles, and (b) 15 particles, volume content was varied.

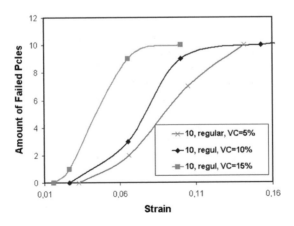

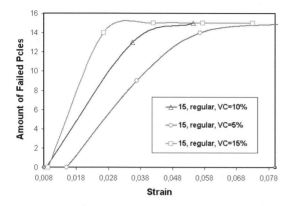

Fig. 2.37 Amount of failed particles in the box plotted versus the far-field applied strain: (a) cells with 10 particles; (b) 15 particles (varied volume content).

Effect of the clustering and gradient distribution of particles

At this stage of the work, the effects of particles arrangement and localization on the deformation and damage evolution in the composite were considered.

Fig. 2.38 shows the random (a), regular (b), clustered (c), and highly gradient (d) arrangements of the particles. Two types of the gradient particle arrangements were considered: an arrangement of particle with the vector of gradient (from low particle concentration region to a high particle concentration region) coinciding with the loading direction (called in the following a "gradient Y" microstructure), and a microstructure with the gradient vector perpendicular to the loading vector (called in the following "gradient Z" microstructure). The standard deviations of the normal distribution of the Y or Z coordinates of the particle centers (for the Y and Z gradient microstructures, respectively) were taken 2 mm, what ensured rather high degree of gradient. The same standard deviations were taken for the clustered particle arrangements. Fig. 2.39 shows the tensile stress–strain curves for the random, regular and gradient microstructures (for 10 particles, volume content of SiC 5%, Fig. 2.39(a)) and for the random, regular, clustered and gradient microstructures (for 15 particles, volume content of SiC 5% and 10%, Fig. 2.39(b)–(d)). Fig. 2.40 shows the amount of failed particles in the box plotted versus the far-field applied strain, for the same microstructures (15 particles, volume content of SiC 5% and 10%). The error bars show the deviations of the stress and the amount of failed particles from averaged values in 3–5 simulations. The deviations fall in the range of 2% for the random microstructures, and are even less for the gradient and clustered microstructures (0.5–0.9%).

It can be seen from Fig. 2.39, that the particle arrangement hardly influences the effective response of the material in elastic area or at small plastic deformation. The influence of the type of particle arrangement on the effective response of the material becomes significant only at the load at which the particles begin to fail (compare the Figs. 2.39 and 2.40). However, after the particle failure begins, the effect of particle arrangement increases with increasing the applied load. (One should note here the difference with the case when only amount of particles and neither their content nor arrangement vary: in this case, the difference becomes sufficiently large only when most particles fail, see Fig. 2.35; in the case of the different particle arrangements, the influence of the arrangement becomes strong when a first particle fail already.)

After the first particle fail, the flow stress of the composite and the strain hardening coefficient increase with varying the particle arrangement in the following order: gradient<random<clustered<regular microstructure (see Fig. 2.39).

In order to analyze the effect of particle arrangement on the strain hardening quantitatively, we determined the stress hardening coefficients for the stress–strain curves shown in Fig. 2.39(c) and (d). The strain hardening coefficients were calculated as the power in the power like equation for the true stress–true strain curves, using regression analysis. table 2.5 shows the strain hardening coefficients ðnÞ for all the considered curves. For all levels of the volume content, the particle failure rate for 2.3 times higher for the cells with 15 particles, than for the cells with 10 particles: 18; . . . ; 29 particle/mm (for the cell with 15 particles and regular particle arrangement) and 45; . . . ; 90 particle/mm (for the cell with 10 particles).

One can see from figure that the critical applied strain decreases in the following order: gradient (10%)>gradient (5%)>random (5%)>regular (5%), random (10%)>cluster (5%) and regular (10%)>clustered (10%) (the volume contents of the SiC phase are given in brackets).

The strength and damage resistance of a composite with a gradient microstructure strongly depends on the orientation of the gradient in relation to the direction of loading. In the case of the "gradient Y" microstructure, the rate of particle failure is very low (about 6.35 particle/ mm) and the particle failure begins at relatively high displacement loading, 0.2 mm. In the case of the "gradient Z" microstructure, the rate of particle failure is the same as for random microstructures. One should note that the highly gradient distributions of particles, considered here, constitute just one snapshot of many possible arrangements. Apparently, if the arrangement of particles changes from the highly gradient arrangement, considered here, to the arrangements with more slow gradients, the properties of the material will be changed. This will be the subject of further investigations.

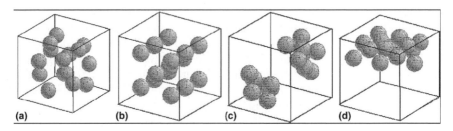

Fig. 2.38 Random (a), regular (b), clustered (c) and highly gradient (d) arrangements of the particles.

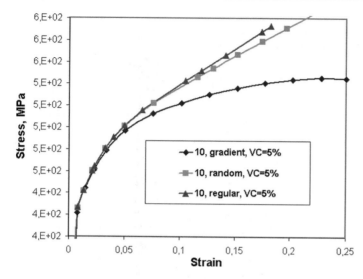

(a)

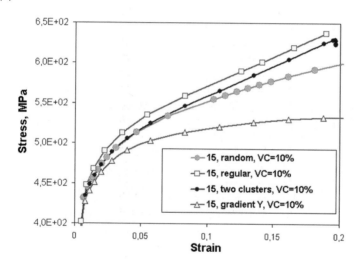

(b)

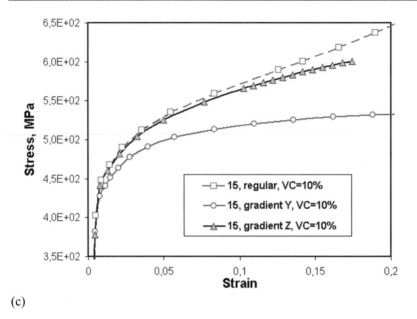

(c)

Fig. 2.39 Stress-strain curves for the unit cells with different particle arrangements: (a) 10 particles, VC=5%, (b, c) 15 particles, VC=10% microstructures

To characterize quantitatively the arrangements of particles and the degree of localization in different microstructures, we determined the nearest neighbor distances between the particle centers (NND), the nearest-neighbor index (NNI, ratio of observed to expected nearest-neighbor distance) and the statistical entropy of the nearest neighbor distance (SENND). The accurate determination of these parameters requires a much larger of particles than those generated to compute the mechanical behavior [7] and those considered in the unit cells above. To overcome this limitation, we followed the idea by Segurado et al. [7] who generated much larger unit cells, than those used in their numerical experiments, using the same algorithms, and determined the statistical parameters of microstructures with the use of the larger cells. They suggested that – the cubic unit cells used for the simulation of the mechanical behavior can be understood as small representative volume elements taken at random from the larger cell.

Table 2.5
Calculated strain hardening coefficient for different particle arrangements

Particle arrangement	Random	Regular	Clustered	Gradient (Y)	Gradient (Z)
N	0.1284	0.193	0.1676	0.1227	0.1543

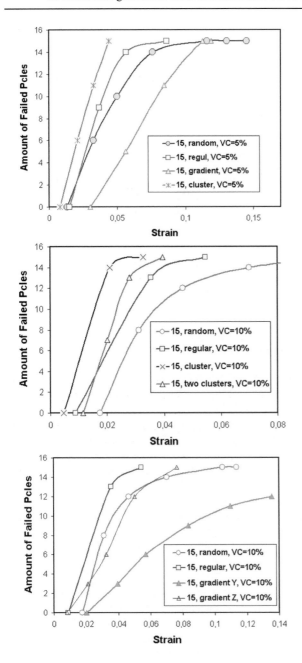

Fig 2.40 Amount of failed particles plotted versus the far field strain: (a) 15 particles, VC=5% , (b, c) 15 particles, VC=10%

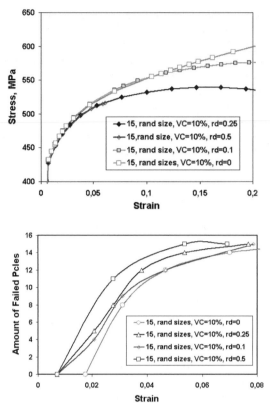

Fig 2.41 Examples of microstructures with particles of randomly distributed sizes (a), and stress-strain curves (b) and the amount of failed particles plotted versus the far-field strain (c) for these microstructures. rd = $\Delta r/r$

Table 2.6
Nearest-neighbor distance between particle centers the considered microstructures.

Microstructure	Average NND, mm	SEPD of NND	NNI
Random	12.69	2.55	1,17
Clustered	2.70	1.26	0.25
Gradient	5.85	2.31	0.54

Abbreviations: SEPD – statistical entropy of the probability distribution, NND- nearest-neighbor distance between particle centers, NNI- nearest neighbor index (ratio of observed nearest neighbor distance to the mean random distance)

Following this idea, we generated unit cells of the size 100x100x100 mm with random, clustered and uniform particle arrangements using the same algorithms as for the simulated cells. The values of NND, NNI and the entropy of NND, deter-mined from the cells, are given in table 2.6. The scattering of the distribution of NND, characterized by the statistical entropy of the distribution, decreases with decreasing the average value of NND.

One can see that the generated random microstructures is really close to the ideal random (NNI = 1.17 versus 1.0 in the ideal case) and that the degree of clustering in the generated gradient and clustered microstructures is rather high.

From the simulations presented in this paragraph, one may draw the following conclusions. The arrangement of particles influences first of all the strain hardening rate, and the damage behavior of composites. This is a remarkable difference from the effect of the volume content of inclusions, which influences the flow stress, not just its slope, i.e., strain hardening rate.

The regular particle arrangement ensures highest flow stress of a composite, especially, after the particle failure begins. The discrepancy between our results and results by Segurado et al. [7] in this point is discussed above. The clustered particle arrangement leads to a very high damage growth rate, and to a low critical applied strain. However, no negative effect of particle clustering on the effective response of the composite was noted. This conclusion is in agreement with the numerical conclusions by Segurado et al. [7] ("the increase in strength due to clustering is almost negligible for the matrix and reinforcement properties typically found in metal matrix composites").

Effect of the variations of particle sizes on the damage evolution

At this stage of the work, the effects of scattering of particle sizes and local strength on the effective response and damage behavior of the composite were investigated.

In order to vary the particle sizes, microstructures with randomly arranged particles of random sizes were generated. The radii of particles were assumed to follow the normal probability distribution law. The degree of scattering of particle sizes, which was characterized by a standard deviation of the normal distribution law, was varied at the level of 0.1, 0.25 and 0.5 of the average particle radius (which was 1.1676 mm, for the cell with 15 particles and 10% of volume content). In order to keep the volume content of particles constant, the randomly distributed radii of particles were normalized. Table 2.7 gives the maximum and minimum sizes of particles for the considered values of the standard deviations. Fig. 2.53(a) shows the arrangements of the particles for different deviations of the probability distribution of the particle sizes. Fig. 2.53(b) and (c) shows the tensile stress–strain curves (b) and the amount of failed particles (c) plotted versus the far-field strain, for these microstructures.

One can see from Fig. 2.53(a) that the variations of the particle sizes cause a strong decrease in the strain hardening rate of the composite during the elasto-plastic deformation with damage. The differences in the effective responses of the composites with different scattering of particle sizes are negligible in the elastic region, but become rather large when the particles begin to fail, and increase with increasing the density of failed particles. From Fig. 2.53(b) it can be seen that the damage evolution in the particles begins at some lower applied strain when the particle sizes vary (0.069 versus 0.175 mm, when the particle radii are constant). Also, the critical applied strain is about 22% lower for the random particle sizes with the standard deviation 0.5 r, and about 60% lower for the random particle sizes with the standard deviation 0.25r, than for the homogeneous particle radii. The difference between the cases of the constant particle radii and the randomly distributed with the deviation 0.1r is negligible, but becomes rather large for the deviations of 0.25 and 0.5r. Therefore, the scattering of particle sizes leads generally to the quicker and earlier damage growth in the composites.

Table 2.7
Maximum and minimum sizes for considered probability distributions of the particle radii

Standard deviation of radii/mean radius	Value of the standard deviation of r	Maximum particle radius	Minimum particle radius
$\Delta r/r$	Δr	r_{max}	r_{min}
0.10	0.1168	1.6595	0.7021
0.25	0.2919	1.8506	0.6282
0.50	0.5838	1.9951	0.1582

Discussion

Fig. 2.41(a) shows the values of the flow stresses, corresponding to the applied displacement u = 0.15 mm, for all the unit cells with 15 particles and volume content 10% as a column charts. For comparison purposes, the values of flow stresses for the regular particle arrangements with volume contents 5% and 15% are shown.

The column charts of the critical applied strain, shown in Fig. 2.41(b), illustrates the effect of the particle arrangement on the damage growth in the composites. Comparing Fig. 2.41(a) and (b), one may see the general tendency: the higher is the stiffness and the flow stress of a composite, the lower critical failure strain should be expected. (There are however some deviations from this relation.). Fig. 2.43 shows the critical applied strain (at which the linear ''quick growth'' part of the curve of the fraction of failed particles versus far-field strain goes into ''asymptotic'' part of these curves) plotted versus the flow stress of the composite at the displacement 0.15 mm for all the cells with 15 particles. From table 2.6 and Fig. 2.41, one can see that there is no monotonic relation between parameters of particle clustering and the effective response or rate of particle failure in the composites.

Considering the ranking of the microstructures in Fig. 2.41 one can see that the gradient microstructures demonstrate a very high damage resistance as compared with the isotropic (random, uniform and clustered) microstructures. The isotropic (random, uniform or clustered) microstructures are grouped in the left part (low critical strain) of the figure. From all the isotropic microstructures, the clustered microstructure ensures the lowest damage resistance.

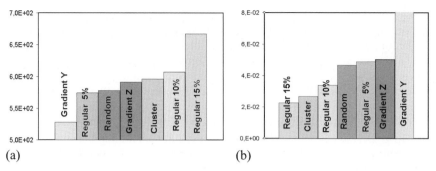

(a)   (b)

Fig 2.42 Flow stress (at the displacement load u = 0.15) (a) and critical far-field applied strain (b) for different particle arrangements. VC = 10% (if not marked)

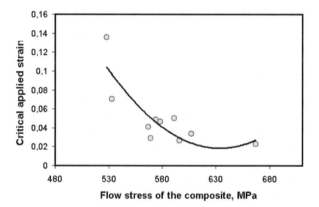

Fig. 2.43 Critical applied strain plotted versus the flow stress of the composite (at the far-field strain 0.15 mm) for all the cells with 15 particles and VC = 10%.

In this relation, the big different between the mechanical behavior of the gradient Y (gradient particle arrangement with a gradient vector coinciding with the loading vector) and gradient Z (gradient particle arrangement with a gradient vector normal to the loading vector) composites is of interest. Both microstructures (gradient Z and gradient Y) show high damage resistances, but the flow stress of these structures differs. The effect of the orientation of the gradient vector on the flow stress and damage resistance of the composite can be explained by using an (over-simplified) illustrative representation of the gradient material as a bilayer material, consisting of a plastic layer (i.e., the part of the composite with a low content of particles) and a stiff layer (i.e., the part of composite with a very high content of particles). If the material is loaded along to the gradient direction, the stiffness of the material is controlled by the stiffness of the high particle density region (which is rather firm, and, on the other hand, allows quick failure of many particles due to the high local deformation, according to the results of paragraph 3.3). Therefore, the stiffness of the gradient Z composite is quite high, and the particle failure goes rather quickly. In the case of gradient Y, the plastic flow of the material is controlled by the lower sections of the specimen with the small density of particles. Therefore, the flow stress of such a material is quite low (close to the flow stress of the pure matrix).The region of high particle density does not form a load bearing construction, and is therefore not subject to a quick failure of particles. One can draw a conclusion that if the regions of high particle density are arranged in such a way that they form a bearing construction (gradient Z structure) or play a role of a "super-reinforcement" (clustered structure), that leads to a quick failure of many particles in the regions. Otherwise, if the high particle density regions does not form a load-bearing construction (gradient Y microstructure), the flow stress of the material is relatively low (since the plastic flow is controlled by the regions of low particle density), yet, the intensity of damage failure is relatively low as well.

Conclusions

Numerical analysis of the effect of microstructure, arrangement and volume content of hard damageable inclusions in plastic matrix on the deformation and damage growth has been carried out. On the basis of the numerical testing of different microstructures, the following conclusions have been drawn.

The stiffness and flow stress of the composites can be controlled both by the volume content of the hard particles, by the degree of localization (clustering) of their arrangement, and by shape and orientation of the regions of high particle density.

Flow stress of composite increases with increasing the volume content of particles. An increase in the volume content by 5% leads to the increase of the flow stress by approximately 4%. The amount of particles (at the same volume content) influences the effective response of the material only at rather high stage of particle failure, when many particles failed already. In this case, the flow stress is higher for a composite with a higher amount of particles. The higher the volume content of particles, the lower is the far-field applied strain, at which the most of the particles fail. An increase of the volume content of particles by 5% leads to the decrease of the load at which the total failure of particles is observed by 4...5%.

The particle arrangement does not influence the effective response of the material in elastic area or at small plastic deformation. The effect of the particle arrangement on the effective response of the material becomes significant only at the load at which the particles begin to fail. After the first particle fail, the flow stress of the composite and the strain hardening coefficient increase with varying the particle arrangement in the following order: gradient < random < clustered < regular microstructure. The applied strain, at which the most particles fail, decreases in the following order: gradient > random > regular > clustered.

The variations of the particle sizes causes very strong decrease in the strain hardening rate and leads to the quicker and earlier damage growth in the composites. The differences in the effective responses of the composites with different degree of scattering of particle sizes are negligible in the elastic region, but become rather large when the particles begin to fail, and the differences increases with increasing the density of failed particles.

References

[1] Mishnaevsky Jr L, Schmauder S. (2001), Appl Mech Rev. 54(1), pp. 49.– 69.
[2] Han W, Eckschlager A, Böhm HJ. (2001), Compos Sci Technol, 61, pp. 1581–90.
[3] Eckschlager A, Böhm HJ, Han W. (2002), Comput Mater Sci , 25, pp. 85– 91.
[4] Böhm HJ, Han W. (2001), Model Simul Mater Sci Eng , 9, pp. 47–65.
[5] Jung I. et al. (1996), Mechanical behaviour of multiphase materials numerical simula-tions and experimental comparisons. In: Proc. IUTAM Symposium on Micromechan-ics of Plasticity and Damage in Multiphase Materials. Kluwer, pp. 99–106.
[6] Eckschlager A, Böhm HJ (2002), 3D Finite element unit cell study of particle failure in brittle particle reinforced ductile matrix composites. In: Mang HA, Rammerstorfer FG, Eberhardsteiner J, (Eds.) Proc. 5th World Congress Compuational Mechanics (WCCM V), Vienna.
[7] Segurado J, González C, LLorca, J. Acta Mater. 51(8), pp. 2355– 60.
[8] Geni M, Kikuchi M. Acta Mater (1998), 46(9), pp. 3125–33; Key Eng Mater 145–149, pp. 895–900
[9] Li M, Ghosh S, Richmond O, Weiland H, Rouns TN. (1999), Mater Sci Eng A A265, pp. 153–73; Mater Sci Eng A ;A266, pp. 221–40.
[10] PavanaChand Ch, Ghosh S. (1998), A 3D voronoi cell finite element model for parti-culate reinforced composites. In: Atluri SN, O'Donoghue PE, editors. Modeling and simulation based engineering. Tech Science Press; pp. 1450–5.
[11] Li M. et al. (1998), Mater Character 41(2–3), pp. 81–95.
[12] Wulf J, Steinkopff T, Fischmeister HF. (1998), Acta Mater, 44(5), pp. 1765–79.
[13] Mishnaevsky Jr L. et al. (1999), Comput Mater Sci 16(1–4), pp. 133–43.
[14] Lippmann N. et al. (1997), Comput Mater Sci;9, pp. 28–35
[15] Zohdi TI, Wriggers P. (2001) Comp Meth Appl Mech Eng;190(22–23), pp. 2803–23.
[16] Zohdi TI, Wriggers P, Huet C. (2001) Comp Meth Appl Mech Eng, 190 (43–44), pp. 5639–56
[17] Zohdi TI. Phil Transact R Soc (2003) Math Phys Eng Sci, 361(1806), pp. 1021–43.
[18] Mishnaevsky Jr L, Lippmann N, Schmauder S. (2003), Int J Fract, 120(4), pp. 581–600
[19] Mishnaevsky Jr L, Lippmann N, Schmauder S. (2003), Zeitschrift f Metallkunde, 94(6), pp. 676–81
[20] Mishnaevsky L, Weber U, Schmauder S. (2004), Int J Fract, 125, pp. 33–50.
[21] Baptiste D. (1999) Damage behaviour of composites'. Cours de formation, Jesi, Italie.
[22] Hu G, Guo G, Baptiste D. (1998), Comp Mater Sci, 9, pp. 420–30.
[23] Derrien K. et al. (1999), Int J Plast 15, pp. 667–85
[24] Mishnaevsky Jr. L, Derrien K, Baptiste D. Compos Sci Technol.
[25] Kobayashi T, et al. Numerical analysis of crack initiation and propagation behavior in aluminum matrix. Composites. Available from: http://alroom.tutpse.tut.ac.jp/efpm/English/Researchsubjects/Numericalanalysis.htm.
[26] Böhm HJ (1998), A short introduction to basic aspects of continuum micromechanics. Vienna: TU Wien;. p. 102.
[27] Thompson J, Soni B, Weatherhill N (1999), Handbook of grid generation. Boca Raton, CRC Press.
[28] Soppa E. (2003) Personal Communication.
[29] Mishnaevsky Jr L. (1998) Damage and fracture of heterogeneous materials. Rotter-dam: Balkema;. p. 230

2.3.2 Step-by-step packing approach to the 3D microstructural model generation and quasi-static analysis of elasto-plastic behavior of composites[6]

In the last few decades great progress has been made in mechanics and physics of solids, which provides new knowledge of material behavior under loading and substantially promotes our understanding of physical mechanisms of deformation and fracture. Over the recent years a new scientific trend – physical mesomechanics has been developed on the basis of a concept of multiscale nature of deformation and fracture processes. In the frame of mesomechanical methodology [14, 15, 17, 19, 28], a solid under loading is considered as a complex system of structural levels – micro, meso and macro, with deformation going on different scale levels in a self-consistent way. For instance, individual dislocations observed at the initial stage of loading tend to organize into cellular and band structures which, in turn, go on to form mesoscale shear bands. Plastic strain localization at the mesoscale level precedes macroscopic neck formation and fracture of the specimen.

A great body of experimental and theoretical data, e.g. [4, 11, 13-15, 17-19, 28, 29, 33], indicates that stress and strain distributions on the mesolevel are characterized by essential non-homogeneity attributed to the structural effects. Although an average response of a representative mesovolume assumes to agree with the experimental stress-strain curve which represents the macroscopic behavior of the material, local values of stresses and strains on the mesoscale can deviate widely from the average values [4, 11, 13-15, 19, 28, 29]. It is, therefore, a challenge to estimate stress-strain fields on the mesoscale level. In the context of this task a numerical simulation shows good promise as a research tool in addition to experimental methods.

A series of computational works [22, 24-27] has been devoted to the numerical simulation of deformation and fracture processes in heterogeneous materials with explicit consideration of their three-dimensional microstructure. In order to introduce in calculations a 3D-structure a method of a step-by-step packing (SSP) to design artificial three-dimensional structures similar to real ones by geometrical and statistical characteristics of their constituents has been developed in [22, 26].

The method proposed has been applied in [22, 24-27] to calculations of several test examples of polycrystalline and composite structures without providing a comprehensive analysis of their stress-strain behavior. First test calculations have been performed in [22, 25-27] for polycrystalline aluminum structures subjected to tension [22, 25] and shock wave loading [26, 27], using the finite-difference (FD) method [35].

In [25] the SSP-method has been applied to design a 3D MMC-structure consisting of Al(6061)-matrix and 10% of Al_2O_3-inclusions in a 50×50×50 cubic grid,

[6] Reprinted from V.A. Romanova, E. Soppa, S. Schmauder, R.R. Balokhonov, "Mesomechanical analysis of the Elasto-Plastic behavior of a 3D composite-structure under tension", Comput. Mech. 36, pp. 475-483 (2005) with kind permission from Springer

with using a procedure of excluding extra phases to obtain inclusions of uneven shape. FDM-calculations have been performed for the MMC-structure subjected to a complex quasistatic loading so that tensile displacements were given along X-direction, with two of the cube lateral faces being free of external forces and the rest three faces treated as symmetry planes. Influence of the composite structure and loading conditions on distributions of equivalent stresses and equivalent plastic strains on the surface and in the volume has been studied at specimen elongation of 0.24%. A comparative analysis of equivalent plastic strain distributions in the 3D MMC-structure and corresponding 2D-structures calculated in terms of plane strain conditions has been done in [24, 25]

The present work has its goal to examine in detail the stress-strain behavior of a MMC-structure subjected to quasistatic tension. In the frame of this contribution, the following tasks not addressed before in [22, 24-27] have been solved:

1. The SSP-procedure applied in [25] to design a two-phase structure consisting of Al(6061)-matrix and Al_2O_3-inclusions is further modified to obtain inclusions more similar to real those in their geometrical and statistical characteristics. The computer-aided design of the MMC-structure is briefly considered in Section 2. Material constants and model parameters used in calculations as well as schematics of load application are given in Section 3.
2. The mechanical problem how far the dynamic formulation can be used for the description of quasistatic problems has been solved, using both the finite element (FE) [7] and finite-difference (FD) [35, 36] methods. Special attention has been given to the comparison of FE- and FD-calculations in order to test the explicit code applicability in solving the quasistatic problem.
3. Evolution of equivalent plastic strain during tensile loading has been investigated with a comparison of meso- and macro-scale processes.
4. The analysis of stress and strain distribution in the bulk of the specimen has been provided on the base of the mesomechanical concept, with special attention paid to the quantitative estimation of local characteristics on the mesoscale level. Particular emphasis has been paid on the investigation of individual contributions from different components of the stress and strain tensors to local and global response of the material.

Discussion of the computational results is presented below.

Microstructure set-up

The incorporation of a material structure, i.e. the distribution of different material properties throughout the volume, is a non-trivial problem in the case of three dimensional simulations. An even more complicated task is to introduce explicitly a *real* 3D-structure since it requires the information about structural patterns not only on the surface but also in the bulk of the specimen. Experimental techniques [3, 9] which could provide layer-by-layer structural images are, as a rule, rather expensive and laborious. In the context of aforesaid, computer-aided modeling is considered to be a reasonable tool to design artificial 3D-structures similar to real ones by characteristics of their structural components, i.e. shape, size, volume content, spatial distribution etc.

In the general case, the design of a 3D-structure is reduced to the problem of packing a finite volume with structure elements without gaps and overlapping. Over the recent years, several methods of computer-aided simulation of microscopic heterogeneities have been developed, including Potts Monte Carlo Method [1], Voronoi tessellation [6], cellular automata [20], pseudo-front tracking method [8] and the phase-field approach [10] etc.. Some of them use a geometrical procedure to generate regular or irregular structures and others are based on certain physical principles and thermo-dynamic formulations. All of these were originally applied in two dimensional simulations and their main goal was to obtain reasonable conclusions regarding the kinetic and statistical aspects of a 2D grain growth. Although in recent works [21] some of these methods were successfully applied in a 3D case, realization of these techniques in three dimensions is for the most part a complicated task calling for optimization of computational algorithms since the memory and processing time requirements needed to make calculations increase dramatically with the increase of spatial dimensions.

In this paper, in order to design the composite structure whose mechanical behaviour will be further numerically investigated, use was made of a step-by-step packing (SSP) algorithm we proposed in [22, 26]. In contrast to the methods listed above [1, 6, 8, 10, 20, 21], this technique disregards physical and thermodynamical aspects of the microstructure evolution such as grain growth, phase coalescence, recrystallization and so on. Its goal is only to design a material structure similar to a real one, whose mechanical behaviour will be further numerically investigated.

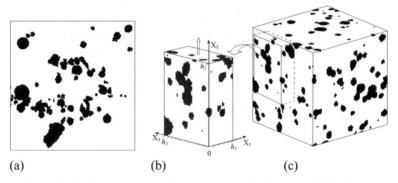

(a)   (b)   (c)

Fig. 2.44 (a) Binary image of the experimentally determined MMC-microstructure obtained by Dr. G. Fischer/ Universität Dortmund, Germany. (b) Schematics of load application to the MMC test-volume cut out from the SSP-designed composite structure (c).

The SSP-algorithm includes the following steps:
1.   The volume to be packed with the structure is discretized with the computational grid and three-dimensional coordinates are defined for each discrete point. Since the design of a 3D-structure precedes the task of

simulating its mechanical behaviour, the test-volume geometry is defined from the conditions of an appropriate mechanical problem and the discretization parameters are dictated by the numerical method to be further applied.

2. Each discrete point of the volume is assigned a so-called structural index (SI) corresponding to a certain phase. Initial conditions imply that all points possess the same SI with the exception of those treated as nuclei of new phases.

3. The discrete volume is filled with structural elements in a step-by-step fashion. At each step in the processing time, the volumes surrounding the nuclei are incremented by preset values in accordance with given law of their growth. Such a procedure repeats until the volume fraction of the growing structural elements has reached the preset magnitude. The initial distribution of nuclei, the law of volume increments and the growth rate can be defined from the analysis of experimental data.

In this work the SSP-technique has been applied to design a MMC-structure composed of Al(6061)-matrix and vol.10% Al_2O_3-inclusions, with the microstructure parameters of a real composite material [30-32]. The experimental data [30-32] indicate that Al_2O_3-particles whose size is varied from 5 to 50 μm are characterized by rather irregular shape, fig. 2.44(a). It has been found that an appropriate microstructure model can be provided via the design of a two-phase composition with the predominant growth of one of its constituents. Initially the nuclei of two sorts were randomly distributed throughout the volume discretized preliminary by a cubic grid of 100×100×100 elements, with the nuclei of both sorts in the ratio of 1:3. Although both phases developed simultaneously in accordance with the spherical law of homogeneous growth in all three spatial directions [26], the first one grew three times faster than the second one. The second phase which will be further treated as Al_2O_3 - particles stops to grow as soon as its volume content reaches the preset value of 10%, after that all unfilled places of the volume are assumed to be matrix in addition to those already possessing the matrix SI. The resulting structure contains 10% of inclusions of different sizes and shapes, fig.2.44(c).

Material properties and loading conditions

Apparently, the finer the computational grid, the better the agreement between statistical characteristics of the experimental and designed structures, provided that the SSP-laws applied in the simulation are correctly defined. This is the reason to design at first a reference structure in a sufficiently detailed computational grid of 100×100×100, fig. 2.44(c). Note, however, the numerical solution of the three-dimensional mechanical problem places more stringent requirements upon the computer memory and processing time in comparison to those needed for the structure design. In addition, the memory requirements are several times higher when using the implicit solver [7]. Since this paper has its goal to compare numerical solutions obtained by implicit [7] and explicit [35] methods, size of the test-structure has been chosen so as to satisfy the memory requirements imposed by the implicit code [7]. In this connection, further calculations have been performed for a smaller test-volume of $55\times41\times22$ elements, as a cut out from the reference structure as presented schematically in fig. 2.44(b-c). From the macroscopic point of view this test-volume is not representative since it contains only several inclusions whose total volume is about vol. 23%, with particle content in different planes varying from vol. 10 to 30 %. The reason for which this structure was chosen to be investigated is a sufficient amount of computational elements contained in the inclusions, which is important in view of the outlined goals to study non-uniform stress-strain patterns resulting from the structural heterogeneities.

Three-dimensional calculations of the MMC-structure behavior under quasistatic tension have been carried out, using both the finite-element (FE) [7] and the finite-difference (FD) [35, 36] methods. Referring to the schematics of loading, fig. 2.44(b), tension is applied to the surface $x_2 = h_2$ in the positive direction of the X_2-axis, while the planes $x_1 = 0$, $x_1 = h_2$, $x_3 = 0$ and $x_3 = h_3$ are treated as free surfaces, and the movement of the plane $x_2 = 0$ is fixed parallel to the X_2-axis. The lengths of the test-volume along the X_1-, X_2- and X_3-axes are 22, 55 and 41 microns, respectively.

Table 2.8
Material constants and model parameters used in the calculations.

Property	Al(6061)-matrix	Al$_2$O$_3$-particles
Density, g/cm^3	2.7	3.99
Shear modulus, GPa	27.7	156.0
Bulk modulus, GPa	72.8	226.0
Young modulus, GPa	68.3	380.0
σ_0, MPa	105.0	-
σ_{max}, MPa	170.0	-
ε_0	0.048	-

According to the experimental data [30-32], Al$_2$O$_3$-particles under loading possess elastic behaviour, whereas the matrix undergoes elasto-plastic deformation. Since at the elongation of 0.5% crack nucleation and development have not been registered in the experiments except of already existing cracks and voids after extrusion [30], fracture processes were disregarded from consideration. In order to describe the strain hardening in the matrix we used the fitting function of Voce type as proposed in [32]:

$$\sigma_{eq} = \sigma_0 + (\sigma_{max} - \sigma_0)\left(1 - \exp\left(-\frac{\varepsilon_{eq}^p}{\varepsilon_0}\right)\right) \qquad (2.16)$$

Here σ_{eq} and ε_{eq} are the equivalent stress and the equivalent plastic strain, σ_{max} is the saturation stress, σ_0 and ε_0 are the equivalent stress and strain corresponding to the beginning of plastic yielding. Mechanical properties of matrix and inclusions as well as fitting constants are given in table 2.8.

Mesomechanical analysis and discussion of computational results

On application of explicit code in solving quasi-static problems

Although recent advances in computer technologies extend considerably the range of mechanical and physical problems feasible to be solved, the memory and processing time requirements remain to be a significant limiting factor in the case of three-dimensional calculations. Thus, the optimization of computational methods and numerical codes as well as simplification of mathematical models continue to be topical tasks in modern computational mechanics and materials science. These problems become even more acute for models explicitly incorporating the internal structure since it requires high resolution of computational grid for the structure elements to be described with an appropriate number of details.

From the viewpoint of memory requirements, explicit numerical codes are more efficient in comparison with implicit ones. Their use, however, is strongly limited by a stability condition which imposes a restriction on the time step to be less than that needed for the elastic wave to traverse one spatial interval of the computational grid. This almost unfailingly leads to a considerable increase in computational time when calculating quasi-static problems.

In spite of this disadvantage, explicit codes have been successfully used to solve quasistatic problems for both homogeneous and heterogeneous materials [4, 11-14, 23, 36, 37]. In [37] an explicit finite-difference method earlier developed in [34] for calculations of wave propagation in elasto-plastic media was applied in 2D- and 3D-simulations of a homogeneous aluminium plate under tension. Later in a series of computational works [4, 11-14, 23] this method was adapted to problems of heterogeneous materials with explicit consideration of their microstructure and applied in 2D-calculations of elasto-plastic deformation and fracture under quasi-static loading. It has been shown for materials non-sensitive to strain rate effects [4, 11, 23] that results of dynamic and static models well agree, provided that the loading rate is slowly incremented to minimize wave phenomena.

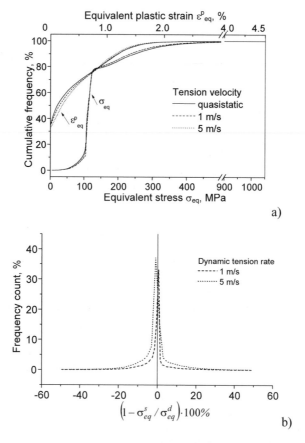

Fig. 2.45 Comparison of quasistatic(FEM) and dynamic (FD) calculations (true effective strain is 0.5%): a) cumulative frequency curves of equivalent stresses and plastic strains; b) normalized frequency count of deviation of quasistatic (σ_{eq}^{s}) and dynamic (σ_{eq}^{d}) equivalent stress fields. Note, the FD and FE methods give almost exact the same results.

The test-volume, fig. 2.44(b), was loaded by applying tension velocity to one of its ends with the other being fixed along the tensile axis. In order to reduce wave effects attributed to the dynamic model itself, the tension rate was smoothly incremented up to its constant value for the waves initiated at the loaded surface to traverse the test-volume length several times at each step of loading.

Presented in fig. 2.45 are the cumulative frequencies of equivalent stresses and plastic strains obtained in FEM- and FDM-calculations and the frequency count of difference between static and dynamic stress fields. The results demonstrate reasonable agreement at a tension rate of 1 m/s – the difference in dynamic and static stress-strain distributions is less than 10 % for most parts of the test-volume. Regarding the discrepancy of 20÷50% observed in several local regions (≈ vol.

0.7%), the comparison of dynamic response at higher and lower tension rates leads us to the assumption that it is not associated with the strain rate effects but rather attributed to numerical codes themselves since both dynamic fields exhibit the same deviation from the static one. Presumably, the reason of this disagreement is a reduction of the dynamic solution accuracy in the areas of high stress-strain gradients. In order to avoid hour-glass distortions of the computational grid in the implicit calculations use was made of an artificial viscosity [36] which can become considerable in the areas of high stress-strain gradients, reducing the accuracy. This conclusion has been additionally confirmed by a layer-by-layer analysis of dynamic and static stress-strain patterns, which has shown that the higher the stress-strain gradients in local areas, the more the disagreement between static and dynamic results. Conceivably, this disagreement can be minimised by smoothing of structure interfaces or mesh refinement.

At a loading rate of 5 m/s in the FD-simulation the deviation of dynamic and static stress-strain fields increases due to the strain hardening in matrix – the higher the loading rate, the weaker the plastic strain localisation effects. Such a disagreement should be taken into account when estimating material quasistatic response, based on the results of dynamic calculations.

In the light of aforesaid, we come to the conclusion that the dynamic code is applicable to simulations of a quasistatic response of the MMC-structure within the range of loading rates below 1 m/s, provided that the rate of loading smoothly increases up to its constant value. The use of the explicit code is expected to allow one to make 3D-calculations for a representative MMC-structure with a high resolution of the computational grid, with essentially reduced computer memory requirements.

It should be noted, however, that this conclusion cannot be extended to the case of other materials and loading conditions without additional calculations. Each kind of structure and method of loading calls for special tests in order to determine the range of strain rates in which dynamic effects become insignificant in regard to the stress-strain behaviour at the meso- and macro-scale levels.

Analysis of plastic strain evolution

In the 3D case, along with the complexity of the numerical solution, visualisation and analysis of the results also become more difficult. Due to its heterogeneity, every layer of the test-volume exhibits an individual stress-strain pattern that, in addition, evolves in time. In this paper we, therefore, give only some representative illustrations of plastic strain patterns in planes selected, with the conclusions drawn for the general case.

A step-by-step and layer-by-layer analysis of the plastic strain distribution has shown that first plastic shears take place on the test-volume surface. It is well-known [14-17, 19, 28] that geometrical special features of the specimen itself, such as corner points, notches, free surfaces and so on, result in strong stress concentration on the macroscopic level. In [22] we have shown for the case of a 3D polycrystalline material that the equivalent stress on the surface is higher than that in the bulk of the specimen.

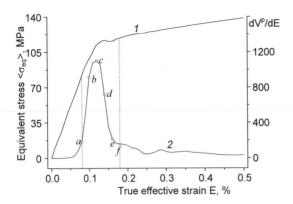

Fig. 2.46 Average equivalent stress (curve 1) and derivative of matrix volume V^P involved into plastic deformation (curve 2) vs. true effective strain.

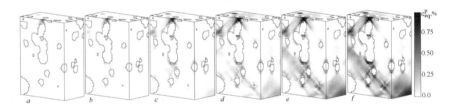

Fig. 2.47 Plastic strain patterns on the surface at different elongations of the test-volume (refer to fig. 2.46).

The reason is that normal components of the stress tensor on the surface are known to be zero, whereas in the volume all stresses make their contribution to the equivalent stress, reducing its value.

This conclusion, however, cannot be extended to the general case of composite materials. Due to considerable differences in elastic properties of matrix and inclusions, incompatible deformation and, as a result, stress concentrations in the vicinity of interfaces can result in higher stress concentrations in some internal regions than that caused by the surface effects, which can lead to crack origination not on the surface but in the bulk of the material [2]. This situation becomes even more unpredictable in the case of inclusions of an irregular shape. This was the reason, therefore, to check whether this conclusion holds for the MMC-structure under study.

Plotted in fig. 2.46 are the average equivalent stress and the increment of the matrix volume plastically deformed vs. true effective strain of the specimen. Here and below the equivalent stress σ_{eq} and its average value $\langle\sigma_{eq}\rangle$ are calculated as follows:

$$\sigma_{eq} = \sqrt{\frac{3}{2}\sigma_{ij}\sigma_{ij}} \qquad (2.17)$$

$$\langle\sigma_{eq}\rangle = \frac{1}{V}\int_V \sigma_{eq}dV \qquad (2.18)$$

where σ_{ij} are the stress tensor components, σ_{eq} is the equivalent stress, and V is the computational volume, with the i- and j-indices varied within the range of 1 to 3. Note, in the general case of a 3D heterogeneous material the average equivalent stress calculated by eq. (2.18) can deviate from both true and nominal stresses as obtained in experiments and cannot be directly associated with the experimental stress-strain curve without additional verification.

Refer to fig. 2.46, due to couple effects of tensile loading and strain hardening, plastic strain accumulation in the places already involved into yielding and origination of new plastic shears in elastically-deformed areas go alternately. Stress concentration in the vicinity of interfaces takes place from the very beginning of loading, which gives rise to plastic shear nucleation in local areas of matrix material at true effective strain of 0.004% which is $\approx$2.7 times less than the preset value of the yield strain ε_0 (see table 2.8). This phenomenon qualitatively agrees with the experimental findings in the field of microplasticity e.g.[5].

From the appearance of the first plastic shears and up to the true effective strain of 0.12% the matrix volume undergoing plastic deformation rapidly increases, which indicates a primary role of the nucleation of new plastic shears in this region. After the plastically deformed volume reaches $\approx$33% of the matrix, the rate of shear nucleation in new regions drastically decreases, that corresponds to the beginning of the strain hardening part in the average stress – true effective strain curve. The evolution of plastic strain patterns on the surface at this stage of loading is illustrated in fig. 2.47.

After plastic deformation covers $\approx$ 63% of the matrix, the curve 2 flattens out, which indicates that new regions are no longer involved into plastic yielding but plastic deformation continues to develop in the shear bands already formed. Summarizing, the formation of plastic strain localization areas is mostly completed at the stage associated with the linear part of the average stress – true effective strain curve, whereas intensive development of plastic deformation in the already formed shear bands corresponds to the stage of macroscopic strain hardening.

Stress and strain distributions on the mesoscale level

Plastic strain patterns in different planes of the 3D test-volume are mainly controlled by both microstructure distribution and orientation of the plane relative to the axis of tension. Let us consider plastic strain localization in specimen cross-sections located parallel (fig. 2.48) and perpendicular (fig. 2.49) to the axis of elongation.

From the macroscopic point of view, the parallel- and perpendicular-oriented planes are in different conditions of loading. In the first case, fig. 2.48, planes $x_1 =$ const undergo tension along the X_2-axis and compression along X_3, which results in the formation of a system of shear bands in the matrix material.

First, plastic shears nucleate near interfaces in the regions of high stress concentration which is attributed to both incompatible deformation in the places of matrix and inclusion contact and uneven contours of the interfaces. Orientation of the shear bands at an angle of $\approx 45°$ to the axis of elongation is primarily dictated by the loading conditions.

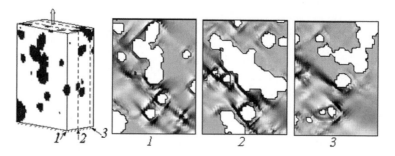

Fig. 2.48 Shaded relief maps of equivalent plastic strains in the selected planes parallel to the axis of tension. True effective strain is 0.5%.

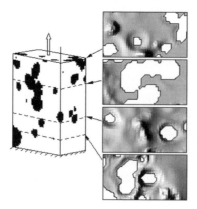

Fig. 2.49 Shaded relief maps of equivalent plastic strains in planes selected perpendicular to the axis of tension. True effective strain is 0.5%.

In planes x_2 = const placed perpendicular to the axis of tension, fig. 2.49, the material undergoes two-axial compression, which prevents shear band formation but gives rise to the plastic strain localisation around inclusions, becoming especially pronounced in the vicinity of their boundaries.

Let us analyse in more details a contribution from different components of stress and plastic strain tensors to the stress and strain patterns. Presented in fig. 2.50 are the frequency count curves of stress and strain tensor components at a true effective strain of 0.5%. The frequency distribution of stress tensor components, fig. 2.50(a), calculated with steps of 5 MPa, was normalized as (q_iN_c) 100%, where q_i is the number of computational elements within the i-th interval, and N_c is the total number of grid cells. Excepting σ_{22} which represents the material response in the direction of tension, all components of the stress tensor demonstrate a symmetric distribution, with the axis of symmetry along $\sigma_{ij} = 0$ and the frequency peaks on this line. Since the matrix takes up 67% of the total volume, it is reasonable to suggest that the stress value corresponding to the highest frequency correlates to a greater extent with the average stress in matrix.

In order to describe the plastic behaviour we used the von Mises yield criterion [7, 34, 37], according to which the material response to loading becomes plastic, provided that the second invariant σ_{eq} of the stress tensor reaches its critical value. It is interesting, therefore, to analyse the contribution of different stress components to the value of equivalent stresses calculated by Eqn. (2.17). From the analysis of the curves plotted in fig. 2.50(a), σ_{22} gives the most significant contribution to the magnitude of σ_{eq}, with the frequency peak corresponding to ≈107 MPa. Note, the stress value is $\approx23\%$ lower than that representing the average material response (point a in fig. 2.46).

The difference in these values well correlates with the volume content of Al_2O_3-particles, which suggests that the maximum frequency corresponds to the stress field in the matrix, whereas the average equivalent stresses are mostly determined by the stress fields in particles. In addition, a contribution from the other stress components to the material average response also appears to be considerable. Even though their values in matrix are close to zero, they demonstrate high stress concentration inside the inclusions and especially in the vicinity of interfaces.

Frequency curves for components of the plastic strain tensor in the matrix, calculated with steps of 0.01%, are plotted in fig. 2.50(b). These curves demonstrate the presence of places undergoing tension, compression, shear and their combination as well. Refer to the extreme and average stress and strain magnitudes given in table 2.9, all components of stress and plastic strain tensors vary in a wide range including positive and negative values. Taking into account that the negative diagonal components correspond to tension and positive ones describe compression along appropriate directions, it is interesting to mark that even σ_{22} and ε_{22}^p take negative values in several local points, which means that these areas are under compression in spite of the external tension applied in the same direction.

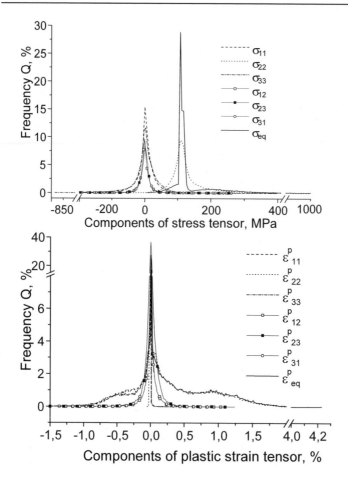

Fig. 2.50 Frequency count of stress tensor components (a) and plastic strain tensor components (b), obtained in FD-calculations (true effective strain is 0.5%).

This is a prominent illustration of specific effects attributed to a three-dimensional heterogeneity. One of the discussed topics of computational mechanics is whether two-dimensional continuum models are capable to provide a reasonable description when modelling deformation and fracture in heterogeneous materials with an explicit consideration of their microstructure and, if so, what are the frames of their applicability. Summarizing our previous experience in the field of 2D- and 3D calculations of heterogeneous media [22, 24-27], we come to the conclusion that each kind of structure and condition of loading requires special consideration to answer the question.

Although 2D-calculations are not presented in this paper, we can make some conclusions regarding this topic, based on the 3D-analysis of the stress and strain tensors. In both cases of plane strain and plane stress conditions the models are reduced to a two-dimensional problem under the assumption that certain components of the stress and strain tensors are equal to zero all over the volume. In the case of the 3D MMC-structure subjected to tension as preset above, all components of stress and plastic strain tensors in the vicinity of interfaces and in the bands of localised deformation, respectively, deviate widely from zero and, thus, make their contribution to the local values of the stress and strain tensors (see table 2.9). Due to their small relative volume, the regions demonstrating the highest values of stresses and strains do not make a considerable contribution into the average material response. So in the case of macroscopic simulation, one- and two-dimensional models provide quite satisfactory results and the heterogeneous fields on the mesolevel can be ignored without serious consequences. It should be noted, however, that the correct estimation of local characteristics takes on great significance for materials whose deformation and fracture behaviour is strongly determined by mesoscale processes and then structural effects have to be taken into account.

Table 2.9
Average and extreme values of stress and plastic strain tensor components at true effective strain 0,5%.

ε_{ij}^{p}	$\left\langle \varepsilon_{ij}^{p} \right\rangle$,	$\varepsilon_{ij\,min}^{p}$,	$\varepsilon_{ij\,max}^{p}$,	σ_{ij}	$\left\langle \sigma_{ij} \right\rangle$,	$\sigma_{ij\,min}$,	$\sigma_{ij\,max}$,
	%	%	%		MPa	MPa	MPa
ε_{11}^{p}	-0.234	-2.361	0.303	σ_{11}	-0.004	-695.1	387.0
ε_{22}^{p}	0.493	-0.058	4.131	σ_{22}	127.5	-324.1	1017.6
ε_{33}^{p}	-0.246	-2.476	0.396	σ_{33}	-0.01	-844.7	396.5
ε_{12}^{p}	-0.0007	-1.365	1.056	σ_{12}	-0.05	-193.5	265.9
ε_{31}^{p}	-0.0086	-1.161	0.873	σ_{31}	0.01	-147.1	283.1
ε_{23}^{p}	0.003	-1.763	1.238	σ_{23}	-0.06	-197.3	272.7
ε_{eq}^{p}	0.53	0.0	4.237	σ_{eq}	138.4	8.4	1017.3

Conclusions

Numerical investigation of the elasto-plastic behavior of a composite material under loading has been carried out, with taking into account its 3D structure. The analysis of stress and strain distribution in the bulk of the specimen has been provided in the frame of the mesomechanical concept, with special attention paid to the estimation of local characteristics on the mesoscale level. Deformation patterns in planes oriented parallel and perpendicular to the axis of tension have been studied. Particular emphasis has been paid on the estimation of individual contributions from different components of the stress and strain tensors to local and global response of the material. It has been shown that, due to its structural heterogeneity, material on the mesoscale level exhibits a complex stress-strain behaviour, with the stresses and strains in local areas (e.g. in the vicinity of interfaces) deviating from their average level by several orders of magnitude.

To introduce in calculations the three-dimensional MMC-structure, use was made of a SSP-algorithm proposed in [22, 26]. The benefit of this method is, undoubtedly, its simple computer-aided realization with the minimal requirements of the memory and computational time. The disadvantage is that the application of an artificial structure makes impossible a direct comparison of computational and experimental results on the mesoscale level. However, this does not exclude the possibility of an indirect verification of this type of modelling. In the assumption that macroscopic responses of a real material and its artificial model should coincide when considering a representative volume, their comparison might be a proper test to verify computational results.

References

[1] Anderson MP, Srolovitz DJ, Crest GS, Sahni PS (1984) Monte Carlo simulation of grain growth in textured metals. Acta metal. 32, pp. 783-789.

[2] Borbély A, Biermenn H, Hartmann O, Buffière JY (2003) The influence of the free surface on the fracture of alumina particles in an Al-Al$_2$O$_3$ metal-matrix composite. Comput. Mater. Sci. 26, pp. 183-188.

[3] Buffiere J, Maire E, Cloetens P, Lormand G, Fougères R (1999) Characterization of internal damage in a MMC using X-ray synchrotron phase contrast microtomography. Acta mater. 47, pp. 1613-1625.

[4] Cherepanov OI (2003) Numerical solution of some quasisatic problems in mesomechanics. SB RAS Publishing, Novosibirsk [in Russian].

[5] Dudarev EF (1998) Microplastic deformation and yield stress of polycrystals. Tomsk University Press, Tomsk [in Russian].

[6] Ghosh S, Nowak Z, Lee K (1997) Quantitative characterization and modeling of composite microstructures by Voronoi cells. Acta Mater. 45, pp. 2215-2237.

[7] Hibbit, Karlson & Sorensen Inc.: ABAQUS 6.2-1, Pawtucket, RI, USA, (ABACOM Software GmbH, Aachen).

[8] Jacot A, Rappaz M (2002) A pseudo-front tracking technique for the modelling of solidification microstructures in multi-component alloys. Acta Mater. 50, pp. 1902-1926.

[9] Jensen DJ (2002) Microstructural characterization in 3 dimensions. In: Pyrz R, Schjødt-Thomsen J, Rauhe JC, Thomsen T, Jensen LR (eds) New Challenges in Mesomechanics (Mesomechanics'2002), Proceedings of International Conference. Aalborg University, Aalborg, pp 541-547.

[10] Krill CE, Chen L-Q (2002) Computer simulation of 3-D grain growth using a phase-field model. Acta Mater. 50, pp. 3057-3073.

[11] Makarov PV (2000) Localized deformation and fracture of polycrystals at mesolevel. Theor. Appl. Fract. Mech. 33, pp. 23-30.

[12] Makarov PV, Schmauder S, Cherepanov OI, Smolin IYu, Romanova VA, Balokhonov RR, Saraev DYu, Soppa E, Kizler P, Fischer G, Hu S, Ludwig M (2001) Simulation of elastic plastic deformation and fracture of materials at micro-, meso- and macrolevels. Theor. Appl. Frac. Mech. 37, pp. 183-244.

[13] Makarov PV, Smolin IY, Prokopinsky IP, and Stefanov YuP (1999) Modeling of development of localized plastic deformation and prefracture stage in mesovolumes of heterogeneous media. Int J. Fract. 100, pp. 121-131.

[14] Panin VE (Ed) (1998) Physical mesomechanics of heterogeneous media and computer-aided design of materials, Cambridge International Science Publishing, Cambridge

[15] Panin VE (2003) Physical mesomechanics – a new paradigm at the interface between physics and mechanics of solids. Physical Mesomechanics. 6, pp. 9-36.

[16] Panin AV, Klimenov VA, Abramovskaya NL, Son AA (2000) Defect flow nucleation and development on the surface of a deformed solid. Physical Mesomechanics. 3, pp. 83-92.

[17] Panin VE (1998) Foundations of physical mesomechanics. Physical Mesomechanics. 1, pp. 5-22.

[18] Pleshanov VS, Kibitkin VV, Panin VE (1999) Optico-television estimation of fracture behavior and crack-resistant characteristics on the mesoscale for polycrystals under cyclic loading. Physical Mesomechanics. 2, pp. 87-90.

[19] Pyrz R, Schjødt-Thomsen J, Rauhe JC, Thomsen T, Jensen LR (Eds) (2002) New Challenges in Mesomechanics. Proceedings of Int. Conference. Aalborg University, Aalborg.

[20] Raabe D (2002) Cellular automata in materials science with particular reference to recrystallization simulation. Annual Review of Materials Research. 32, pp. 53-76.

[21] Raabe D, Roters F, Barlat F, Long-Qing Chen (Eds) (2004) Continuum scale simulation of engineering materials. Wiley-VCH Verlag GmbH & Co. KGaA.

[22] Romanova V, Balokhonov R, Karpenko N (2004) Numerical simulation of material behavior with explicit consideration for three-dimensional structural heterogeneity. Physical Mesomechanics. 7, pp. 71-79 [in Russian].

[23] Romanova VA, Balokhonov RR, Makarov PV, Smolin IYu (2000) Numerical modeling of the behavior of a relaxing medium with an inhomogeneous structure under dynamic loading. Chem. Phys. Reports. 18, pp. 2191-2203.

[24] Romanova V, Balokhonov R (2004) Numerical investigation of mechanical behavior of metal matrix composite Al/Al2O3 with explicit consideration for its three-dimensional structure. Physical mesomechanics. 7, pp. 27-31 [in Russian].

[25] Romanova V, Balokhonov R (2004) Numerical simulation of elastoplastic deformation of artificial 3D-structures. In: Sih G, Kermanidis Th, Pantelakis Sp (eds) Multiscaling and Applied Science. Proceedings of International Conference "Mesomechanics-2004". University of Patras, pp 266-272.

[26] Romanova V, Balokhonov R, Makarov P, Schmauder S, Soppa E (2003) Simulation of elasto-plastic behaviour of an artificial 3D-structure under dynamic loading. Comput. Mater. Sci. 28, pp. 518-528.

[27] Romanova V, Balokhonov R, Soppa E, Schmauder S, Makarov P (2003) Simulation for elasto-plastic behavior of artificial 3D-structure under shock wave loading. J. Phys. IV France. 110, pp. 251-256.

[28] Sih G. (ed) (2000) Role of Mechanics for Development of Science and Technology. Proceedings of Int. Conference. Xi'an Jiaotong University.

[29] Sih G, Kermanidis Th, Pantelakis Sp (eds) (2004) Multiscaling and Applied Science. Proceedings of International Conference "Mesomechanics-2004". University of Patras.

[30] Soppa E, Schmauder S, Fischer G (2004) Particle cracking and debonding criteria in Al/Al$_2$O$_3$ composites. In: Sih G, Kermanidis Th, Pantelakis Sp (eds) Multiscaling and Applied Science. Proceedings of International Conference "Mesomechanics-2004". University of Patras, pp 312-317.

[31] Soppa E, Schmauder S, Fischer G, Brollo J, Weber U (2003) Deformation and damage in Al/Al$_2$O$_3$. Comput. Mater. Sci. 28, pp. 574-586.

[32] Soppa E, Schmauder S, Fischer G, Thesing J, Ritter R (1999) Influence of the micro-structure on the deformation behaviour of metal–matrix composites. Comput. Mater. Sci. 16, pp. 323-332.

[33] Toyooka S, Widiastuti R, Zhang Q, Kato H (2001) Dynamic observation of localized pulsation generated in the plastic deformation process by electronic speckle pattern interferometry. Jpn. Appl. Phys. 40, pp. 873-876.

[34] Wilkins M (1967) Calculation of elasto-plastic flow. In: Older O, Fernbach S, Rotenberg M (eds) Numerical methods in hydrodynamics. Mir, Moscow, pp 212–263.

[35] Wilkins M, French S, Sorem M (1975) Finite-difference scheme of 3D-spatial and time-dependant problems. In: Numerical methods in fluid dynamics. Mir, Moscow, pp 115-119.

[36] Wilkins M (1980) Use of artificial viscosity in multidimensional fluid dynamic calculations. Journal of Computational Physics 36, pp. 281-303.

[37] Wilkins M, Guinan M (1976) Plane stress calculations with a two dimensional elastic-plastic computer program. Preprint UCRL-77251, University of California, Lawrence Livermore Laboratory.

Chapter 3: Simulation of Damage and Fracture

The purpose of this chapter is to describe methods of computational modelling of damage and fracture in heterogeneous materials, and to demonstrate their applications for the analysis of the microstructure-strength and microstructure-fracture resistance relationships of different materials.

Micromechanical aspects of crack extension in multiphase materials are essential in understanding fracture characteristics on a macroscopic scale. Typically, hard second-phase particles fracture or decohere in the course of externally applied loading before the matrix between these particles fails and macroscopic crack advance is observed. The section 3.1 discusses some of these phenomena and methods of their modelling for the ductile/brittle systems WC/Co, Al/SiC and Al/Si.

In section 3.2, several methods of modelling of ductile fracture of composites are presented. An approach to the damage and fracture modelling, based on the automatic elimination of elements based on criteria such as critical plastic strain or stress triaxiality, is introduced. The method is applied to shear band failure and cavity nucleation. Further, an elastic-plastic finite element analysis of the crack-tip field in a WC-Co alloy is presented. A model in which a Co-phase was embedded at the crack-tip in an elastic solid was employed, and Gurson's constitutive equations for a porous plastic material were used for the Co-phase in order to take into account the nucleation and growth of microvoids. Effects of the shape of Co-phase and the stress state (plane stress or plane strain) on the distributions of hoop stress, hydrostatic stress and microvoid volume fraction are discussed based on the computational results. The process of ductile fracture under constraint of deformation is also discussed in this section.

In order to model the crack growth in a WC/Co hard metals, a micromechanical model, consisting of a unit cell with a cobalt island in a carbide environment, embedded in a composite surrounding, is developed. The energy release rate is calculated for a crack propagating along the symmetry plane of the model on a microscopic scale. The cobalt phase influences the crack driving force in an important way. The energy release rate of a crack approaching the cobalt phase increases, while it decreases rapidly for the crack propagating towards the center of the cobalt island. Parametric studies are carried out to determine the influence of different cobalt inclusion shapes and cobalt volume fractions on the energy release rate. Moreover, the energy release rate is calculated for a unit cell with two square cobalt inclusions and compared to crack propagation in a computational cell with a single inclusion

In section 3.3, numerical simulations of crack initiation and growth in real microstructures of materials with the use of multiphase finite elements (MPFE) and the element elimination technique (EET), are employed for the simulation of crack growth in idealized quasi-real microstructures (net-like, band-like and random distributions of primary carbides in the steels). On the basis of a comparison of fracture resistances of different microstructures, recommendations to the improvement of the fracture toughness of steels are developed. The fracture toughness and the fractal dimension of a fracture surface are determined numerically for each microstructure. It is shown that the fracture resistance of the steels with finer microstructures is sufficiently higher than that for coarse microstructures. Three main mechanisms of increasing fracture toughness of steels by varying the carbide distribution are identified: crack deflection by carbide layers perpendicular to the initial crack direction, crack growth along the network of carbides and crack branching caused by damage initiation at random sites.

Further, a systematic computational study of the effect of microstructures of materials reinforced with brittle hard particles on their fracture behavior and toughness is presented. Crack growth in particle-reinforced materials (here, in high speed steels) with various artificially designed arrangements of brittle inclusions is simulated using microstructure-based finite element meshes and an element elimination method. Along with simple microstructures, layered and clustered arrangements, with different inclusion sizes and orientations, have been considered. Crack paths, force-displacement curves, fracture toughness and fractal dimension of fracture surfaces are determined numerically for each microstructure of the materials. It is demonstrated that extensive crack deviations from the initial cracking directions and an increase in fracture toughness can most efficiently be achieved by using complex microstructures, such as alternated layers of fine and coarse inclusions.

In the last section of this Chapter, the interface fracture for metal/ceramic compounds is considered. The influence of the plastic properties of the metal part on the interface strength and on the energy release rate is examined.

3.1 Crack Growth in Multiphase Materials[1]

3.1.1 Failure Phenomena and Criteria for Crack Extension

In this subsection, attention is focused on two-phase materials consisting of a brittle and a ductile phase, as exemplarily shown in Fig. 3.1 for the case of an Al/SiC (20vol. %) metal matrix composite (Schmauder *et al.* 1996). Principally, three failure modes have to be taken into account when cracking on the microstructural level is considered: particle cracking, interface debonding, and matrix failure. In Fig. 3.1, matrix failure is the relevant cracking mode while only few particles are cracking or decohere.

In most cases, cracking of hard brittle (e.g., ceramic) particles follows a normal stress criterion (e.g., Lippmann *et al.* 1996), similarly as interface debonding. Such critical parameters depend on microstructural details as particle size and shape. It is well accepted to obtain these parameters by inverse modeling where these stresses are calculated for the respective microstructural event at failure load (Lippmann *et al.* 1996).

In order to simulate ductile failure of metallic phases, damage models have to be applied. Among them, Gurson-type damage models are the most prominent ones in which ductile failure is approximated by following the micromechanical phenomena of nucleation, growth, and coalescence of voids. These types of models have been proven to succeed in the case of macroscopic failure description of ductile metals. However, they were found to be of limited value in ca se of microstructural crack advance in the metallic phase of brittle/ductile composites.

Another, more promising approach is to apply local failure criteria for crack propagation in the metallic matrix phase of a microstructure as in Fig. 3.1.

Amongst these failure criteria, critical normal ($\sigma_{n,c}$) or equivalent stresses ($\sigma_{v,c}$) as well as critical plastic strains ($\varepsilon_{pl,c}$) or triaxialities ($\eta_c = (\sigma_H/\sigma_v)_c$ where σ_H = hydrostatic stresses) and mixed criteria (e.g., $\varepsilon_{pl} \geq \varepsilon_{pl,c}$ and $\sigma_H \geq \sigma_{H,c}$) have been proposed and tested. It was found that all of these criteria failed to reproduce the crack path in the matrix, especially in the vicinity of the second-phase particles. However, all of the above mentioned criteria for failure prediction in the matrix are suitable to predict void formation during sintering or crack propagation.

A successful way to model ductile failure in a microstructure is based on the work of Rice and Tracey (1969) and Hancock and Mackenzie (1976). The damage parameter was originally written as:

[1] Partially reprinted from S. Schmauder, "Crack Growth in Multiphase Materials", Encyclopedia of Materials: Science and Technology, Elsevier Science Ltd., pp. 1735-1741 (2001), with kind permission from Elsevier

$$D = \alpha \int_0^{\varepsilon_{pl}} e^{3\eta/2} d\varepsilon_{pl} \qquad (3.1)$$

where α is a constant and failure initiation takes place at a critical damage parameter value of D_c. This damage parameter takes into account the complete failure history and evaluates the triaxiality as well as the plastic straining in a multiplicative manner. It may thus be seen as an energetic criterion. Recently, failure curves have been derived for a number of materials and expressed as (Arndt *et al.* 1996). These curves separate the areas of failure ($\varepsilon_{pl} \geq \varepsilon_{pl,c}$), from those where no failure ($\varepsilon_{pl} < \varepsilon_{pl,c}$) occurs and were recently extended to a three-parameter approach. In Eqn. (3.2) A and B are material dependent parameters. The following damage parameter (Hönle *et al.* 1998). is an extension of Eqn. (3.1) and consistent with Eqn. (3.2). By this definition, failure will occur when $D \geq D_c = 1$.

$$\varepsilon_{pl,c} = Ae^{-B\eta_c} \qquad (3.2)$$

$$D = \frac{1}{A} \int_0^{\varepsilon_{pl}} e^{B\eta} d\varepsilon_{pl} \qquad (3.3)$$

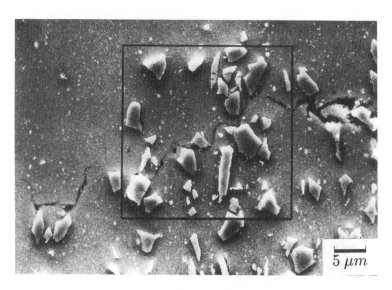

Fig. 3.1 Crack path of an Al/SiC (20vol.%) metal matrix composite which is used for the simulation (the crack approaches from the right-hand side where the main crack tip is visible; the most advanced crack tip is seen on the left-hand side of this cross-section).

3.1.2 Micromechanics of Deformation in Multiphase Materials

Deformation analyses of multiphase materials have been performed numerously in the late twentieth century for inclusion type of microstructures as well as for inter-penetrating microstructures. From these analyses it was found that strengthening as well as failure initiation in the matrix are strongly dependent on volume fraction and arrangement of the particles and only to a minor degree on the shape of the particles. While shear bands in regularly arranged inclusion-type microstructures propagate throughout the material they are of finite length in real inclusion-type microstructures and in interpenetrating microstructures. However, strong localized shear is present in the vicinity of particulates, independent of whether particle/matrix interfaces are debonded or not. Plastic straining in combination with stress triaxialities is, therefore, elementary for the failure process in the matrix.

Ductile/Brittle Composites

In this section, three examples of crack extension in multiphase materials are presented for inclusion-type microstructures with a ductile metal matrix and hard Si or SiC particles as well as for interpenetrating WC/Co microstructures.

Al/SiC

In order to model multi phase materials, multiphase elements have been introduced. This method is based on the approach that Gaussian points of one element may belong to different phases. Thus, the FE-mesh is independent of the phase structure and the preprocessing is simplified. The crack path is modeled by the elimination of elements exceeding a predefined failure criterion.

Fig 3.1 shows a crack path in an Al/SiC (20vol. %) metal matrix composite (MMC) which is characterized by the failure of the Al-matrix (Schmauder *et al.* 1996). Fracture of SiC-particles occurs rarely, depending on the length/height relation and on the size of particles. Nucleation of pores and decohesion of SiC-particles from the matrix is observed in front of the main crack tip. The fracture surface is characterized by dimples as a result of the failure in the Al-matrix. Thus, modelling the fracture behavior of Al/SiC MMCs can be reduced to the simulation of ductile crack propagation in the Al-matrix. Particle fracture is not considered.

In the present study, Rice and Tracey's damage parameter (Eqn. (3.1)) is applied to areal microstructure for the first time. Figure 3.2 gives an overview of the thus calculated crack path. Using the damage parameter, the crack path is fairly well simulated in comparison to the experimental results. In agreement with the experimental observations the nucleation, growth, and coalescence of pores was modeled. The stages of ductile failure as well as the whole crack path are modeled accurately.

Al/Si

The mechanical properties of subeutectic AlSi-cast alloys are critically determined by the microstructure of the Si-eutectic (Lippmann *et al.* 1996). With respect to the optimization of the mechanical properties, it is necessary to obtain an understanding of the failure mechanisms on a microscopic scale. Modeling of real microstructures with the finite element method permits the simulation of several phases as well as external notches.

In-situ tensile testing in a scanning electron microscope (SEM) is a simple method to investigate accurately the course of events during the failure of such materials. The concentration of stress in a confined region is necessary to study all details of crack initiation at the materials surface. Consequently, notched specimens with an elastic stress concentration factor and a slender ligament are used.

Material properties of the compound as well as of the individual microstructural constituents are needed as input parameters for the simulation. The stress strain diagram for the AlSi-cast alloy was experimentally determined. A linear-elastic behavior with a Young's modulus of 116GPa is assumed for the Si particles in the FE-model. The constitutive relation of the matrix is introduced in the calculations with the following functional dependence where σ_0 is the yield stress (200MPa), E is Young's modulus (70 GPa), and n is the hardening exponent (0.25). Initial deviations from the linear-elastic behaviour in the stress-strain diagram are related to the brittle fracture of Si-particles. Especially, in the notch region, a strong plastification of the matrix in conjunction with particle failure is observed and the Si-particles appear to be the "weak link" in the microstructure.

Assuming a normal stress criterion for brittle fracture, a failure stress of about 310 MPa is determined by inverse modeling and used as the failure criterion for the Si-particles in the FE-simulation of the crack initiation. Using multiphase finite elements, a cell representative of the two-phase microstructure is embedded in the central region of a two-dimensional model of the in-situ tensile specimen with 15256 triangular elements.

The different stages of crack initiation are simulated with a two-criterion model. Elements, which are assigned to the Si-particles, are eliminated on the basis of a normal stress criterion at fracture stress.

$$\sigma = \sigma_0 \left(\frac{E\varepsilon}{\sigma_0} \right)^n \qquad (3.4)$$

Rice and Tracey's void growth model is used as failure criterion for the subcritical crack growth in the matrix (Lippmann *et al. 1996*).

In the course of the calculation, first cracks are formed in Si-particles in the notch ground region at a total strain of about 0.2%. In agreement with the experimental observations a large amount of Si particles break before matrix failure occurs. At the tip of the cracks, which are formed by consecutive particle cracking, the damage parameter locally exceeds the critical value at higher total strains (0.27%). In the model, this is the onset of crack propagation into the matrix by further element elimination.

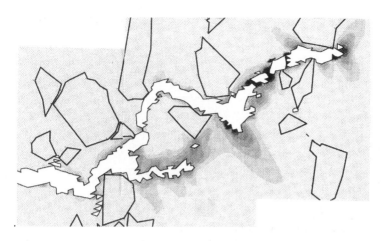

Fig 3.2 Calculated crack path in Al/SiC (20vol.%) (inside the frame in Fig. 3.1).

A comparison between experiments and calculations shows good agreement with respect to the crack pattern (Fig. 3.3) as well as the crack initiation strain of approximately 0.18 %.

WC/Co Hardmetals

When investigating deformations and failure of composites it is worth noting that may exist between the surface and the bulk of the material. The numerical analysis of a microstructural area in *WC/Co* under external tensile loading, which is embedded in a composite surrounding to account for correct boundary conditions, provides detailed information about different local plastifications. While plane stress conditions prevail at the surface, the interior is represented by plane strain assumptions.

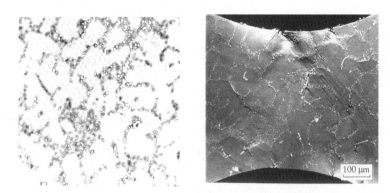

Fig 3.3 Crack path in AlSi with globular microstructure-comparison of (a) calculation and (b) experiment.

The plastic zone at the crack tip is different with respect to its shape and size. Voids can form underneath the materials surface under high hydrostatic constraint. Triaxial stresses are a valid indicator for this constraint, especially in the plastically deformed area where voids will actually form. First void indications are thus found ahead of the crack tip in agreement with the experimental observation. WC/Co hardmetals are produced in a sintering process. Due to the different thermal expansion coefficients of WC and Co high residual stresses appear in either phase; compressive in WC and mainly tensile in Co which results in tiny sintering voids with a diameter well below one micrometer (Fig. 3.4).

The locations of these sintering voids as well as of voids which arise when extern al load is applied (Fig. 3.5 (a)) can be successfully predicted by applying the S-criterion which takes hydrostatic stresses and triaxiality into consideration. A comparison of experiment and calculation is shown in Fig. 3.5(b) with respect to loading voids

$$S = \sigma_H^2 / \sigma_v = \max \tag{3.5}$$

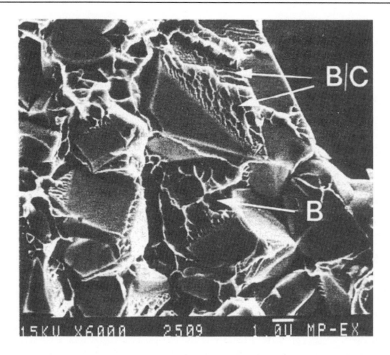

Fig 3.4 Experimentally observed crack path in WC/Co alloy: SEM fractograph ("B", large dimples, corresponding to large pores); in the binder but close to carbide crystals ("BIC", small dimples). Reprinted from H.F. Fischmeister, et al., Modelling Fracture Processes in Metals and Composite Materials, Z. Metallkde. 80, pp. 839-846 (1989)

Unit cells containing an initial void (referring to the sintering porosity) have been recently investigated by crystal plasticity means in the framework of continuum mechanics. When stretched, the void is enlarged and the unit volume is assumed to fail when a critical void volume fraction of $f_c = 0.15 \div 0.25$ is reached. This procedure has been repeated for different Co-crystal orientations and under multiaxial loading conditions. As a result, a damage curve and the parameters A and B of the damage parameter (Eqn. (3.3)) have been derived (Hönle *et al.* 1998):

$$D = \frac{1}{1.04} \int_0^{\varepsilon_{pl,c}} e^{1.12\eta} d\varepsilon_{pl} \qquad (3.6)$$

This criterion was applied to a hardmetal with a high cobalt content where no contact between carbide particles was modeled and crack evolution under external tensile loading was investigated. This study focused on the failure behavior of the ductile Cophase. Brittle fracture in the carbide phase was thus suppressed.

In the simulation, the crack enters the real structure by initiating a void, which starts to grow under increasing load. Further increase of the applied load leads to void initiation in the matrix in front of the crack tip and coalescence with the main crack. Crack propagation is thus found to be a consequence of nucleation, growth, and coalescence of the voids. This numerical study is in agreement with experimental findings on WC/Co hardmetals.

The experimental force-displacement curve for crack propagation in WC/Co depicts the failure behavior of WC/Co hardmetals as quasi-brittle fracture in agreement with the present calculations where highly constraint Co-areas fail preferably in a rather brittle manner: the applied load increases nearly linearly, while it drops immediately when the critical load is reached.

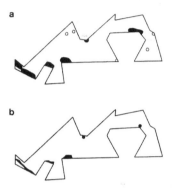

Fig 3.5 (a) Experimentally observed void formation locations in a Co-ligament: loading voids (filled) and sintering voids (open circles), (b) calculated loading void formation.

Actual tendencies in modeling crack extension in multi phase materials include the consideration of crystal plasticity effects, three-dimensional modelling and high temperature effects, and further intensifications of comparisons with experiments. Moreover, multiscale level modeling, size effects as well as the interaction of microcracks are under present investigations.

Conclusions

An overview has been given on the present state of the art in the simulation of crack extension in multiphase materials. Phenomena and criteria for crack extension have been discussed and examples been given for Al/SiC, Al/Si, and WC/Co. The results compare very well with experimental observations.

References

[1] Arndt J, Majedi H, Dahl W (1996), Influence of strain history on ductile failure of steel. J. Phys. IV Coll. C6, C6-23-32.
[2] Hancock J W, Mackenzie AC (1976), On the mechanisms of ductile failure in high-strength steels subjected to multi-axial stress-states. J. Mech. Phys. Sol. 24, pp. 147-169.

[3] Hönle S, Dong M, Mishnaevsky L Jr, Schmauder S 1998 FE simulation of damage evolution and crack growth in twophase materials. In: Bertram A, Forest S, Sidoroff F (eds.), Proc. Euromech-Mecamat, Mechanics of Materials with Intrinsic Length Seale: Physics, Experiments, Modeling and Applications. Otto-von-Guericke-Universität Magdeburg, pp. 189-196.

[4] Lippmann N, Schmauder S, Gumbsch P (1996), Numerical and experimental study of early stages of the failure of AlSi-cast alloys. J. Phys. IV Coll. C6, C6-123-31.

[5] Rice J R, Tracey DM (1969), On the ductile enlargement of voids in triaxial stress fields. J. Mech. Phys. Solids 17, pp. 201-217.

[6] Schmauder S, Wulf J, Steinkopff Th, Fischmeister HF (1996), Micromechanics of plasticity and damage in an Al/Sie metal matrix composite. In: Pineau A, Zaoui A (oos.) IUT AM Symp. Micromechanics of Plasticity and Damage of MultiphaseMaterials. Kluwer, Dordrecht, The Netherlands, pp. 255-262.

3.2 Ductile damage and fracture

3.2.1 *Numerical modelling of damage and fracture in Al/SiC composites: element removal method[2]*

A failure criterion, based on the development of a void in a highly constrained environment (so-called S-criterion) has been successfully employed to the modelling of the formation of discrete voids in WC-Co hard metals [1,2]. However, these FE calculations suffer from the disadvantage that the mesh has to be manipulated manually at each loading step. The present investigation provides an automatic crack propagation algorithm which will simulate the appearance of voids by automatically removing those elements of the FE-net which reach a given failure criterion during loading. The choice of the failure criterion can be adapted to the physical situation and the preference of the user.

Method

The main feature of the present method is to automatically eliminate elements which reach critical loads of global quantities (e.g. critical stress intensities) or local quantities (e.g., critical plastic strain, ε_{pl}, hydrostatic stress, σ_H, or von Mises equivalent stress, σ_v, single components of the stress tensor as well as arbitrary combinations of these quantities). These elements will no longer be considered in subsequent loading steps. In this way the nucleation, growth and coalescence of voids and the growth of ductile cracks can be simulated. At the present stage, triangular elements with a quadratic displacement field are used for three reasons: (i) triangular elements possess the special topological property of filling an arbitrary n-edged polygon without leaving gaps, which is not possible with, e.g. quadrilaterals; this is important in modelling irregularly shaped inclusions; (ii) quadratic displacement fields facilitate the modelling of typical void shapes, in particular circles can be approximated very well; (iii) higher order displacement fields require much greater computation times for a small gain in accuracy. An element fulfilling the failure criterion is not erased from the FE-data base but the stresses in that element are set equal to zero. Elements of zero stress no longer possess physical relevance.

---

[2] Reprinted from J. Wulf, S. Schmauder, H. Fischmeister, "Finite Element Modelling of Crack Propagation in Ductile Fracture", Computational Materials Science 1, pp. 297-301 (1993) with kind permission from Elsevier

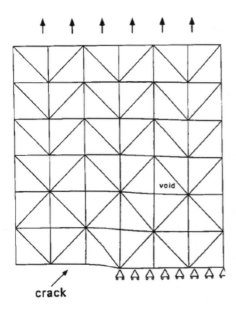

crack

Fig. 3.6 Coarse mesh of a simple crack-tip geometry with triangular void in an elastic-plastic model. Upper part of the symmetric model is shown at 5% overall strain.

The elastic-plastic large-strain FE-program "LARSTRAN" [3] is used for the present purpose. It is applied in a displacement controlled manner. From the displacement field, the strain tensor is calculated. Then the stress tensor and finally the force field are determined. An element whose stresses have been set to zero will carry no forces and will not contribute to further build-up of stresses and will not hamper deformation in the surroundings. To search a solution in the "right direction" the stiffness matrix $K(u_n)$ is corrected by setting Young's modulus of the eliminated elements to zero.

To avoid a singular stiffness matrix, nodes which are no longer associated with any element are removed from the matrix.

The main feature of the Newton-Raphson algorithm used here is to solve an equation of the following form,

$$K(u)u + f = 0 \tag{3.7}$$

where the forces, f, are calculated from the external forces. Equation (3.1) is solved as a problem of fixed points[3]. Setting now

---

[3] A problem of the form $g(x) = 0$ can be reformulated as a problem of fixed points by writing $g(x) + x = x$ and defining a new function $h(x) = g(x) + x$. This leads to $h(x) = x$. In the above case identify $g(x) = K(u)u + f$.

$$\psi\ (u) = K(u)u + f \tag{3.8}$$

leads to an iterative equation for improving the solution u, from the nth iteration

$$u_{n+1} = u_n + \Delta u_n \qquad with \quad u_0 = 0 \quad and \tag{3.9}$$

$$\Delta u_n = -\ K^T\ (u_n)^{-1}\ \psi(u_n) \tag{3.10}$$

where K^T is the tangential stiffness matrix.

Application

To check the limits of stability in this method, we consider a model of a crack with a void (fig. 3.6). The influence of discretization is investigated initially for a very coarse mesh and subsequently for a more realistic, finer mesh. With this example, both the stability of the method and physical effects can be studied, but at the moment attention is focussed on the method as such. Elements are eliminated not at a predetermined loading, but at a critical level of a local criterion. All physical quantities such as strains and stresses are averaged over the Gaussian points in each element. Figure 3.6 shows the elastic-plastic model situation for a crack with a predefined triangular "void". For illustration we chose the criterion $\varepsilon_{pl} \geq \varepsilon^c_{pl} = 10\%$ for the elimination of the element which starts the void. Such a criterion has been supposed for ductile fracture [4]. Numerous other criteria exist for ductile fracture, some of which are based on consideration of stress and strain fields near the crack tip in terms of Gurson's constitutive model of porous solids [5, 6].

Chu and Needleman studied void nucleation effects considering a void nucleation criterion with two parameters σ_H and σ_v [7]. Other analyses of ductile fracture were based on considerations of stress triaxiality, η, [8,9] or of the strain energy density function [10]. Element elimination was used to study particle/matrix debonding in a Gurson type approach [11]. It must be emphasized that the physical relevance of all of these criteria needs further clarification before any one of them can be generalized. Element elimination was used to study particle/ matrix debonding in a Gurson type of approach [11].

Figure 3.7 illustrates the elimination of elements at successive loading steps. At first, elimination occurs only at the crack tip and at some distance from it. Later this second region joins up with the void at the tip. At a total strain of 7% all elements in the shear band are eliminated, separating specimen into two parts. At this stage, the stability of the model is only upheld by the boundary conditions (fixed degrees of freedom of nodes in the x- or y-direction on boundaries), and the simulation ceases to be physically meaningful. In reality, the sample would shear in the direction of the shear band, but this cannot be modelled with the present boundary conditions. However, this simple model demonstrates the stable functioning of the procedure adopted for element elimination.

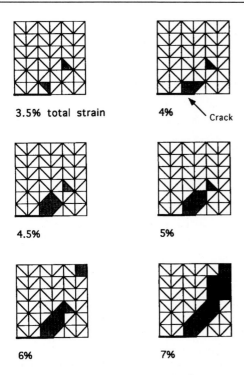

3.5% total strain   4%   Crack

4.5%   5%

6%   7%

Fig. 3.7 Successive element elimination at different states of prescribed overall strain. Eliminated elements are shaded.

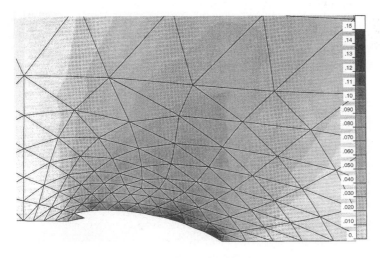

Fig. 3.8 Crack-tip region of fine meshed elastic-plastic model with crack. Element elimination for $\varepsilon_{pl} \geq 15\%$ at given 0.55% global strain. Distribution of plastic strain is shown.

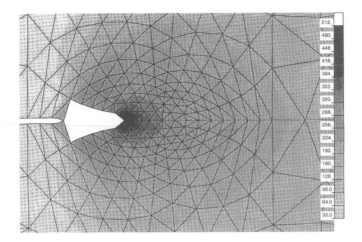

Fig. 3.9 Elastic-plastic structure with crack at total strain of 0.7%. Element elimination for $\sigma^2_H / \sigma_v = 700$MPa. The distribution of σ^2_H / σ_v is shown.

Next, a finer meshed elastic-plastic model is considered which contains a crack tip. As in the previous example, the properties of the material are described by a Voce-law with constants appropriate for aluminium. In this case, a local elimination criterion, $\varepsilon_{pl} \geq \varepsilon^c_{pl} = 15\%$ was adopted. Because of symmetry reasons only one half of the model was analyzed.

In this case we get no shear band failure as with the coarse model, but rather crack growth perpendicular to the loading direction (fig. 3.8).This is because the model geometry does not change significantly with the elimination of an element in this finely meshed model. There is no interaction between the crack tip and the boundary as in the coarse model. Alternatively, we use the failure criterion derived in [2] for the nucleation and growth of isolated voids in homogeneous stress fields.

It prescribes element elimination for $S = \sigma^2_H / \sigma_v \geq S^c$. Arbitrarily choosing $S^c = 700$ MPa and loading the model (fig. 3.9), we obtain a cavity ahead of the crack tip. The shape of the cavity is clearly not realistic because the mesh is still too coarse at the crack tip. This approach, with a finer net but with a manual elimination of elements, has been successfully used to predict crack paths in the two-phase material WC-Co [1].

Concluding remarks

The new technique proposed to simulate crack propagation by automatic element elimination within a commercial finite element program shows promise for application to shear band failure and void nucleation at crack tips in homogeneous materials. In future work, the method will be applied to real microstructures and the results will be compared with experiments on two-phase composites.

References

[1] Spiegler R. and. Fischmeister H.F (1992), Acta Metall. Mater. 40, p. 165.

[2] Spiegler R., Schmauder S. and. Fischmeister H.F. (1992), Int. J. Plast. submitted.

[3] LASSO Ingenieurgesellschaft, Markomannenstr. 11, D-7022 Leinfelden-Echterdingen, Germany.

[4] Brown L.M. and Embury J.D. (1973), in: Proc. 3rd Int. Conf. On Strength of Metals and Alloys, Cambridge, vol. 1, p.164.

[5] Gurson A.L (1975), Ph.D. thesis, Division of Engineering, Brown University, Providence, RI.

[6] Aoki S, Kishimoto K, Takeya A and Sakata M (1984), Int. J. Fract. 24, p. 267.

[7] Chu C.C. and Needleman A. (1980), Eng J. Mat. & Techn. 102, p. 249.

[8] Fischmeister H.F., Schmauder S and Sigl L. (1988), Mat. Sci. & Eng. A105/106, p. 305.

[9] Barnby J.T, Shi Y.W and. Nadkarni A.S (1984), Int. J. Frac. 25, p. 273.

[10] Sih G.C and Madenci E (1983), Eng. Frac. Mech. 18, p. 1159.

[11] Needleman A and. Tvergaard V, J (1987), Mech. Phys. Solids 35, p. 151.

3.2.2 FE Analysis of fracture of WC-Co Alloys: Microvoid growth[4]

The WC-Co hard alloys (hereafter these will be abbreviated as WC-Co) have high hardness as well as high wear resistance, and hence are widely used or will be widely used for machining tools and structural parts of various machines. To improve the strength of the WC-Co, it is necessary to understand the fracture behaviors of these alloys, which consist of WC crystals and Co binder (hereafter these will be referred to as WC and Co, respectively), as shown in Fig. 3.10 [1].

Togo *et al.* [2, 3] performed a finite element analysis of the stress field near a crack tip. The results show that micro cracks are at first nucleated by fracture of WC or debonding between WC and Co, and then ductile fracture of Co occurs followed by crack extension. It is suggested from these results that Co plays an important role for improving the fracture toughness of WC-Co. Sato and Honda [4] have shown by conducting an experiment that the fracture toughness of WC-Co increases with an increase in the volume fraction of Co. Sigl and co-workers [5, 6] measured the actual shape of WC and Co, and performed a finite element calculation for the measured shape of WC and Co to discuss the fracture of Co.

In this study, a finite element analysis is performed to examine the effect of the shape of Co near a crack tip on its fracture behaviors. The Co, which is much softer than the WC, is subjected to the deformation constraint from the surrounding WC crystals. The study of the ductile fracture under these constraints may be important not only in WC-Co, but also for various composite materials.

Numerical Procedures

Let us consider a WC-Co plate with a crack and model (Figs 3.11 and 3.12), where a rectangular or rhombic Co-region is located at the tip of a sharp notch. We assume that the small scale yielding condition is satisfied and the solid surrounding the Co in Figs 3.11 and 3.12 is an elastic body with the average material constants of the WC-Co composite (WC:90wt% + Co:10wt%) as shown in Table 3.1. We consider three different shapes of Co as shown in Fig. 3.11.

We also assume that Co is an elastic-plastic body and employ Gurson's constitutive equation [7] to take account of the effect of nucleation and growth of microvoids.

---

[4] Reprinted from S. Aoki, Y. Moriya, K. Kishimoto, S. Schmauder, "Finite Element Fracture Analysis of WC-Co Alloys", Eng. Fract. Mech. 55, pp. 275-287 (1996) with kind permission from Elsevier

Fig 3.10 Cracked WC-Co hard alloy

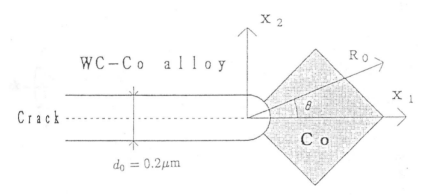

Fig. 3.11 Model of cracked WC-Co alloy for computation.

The yield function is given by

$$\phi = \frac{3}{2}\frac{\sigma_{ij}'\sigma_{ij}'}{\bar{\sigma}_m^2} + 2f\cosh(\Sigma) - [1 + f^2] = 0 \qquad (3.11)$$

$$\Sigma = \frac{1}{2}\frac{\sigma_{kk}}{\bar{\sigma}_m} \qquad (3.12)$$

where σ_{ij} is the macroscopic true stress $\sigma_{ij}' = \sigma_{ij} - \delta_{ij}\sigma_{kk}$ is the stress deviator, δ_{ij} is Kronecker's delta, $\overline{\sigma}_m$ is an equivalent tensile flow stress representing the actual microscopic stress state in the matrix material, and f is the volume fraction of microvoids. The microvoid volume fraction f increases during plastic deformation partly due to the growth of existing microvoids, and partly due to the nucleation of new microvoids:

$$\dot{f} = \dot{f}_{growth} + \dot{f}_{nucleation} \tag{3.13}$$

The increment due to growth is given by

$$\dot{f}_{growth} = (1 - f)D_{kk}^p \tag{3.14}$$

where D_{kk}^p is the plastic part of the macroscopic deformation rate. We employ the following equation proposed by Needleman and Rice [8]:

$$\dot{f}_{nucleation} = F_1 \frac{\sigma_m}{\sigma_0} + F_2 \frac{\dot{\sigma}_{kk}}{3\sigma_0} \tag{3.15}$$

where σ_0 is the tensile yield stress. We assume $F_1 = 0.01$, $F_2 = 0$ and the initial microvoid volume fraction, $f_0 = 0$.

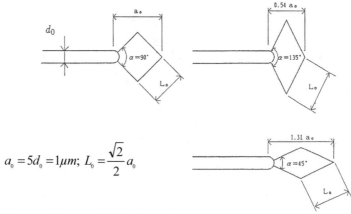

$$a_0 = 5d_0 = 1\mu m; \quad L_0 = \frac{\sqrt{2}}{2}a_0$$

Fig 3.12 Assumed shapes of Co binder

Table 3.1
Material constants

	Youngs modulus E	Poisson's ratio v	Yield stress σ_0	Work-hardening coef. n
WC-Co	600 GPa	0.22		
Co	200 GPa	0.31	500 MPa	4

The stress vs strain curve of matrix material is assumed to be given by:

$$\gamma \begin{cases} \dfrac{\tau}{G} & \tau \le \tau_0 \\[2mm] \dfrac{\tau_0}{G}\left[\dfrac{\tau}{\tau_0}\right]^n & \tau > \tau_0 \end{cases} \tag{3.16}$$

where τ denotes the effective shear stress $(=\sigma_m/\sqrt{3})$, τ_0, the uniaxial yield strength in shear $(=\sigma_{m0}/\sqrt{3})$, γ the total equivalent shear strain, n the strain hardening exponent. $G = E/2\,(1 + v)$ the shear modulus, E the Young's modulus, and v the Poisson's ratio. The material constants. E, v, σ_0 and n of Co are shown in Table 3.1.

The finite element mesh is shown in Fig. 3.13. The displacement given by the Mode-I elastic singular solution, which is characterized by the stress intensity factor K. is applied to the circular boundary far from the crack tip (R = 12,000d, d= depth of notch, Fig.3.11). Finite element calculations based on the finite displacement theory are carried out under the assumption of the plane strain or plane stress condition.

Numerical Results and Discussions

Circumferential and hydrostatic stress

The circumferential stress σ_θ on the positive X-axis is shown in Fig. 3.14, where $K \cong 15 MPa\sqrt{m}$ (= K_{1c}: fracture toughness of WC-Co). The σ_0 in Co increases as the shape of Co becomes flat for the plane stress condition [Fig. 3.14(a)], while the σ_θ in the flat Co-region decreases considerably for the plane strain condition [Fig. 3.14(b)].

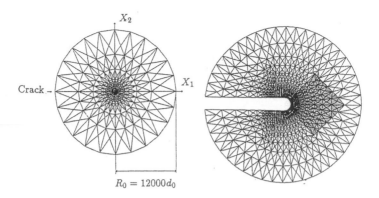

Fig. 3.13 Finite element mesh. (a) Whole mesh. (b) Detailed mesh near the crack tip.

The decrease in σ_θ in the flat Co under the plane strain condition may be attributable to the surrounding elastic solid. However, the detailed description of the reason needs a careful consideration, because if we assume that a strong displacement constraint increases the hydrostatic stress σ_{kk}, and hence the microvoid volume fraction f causing material softening, then it follows that the σ_θ must decrease due to the softening and this is contradictory to the assumption. In fact, the σ_{kk} in the flat Co decreases under the plane strain condition, as shown in Fig. 3.15(b).

This problem is important for understanding the ductile fracture under a displacement constraint and, hence, will be discussed in the following two sections. Hereafter, let us concentrate on the distribution on the X-positive axis under the plane strain condition, and refer to it as the distribution.

Microvoid volume fraction

The distribution of the microvoid volume fraction f is shown in Fig. 3.16, where $K/K_{Ic} = 1$. It is found that f in the flat Co is particularly increased, and this suggests that the large value off is connected with the decrease of σ_θ and σ_{kk} [Figs 3.14(b) and 3.15(b)].

To examine when f in the flat Co becomes high, the distributions off at various stress levels are shown in Fig. 3.17(a) for the flat Co and in Fig. 3.17(b) for the rectangular Co. It is found that f in the flat Co is slightly higher than that in the rectangular Co at $K/K_{Ic} = 0.4$, while great difference appears for $K/K_{Ic} > 0.6$.

The rate of the microvoid volume fraction f was assumed to be equal to the sum of the rates due to nucleation and growth of microvoids, as stated in eq. (3.13) [Section 2]. To examine whether nucleation or growth is more dominant, the increments of f due to nucleation and that due to growth are shown at various stress levels in Fig. 9 and Fig. 3.19, respectively. It is found from these figures that the microvoid growth has a crucial effect on the increase in f in the flat Co, but the microvoid nucleation has almost no effect.

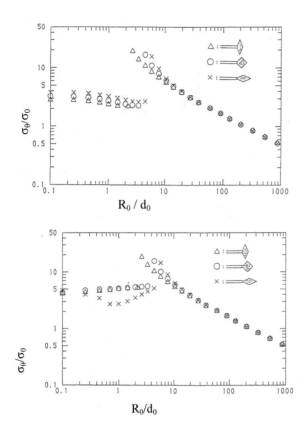

Fig. 3.14 Distribution of a, on X_1 axis at $K = K_{1c}$. (a) Plane stress. (b) Plane strain.

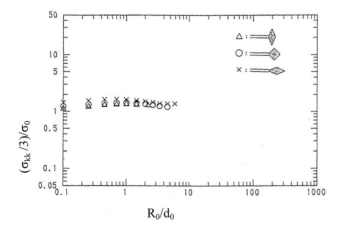

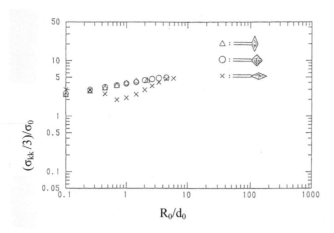

Fig. 3.15 Distribution of $\sigma_{kk}/3$ on X_1 axis at $K = K_{1c}$. (a) Plane stress. (b) Plane strain.

Ductile fracture under deformation constraint

In the last section the growth of microvoids is found to have a great effect on the ductile fracture under a strong deformation constraint. The rate of microvoid volume fraction due to growth $\dot{f}_{growth}$ is proportional to the plastic part of macroscopic deformation rate D^P_{ij} as shown by eq. (3.14), and the D^P_{ij} is related to the hydrostatic stress σ_{kk} through the constitutive equation [8].

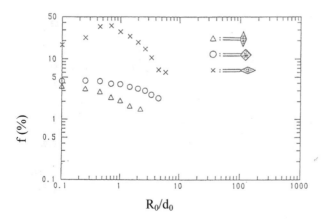

Fig. 3.16 Effect of Co shape on microvoid volume fraction f under plane strain condition.

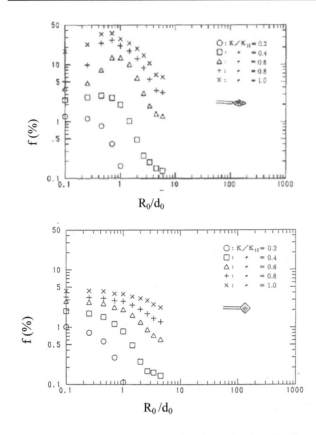

Fig. 3.17 Comparison of microvoid volume fraction f in flat and rectangular Co. (a) Flat Co. (b) Rectangular Co.

The σ_{kk}, in the flat and rectangular Co at several stress levels are shown in Fig. 3.20(a) and Fig. 3.20(b), respectively. There is no such great difference in σ_{kk} ~ between the flat and rectangular Co as in f [Fig. 3.17(a) and Fig. 3.17(b)] or in $\dot{f}$ growth [Fig. 3.19(a) and Fig. 3.19(b)]. Especially for $K/K_{Ic} \leq 0.4$, the σ_{kk} in the rectangular Co is slightly higher than that in the flat Co, although the difference is small. From these numerical results it is suggested that ductile fracture under strong deformation constraint proceeds in the following way.

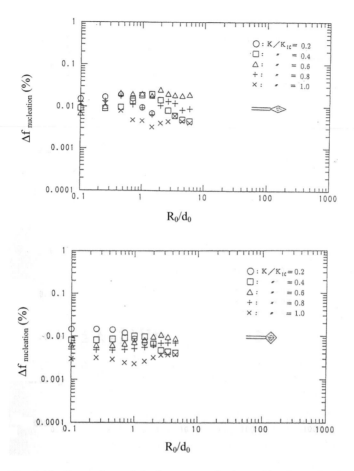

Fig. 3.18 Comparison of the increment of microvoid volume fraction due to nucleation $\Delta \dot{f}_{nucleation}$, in flat and rectangular Co. (a) Flat Co. (b) Rectangular Co.

Let us focus on damage evolution in a single Co-region. At an early stage of deformation, σ_{kk} and hence f are slightly higher under a stronger deformation constraint than under a weaker constraint, because the value of f is small and the material softening is not significant (cf. Fig. 3.17 and Fig. 3.20). Figure 3.21(a) shows schematically this state as two points on the yield surface. It is noted that the normality rule (D^p_{kk}, $\dot{\varepsilon}^p_e$) is normal to the yield surface in Fig. 3.20) holds because F_2 in eq. (3.12) is assumed to be zero [8]. Here $\dot{\varepsilon}^p_e = (2D^{p'}_{ij}D^{p'}_{ij}/3)^{1/2}$ and $D^{p'}_{ij} =$ deviatoric component of the plastic part of macroscopic displacement rate.

Although the two points in Fig. 3.21(a) are close to each other, there is a small difference in direction of the normal to the yield surface between the two points.

This means that D_{kk}^p and hence $\dot{f}$ growth, eq. (3.14)] in the flat Co is slightly higher than in the rectangular Co. The difference in $\dot{f}$ growth is nearly equal to that in f, because the difference of $\dot{f}$ nucleation, is negligible as stated above.

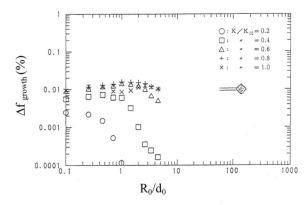

Fig. 3.19 Comparison of the increment of microvoid volume fraction due to growth $\Delta \dot{f}$ growth in flat and rectangular Co. (a) Flat Co. (b) Rectangular Co.

Since the yield surface shrinks more rapidly for greater $\dot{f}$, the difference in the yield surface between the fiat and rectangular Co becomes larger in the next increment of load. During further loading, the state shown in Fig. 3.21(b) is attained, i.e. the result that even though there is almost no difference in hydrostatic stress [Fig. 3.19(a) and Fig. 3.19(b)], there exists a large difference in microvoid volume fraction f [Fig. 3.17(a) and Fig. 3.17(b)] and its increment [Fig. 3.19(a) and Fig. 3.19(b)] for $K/K_{1c} > 0.4$. Figure 3.22 is the calculated result for a point at $X_2 = 0$, $X_1 = 1.0$ d_o at $K/K_{1c} = 0/4$, showing that the state in Fig. 3.21(b) is really attained.

As loading is further increased, the shrinkage of the yield surface of fiat Co becomes prominent. It follows from this that the hydrostatic stress σ_{kk}, begins to decrease after reaching the maximum, while the microvoid volume fraction f continues to increase.

The behaviours shown in Figs 3.17(a) and 3.20(a) for $K/K_{1c} \leq 0.6$ and also the phenomena shown in Fig. 3.16 and Fig. 3.15(b) [and hence Fig. 3.14(b)] may be explained by the above consideration. Furthermore when the microvoid volume fraction f reaches a critical value, macroscopic ductile fracture may evolve.

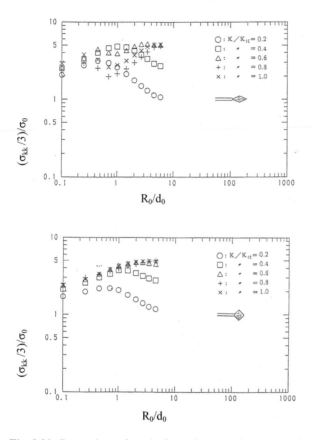

Fig. 3.20 Comparison of σ_{kk}, in flat and rectangular Co. (a) Flat Co. (b) Rectangular Co.

Concluding Remarks

A finite element analysis on the ductile fracture in the Co binder near the crack-tip in a WC-Co hard alloy was performed taking account of the nucleation and growth of microvoids. The ductile fracture behaviors in the Co binder or more generally under strong displacement constraint were discussed based on the numerical results. Microvoids are predicted to glow predominantly in flat Co-regions and increasing hydrostatic stresses accompany nucleation controlled state, while the subsequent growth controlled stage is related to decreasing hydrostatic stresses. In non-fiat Co-regions, microvoid evolution is nucleation and growth controlled. However, it is noted that the discussion in this paper is based on $F_2 = 0$ [eq. (3.15)]. If F_2 is not equal to zero, not only does the normality rule not hold, but also the nucleation may be influenced by the displacement constraint. Further study is necessary for $F_2 \neq 0$.

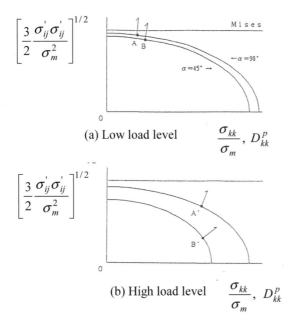

(a) Low load level $\dfrac{\sigma_{kk}}{\sigma_m}$, D_{kk}^p

(b) High load level $\dfrac{\sigma_{kk}}{\sigma_m}$, D_{kk}^p

Fig 3.21 States of stress in Co and change of yield surface (schematic)

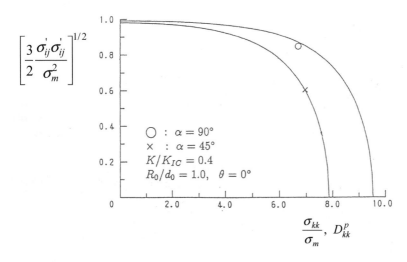

Fig. 3.22 States of stress in Co and yield surface (numerical result).

References

[1] Sigl, L. S. and Exner, H. E. (1987), Experimental study of the mechanics of fracture in WC-Co alloys. Metallurgical Trans., 18A, pp. 1299-1308.

[2] Tohgo, K., Tagawa, T. and Otsuka, A. (1987), Analysis of elastic and elastic-plastic deformation behavior of two-phase sintered materials. Trans. JSME, Ser.A, 53-494, pp. 1870-1878, in Japanese.

[3] Tohgo, K., Tagawa T. and Otsuka, A (1990), Microscopic stress/strain field around a crack tip and fracture process in WC-Co composite. Trans. JSME, Ser.A, 56-532, pp. 2417-2424, in Japanese.

[4] Sato, K. and Honda, H. (1990), Effects of WC grain-size and Co contents on fracture toughness of WC-Co alloys. J. Japanese Soc. for Non-Destructive Inspection, 39-3, pp. 237-242, in Japanese.

[5] Fischmeister, H. F., Schmauder, S. and Sigl, L. S. (1988), Finite element modeling of crack propagation in WC-Co hard metals. Materials Science and Engineering, A105/106, pp. 305-311.

[6] Sigl, L. S. and Schmauder, S. (1988), A finite element study of crack growth in WC-Co. Int. J. Fracture, 36, 305-317. Finite element fracture analysis of WC-Co alloy 287

[7] Ourson, A. L.(1977), Continuum theory of ductile rupture by void nucleation and growth: Part l--Yield criteria and flow rules for porous ductile materials. J. Eng. Mat. and Tech., Trans. ASME, 99, pp. 2-15.

[8] Needleman, A. and Rice, J, R.(1978), Limits to ductility set by plastic flow localization, in Mechanics of Sheet Metal Forming (ed. D. P. Koisten and N. M. Wang), pp. 237-267. Plenum.

3.2.3 Micromechanical simulation of crack growth in WC/Co using embedded unit cells[5]

Coated hard metal inserts contain cracks as a result of the coating process [1-6]. Cobalt enriched gradient zones (Fig. 3.23) [7] underneath the coating prevent crack growth into the tool to a certain extent. Subject of the present paper is the numerical simulation of crack advance and the local material response of the WC/Co hard metal in a fracture process.

Model description

A micromechanical elastic two-dimensional FE-model with elastic material properties has been developed in order to simulate crack growth in WC/Co (Fig. 3.24). The model includes a non-self consistent unit cell with a cobalt island in a carbide environment. The unit cell is embedded in an elastic body with the average material properties of the WC/Co composite, depending on the volume fraction of the material under consideration, according to Fig. 3.24.

10 µm

Fig. 3.23 Cobalt enriched zone (example marked) in a graded WC/Co hard metal.

[5] Reprinted from S. Hönle, S. Schmauder, "Micromechanical Simulation of Crack Growth in WC/Co Using Embedded Unit Cells", Computational Materials Science 13, pp. 56-60 (1998) with kind permission from Elsevier

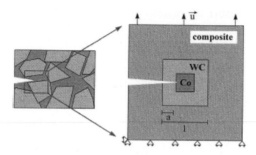

Fig. 3.24 Scheme of an embedded two-dimensional unit cell model (with prescribed displacement $\bar{u}$).

The elastic material properties of the composite material and the constituents are listed in Table 3.2. The energy release rate (ERR) G of a crack propagating along the symmetry plane of the cell is calculated according to the procedure given in [8] for different crack lengths on a microscopic scale. The dimensionless normalized energy release rate G* is calculated as follows [9]

Table 3.2

Material properties for WC/Co composite and constituents	
E_{comp} = 595 Gpa	V_{comp} = 0.22
E_{WC} = 714 GPa	V_{WC} = 0.19
E_{Co} = 211 GPa	V_{Co} = 0.31

with E, Young's modulus of the composite, σ, average stress in the model, h, height of the model. Moreover, the geometry of the cobalt inclusion is varied as well as the cobalt volume fraction and the arrangement of the inclusion.

$$G* = \frac{GE}{\sigma^2 h} \qquad (3.17)$$

Results and discussion

In order to simulate the failure behavior of a WC/Co hard metal, a micromechanical computational cell model was set up to calculate the ERR for a crack propagating through a carbide-cobalt cell (Fig. 3.24). Therefore, the model containing an initial crack was loaded uniaxially with a strain value related to the global fracture toughness K_{1c} = 16.7 MPa m$^{-1/2}$ for WC/Co with a cobalt volume fraction of 16% [10]. The energy release rate was calculated for different crack lengths a in the WC/Co-cell, varying from 0 to l, where l represents the width of the cell.

Inclusion shape

At first, the ERR for a propagating crack was calculated for different cobalt inclusion shapes in a volume element of hard metal with a given cobalt volume fraction of 16% (Fig. 3.25). The cobalt phase influences the crack driving force in an important way. The energy release rate of a crack approaching the cobalt phase increases strongly, while it decreases rapidly for the crack propagating towards the center of the cobalt island (Fig. 3.26). The cobalt phase absorbs a large amount of cracking energy. Thus, such a crack may probably be arrested in the cobalt phase.

All the curves exhibit a similar shape and the maximum values of the elastic energy release rates vary in a range of about 20% (Fig. 3.26). A crack is found to be more attracted by a sharp-edged cobalt inclusion in front of the crack tip.

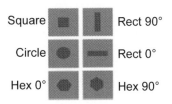

Fig. 3.25 Variation of cobalt inclusion shapes.

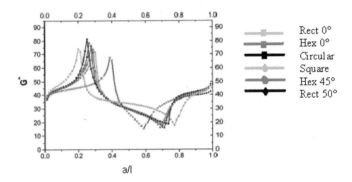

Fig. 3.26 Normalized elastic energy release rate (ERR) for varying cobalt inclusion shapes.

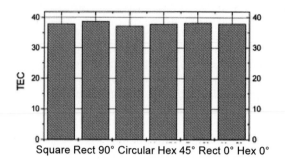

Fig. 3.27 Total energy consumption (TEC) for varying cobalt inclusion shapes.

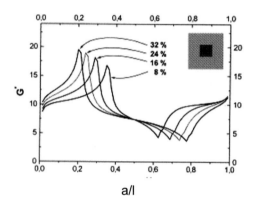

Fig. 3.28 Normalized elastic energy release rate (ERR) for varying volume fractions of a square cobalt inclusion.

This cobalt inclusion arrangement can be related to cobalt inclusions between carbide grains in realistic structures with small "opening angles".The total energy consumptions (TEC) of the cracks, which are determined by integrating the ERR curves; vary in a range of only a few percent for the different inclusion shapes (Fig. 3.27). Thus, the cobalt inclusion shape influences the attraction of the crack tip in the elastic regime, but has a negligible effect on the total elastic energy consumption of the crack propagating through the computational cell.

Inclusion volume fraction

To show the influence of the cobalt volume fraction, a parametric study was carried out using a computational cell with a centered square cobalt inclusion. The volume fraction of the cobalt inclusion was varied from 8% to 32% which is a typical regime for WC/Co hard metals [6].

The calculated energy release rates for varying cobalt volume fractions are illustrated in Fig. 3.28. The maximum value of the normalized energy release rate (G^*_{max}), which gives rise to the attraction of the crack by the cobalt inclusion, is found to be an increasing function with increasing cobalt volume fraction (Fig. 3.29). The total energy consumption (TEC) decreases linearly with increasing cobalt content (Fig. 3.30). Thus, when propagating through a linear-elastic hard metal volume element, the crack is more attracted by higher cobalt content while consuming less energy.

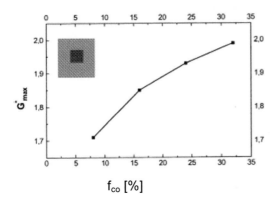

Fig. 3.29 Maximum ERR for varying volume fractions of a square cobalt inclusion.

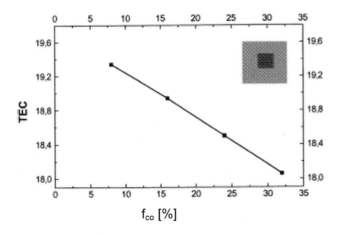

Fig. 3.30 TEC for varying volume fractions of a square cobalt inclusion.

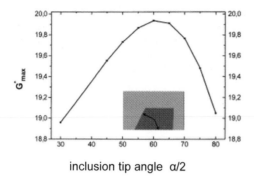

inclusion tip angle α/2

Fig. 3.31 Maximum normalized ERR for varying cobalt inclusion tip angles.

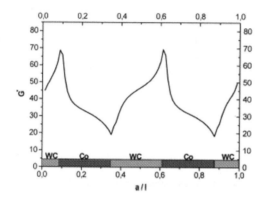

Fig. 3.32 Normalized ERR for two square cobalt inclusions.

Inclusion tip angle

Due to the fact of high maximum values of the ERR related to sharp-edged cobalt inclusions (Fig. 3.26) a parametric study was carried out, using a hexagonal cobalt inclusion with a fixed cobalt volume fraction of 16%. The entrance angle α/2 of the cobalt inclusion was varied from 30° to 80°. As a consequence of the stress-strain distribution at the crack tip the maximum value of the ERR is found at an angle of about 60° which represents a hexagon angle of 120° (Fig. 3.31). Thus, among the inclusions investigated, the crack is most attracted by cobalt inclusions with an inclusion tip angle of 120°.

Multiple inclusions

The energy release rate was calculated for a unit cell with two square cobalt inclusions in the crack plane, where the overall cobalt volume fraction was again set to 16%. According to the results presented in Fig. 3.32 the energy release rate increases while the crack is approaching the first cobalt inclusion and decreases rapidly when propagating through the first inclusion in the same way as in the calculation for a single inclusion. The energy released by the crack in the first inclusion is not influenced by the second cobalt inclusion. The total energy consumption for this crack has the same value as in the case of a single cobalt inclusion with the same cobalt volume fraction.

Conclusions

The following conclusions can be drawn according to the presented results:
- A crack is most attracted by sharp-edged cobalt inclusions in front of the crack tip.
- An inclusion tip angle of the cobalt island of 120° was found to be most beneficial for attracting a crack.
- Unit cells with a higher content of cobalt are more attractive to cracks, although less energy is consumed when the crack propagates through the cell.
- The arrangement of the cobalt inclusions in crack direction has no important influence on the energy consumption of a propagating crack.

References

[1] Kolaska, H. (1992), Pulvermetallurgie der Hartmetalle, Fachverband Pulvermetallurgie.
[2] Schedler, W. (1988), Hartmetall für den Praktiker, VDI Verlag, Düsseldorf.
[3] Schumacher, G (1969), Wendeschneidplatten aus Hartmetall mit TiC Schicht, Techn. Zentalbl. für prakt. Metallbearb. 63, pp. 275-278.
[4] Hintermann, H.E. (1979), Verschleiß- und Korrosionsschutz durch CVD- und PVD-Überzüge, VDI Fortschritt-Berichte 333, pp. 53-67.
[5] Nordgren, A., Jonsson, S. (1992), Combined cracking of TiN, TiC and Al₂O₃-coated cemented carbide during milling of steel. Report No. IM-2901. Swedish Institute for Metals Research, Stockholm.
[6] Nordgren A., Jonsson. S (1994), Residual stress in CVD TiN coatings on unworn and worn cemented carbide and the influence upon crack formation, Report No. IM-3180, Swedish Institute for Metals Research, Stockholm,
[7] Hönle, S., Rohde, J. , Schmauder, S (1996), Meso- and Microscopic modeling of failure behaviour of graded zones in hard metals, Abstracts Junior Euromat'96, DGM, Oberursel, p. 275.
[8] Theilig, H., Wiebe, P., Buchholz, F.G. (1992), Computational simulation of non-coplanar crack growth and experimental verification for a specimen under combined bending and shear loading. Reliability and Structural Integrity of Advanced Materials, vol. 2, Proceedings of Ninth European Conference on Fracture (ECF9), Varna, pp. 789-794.
[9] Rohde, J., Schmauder, S., Bao, G. (1996), Mesoscopic modelling of gradient zones in hard metals, Comp. Mat. Sci. 7, pp. 63-67.
[10] Sigl, L., Exner, H.E. (1987), Experimental study of the mechanics of fracture in WC-Co alloys, Metallurgical Transactions A 18, pp. 1299-1308.

3.3 Damage and fracture of tool steels

3.3.1 Modeling of crack propagation in real and artificial micro-structures of tool steels: simple microstructures[6]

In this part of the work, the crack growth in a real microstructure of high speed steel HS6-5-3 is simulated. The simulations were made for 2D plane strain conditions. In the simulations, multiphase finite elements have been used to simulate the real microstructure of the steel, and the element elimination technique to determine the crack path. The following input data were needed for the simulation: the geometry of the specimen, the real structure of the steel, and the elastic and strength properties of carbides and the matrix.

Short rod specimens are proven to be applicable to study the fatigue behaviour of rather brittle materials. To ensure the possibility of comparison of numerical and experimental investigations, and further extrapolation the results of this work for the cyclic loading of steels, the short rod specimens have been taken for macrosimulations.

Simulation of Crack Initiation and Propagation in Heterogeneous Materials

Methods of modelling crack propagation

Generally, there are two main approaches to the numerical simulation of fracture: a crack can be modeled as a discontinuity between adjacent elements, and a crack may be "smeared" over entire elements [1-5]. These approaches determine also two main methods of FE implementation of fracture models: methods, based mainly on special types of finite elements [6] and methods which assume the specific material properties in the region of possible crack propagation. Table 3.3 presents some often used methods of numerical simulation of crack propagation. A more detailed review of the available methods is given in [4].

---

[6] Reprinted from L. Mishnaevsky Jr., N. Lippmann, S. Schmauder, "Computational modeling of crack propagation in real microstructures of steels and virtual testing of artificially designed materials", Int. Jour. Fract. 120, pp. 581-600 (2003). with kind permission of Springer

Table 3.3
Some methods of FE-simulation of microstructural crack growth

Model	Main ideas	Ref.
Cohesive zone models (CZM)	Crack path is presented as a thin material layer with its own constitutive relation (traction-separation law). The relation is such that with increasing crack opening, the traction reaches a maximum, then decreases and eventually vanishes so that complete decohesion occurs.	[7]
Computational cell methodology (CCM)	Crack propagation is a result of void growth in front of the crack tip. Void growth is confined to a layer, which consists of cubic cells with a void and the thickness of which is equal to the mean distance between inclusions which causes void initiation. When the void volume fraction in a cell reaches some critical level, the cell is removed and therefore the crack grows.	[8,9]
Cell model of material	A generalized formulation of the cell models of crack growth. The material is accepted to consist on cells (which is defined as a „smallest material unit that contains reasonably sufficient information about crack growth in the material"), which is characterized by its size and cohesion-decohesion relation.	[10]
Smeared crack models	A crack is considered as a continuous degradation (reduction of strength/stiffness) along the process zone. The displacement jump is smeared out over some characteristic distance across the crack, which is correlated with the element size. The degradation of individual failure planes is described by the constitutive law. In the fixed crack model (the classical version of the smeared crack model), the degradation is controlled by the maximum tensile stresses only; other versions of the smeared crack model (rotating crack model, multiple fixed model) allow to take into account the variations of crack growth direction during crack propagation, and the formation of secondary cracks.	see review [11], [12]
Embedded crack model (ECM)	Finite elements with an embedded discontinuity line or localization band are used. The discontinuity crosses the element and divides it into two parts. The constitutive model of the element with a discontinuity is given by both the traction-separation law and the stress-strain law.	[5]
Hybrid fracture/ damage approach (HFDA)	The method seeks to combine approaches based on fracture mechanics and continuum damage mechanics. A "super-element" consisting of a singular element (crack tip element) and several variable-node elements (transitional from the singular element to the linear four-node ones), and located just on the tip of the growing crack is used. At the crack tip, the material stiffness is reduced 1000 times in each timestep (the crack propagation is simulated dynamically). As a criterion of crack propagation, the J-integral (over the contour inside the super-element) was used.	[13]
Cohesive surface model (CSM)	The crack path is prescribed as a cohesive (contact) surface. The decohesion criterion is accepted to be controlled by the normal traction transmitted through the cohesive surface.	[3]

Comparing the approaches described in Table 3.3, one can see that the following main problems are confronted by most researchers in the simulation of crack growth: combining advantages of both continuum damage mechanics and fracture mechanics (less mesh and damage localization sensitivity); generalized approach to model void and crack growth [compare the paper by Siegmund et al. [14] who considered interrelations between the model of crack growth based on void evolution and coalescence (modeled with the Gurson-Tvergaard-Needleman approach) and the cohesive zone model (CZM)]; possibility of taking into account the microstructure of material in crack growth modeling (it can be done by combining the unit cell approach with any model of crack growth).

Element Elimination Technique (EET) and Multiphase Element Method (MPFE)

The problem considered in this paper requires an approach which allows to model both fracture and damage evolution using the same criteria, and to take into account the microstructure of the material without prescribing a crack path. In order to solve this problem, the authors applied the element elimination technique (EET) [2], which, by the author's opinion, has all advantages of the numerical methods described in Table 3.3. This technique is described in the section 3.2. The advantages of EET are especially evident in combination with the multiphase element method (MPFE). The main idea of the MPFE is that the different phase properties are assigned to individual integration points in the element (as differentiated from the common approach, when each element of the FE mesh is attributed to one phase and the phase boundaries are supposed to coincide with the edges of finite elements). Therefore, the FE-mesh in this case is independent of the phase arrangement of the material, and one can use relatively simple FE-meshes in order to simulate the deformation in a complex microstructure. Using EET together with MPFE, one can simulate the formation, growth and coalescence of voids or microcracks, and the crack growth in multiphase materials, what enables the study of effects of the microstructure of materials on its fracture. EET is incorporated in the FE code LARSTRAN [15].

FE Simulation of Crack Propagation in Real Structures of a Tool Steel

Mechanical properties of the constituents of investigated tool steels

For the mesomechanical simulation of deformation, damage and fracture in tool steels, the mechanical properties of the steel constituents are required. Generally speaking, it is rather difficult to determine the mechanical properties of all the constituents of materials to be optimized. Table 3.4 gives the mechanical properties of ledeburitic tool steels which have been determined by different authors with the use of different methods [17-23].

Table 3.4

Mechanical properties of the constituents of tool steels

Property	Value	Method of determination and reference
Cold work steel X155CrVMo12		
Young's modulus of primary carbides, GPa	276 (large primary carbides)	hardness tests (Vickers indenter) in metallographically prepared surfaces (after grinding, polishing and etching), with simultaneous observation in build-in microscope, and recording the force-displacement (uploading) curve [19]
Young's modulus of matrix, GPa	232	
Failure stress of carbides, MPa	1826-1840 (carbides about 17-20 μm), or 1520 (carbides ~ 30 μm)	SEM-in-situ experiments on 3-point bending of specimens with inclined notch plus macromechanical simulation of the specimen deformation and mesomechanical simulation of carbide failure [17, 18]
Constitutive law of the matrix	$\sigma_y = 1195 + 1390 [1-\exp(-\varepsilon_{pl}/0.0099)]$	approximation of results of the testing of a "matrix material" which possess the composition identical with this of the matrix and produced by PM means [20]
High speed steel HS6-5-2		
Young's modulus of primary carbides, GPa	286 (carbides M6C), 351 (MC)	hardness tests, with recording force-displacement (uploading) curve [22]
Young's modulus of matrix, GPa	231	
Failure stress of carbides, MPa	1604 or 1840 (carbide bands are oriented along or perpendicular to the load, respectively) (carbides ~10...17 μm)	like in the case of cold work steel, see above
Constitutive law of the matrix	1. $\sigma_y = 1500 + 1101 [1-\exp(-\varepsilon_{pl}/0.00369)]$ 2. $\sigma_y = 1500 + 471 [1-\exp(-\varepsilon_{pl}/0.0073)]$	1. like in the case of cold work steel, see above 2. Assumed that the matrix of the high speed steel behaves like a cold work steel [21]
Poisson's ratio of carbides and matrix	0.19 (carbides), 0.3 (matrix)	[21, 22]

The material properties of the high speed steel as a quasi-homogeneous material (such properties are needed for the macroscopic modelling) were taken as follows: Young's modulus E = 236 MPa and Poisson's ratio ν = 0.3 [21]. The constitutive law of the steels was taken as [21]:

$$\sigma_y = 2200 + 820 \, [1\text{-exp}(-\varepsilon_{pl}/0.002)], \qquad\qquad (3.18)$$

where σ_y – von Mises stress, MPa, ε_{pl} – plastic strain.

Hierarchical modelling: macro-meso transition

Taking advantage of symmetry conditions, the FE model of the short rod specimen with homogeneous material properties was developed. According to the description of the specimens, the diameter was taken to be 12 mm, height 18 mm, notch depth 5.32 mm [16, 17]. Figure 3.33 gives the scheme and boundary conditions of the macroscopic FE model of the short rod specimen.

The scheme of the micromodel and the position of the real microstructure of the steel in the model are shown in Figure 3.34. The displacement of the points in the vertical direction on the plane of symmetry was set to be zero. The point on the symmetry plane of the specimen, which lies on the other end from the notch was fixed in the X-direction as well. The loading was displacement-controlled, and applied in a point at a distance 1.88 mm from the end of the specimen. The loading displacement varied from 0 to 1 mm.

The macromodel was constructed in such a way that one could determine the displacement distribution on the boundaries of an area 300 μm x 500 μm in the vicinity of the notch. This area presented then the mesomodel, which included also a region with a real microstructure of the steel.

The boundary conditions in the mesomodel (the small area 300 x 500 μm near the notch of the short-rod-specimen) were given as vertical displacements. The displacement distribution on the boundaries of the mesoscopic model was determined from the macromodel, and then approximated by a linear function of a loading step in the macromodel and by a linear function of the coordinates of a point as

$$U_y^{micro} = f \, (N_i, X) \qquad\qquad (3.19)$$

where N_i – the number of the loading step, and X – X-coordinate of the point of the boundary of the mesoscopic model. In other words, each loading step in the macromodel, and each point on the boundary of the macroscopic model have had different values of applied displacement. Figure 3.36 shows the correspondence between the displacements in the macromodel and those in different points (notch, middle point and other end) of the micromodel. After approximation of the numerically obtained relationship between U_y^{micro}, N_i and X, given in Figure 3.36, the formula (1) took the form:

$$U_y^{micro} = 0.0002 \, N_i \, (-1.96 \, X +1), \qquad\qquad (3.20)$$

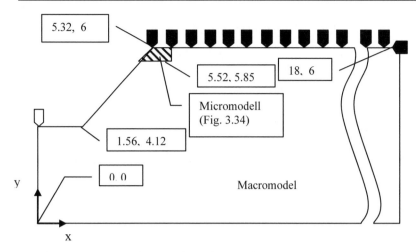

Fig. 3.33 Geometry (in mm) and boundary conditions of the macromodel of the short rod specimen

Thus, whereas the applied displacement Uy in the macromodel presents a concentrated load, the displacement applied to the boundary of the mesomodel is distributed along its upper and lower boundaries.

Mesomechanical model of crack growth in a real microstructure

The 2D model with a real microstructure of high speed steel was placed in an area 100 μm x 100 μm near the notch in the mesomodel. The elastic and elasto-plastic properties of the carbides and matrix have been taken from Table 3.4. With the given properties of the components and the steel, and the real microstructure of the steel, the crack initiation and growth in the steel is simulated. Figure 3.35 gives the metallographic micrograph of the high speed steel HS6-5-2 which was used in the simulations.

As criteria for the element elimination, both the value KTYP and the critical values of failure stress (for carbides) and plastic strain (for matrix) were used. The value KTYP means the number of Gaussian points in an element when it is decided whether this element is to be assigned to the matrix or to a carbide. This value was 3; i.e., if 4 (or more) Gaussian points of an element (which contains 6 points) are lying in the matrix, the element was supposed to be eliminated as a matrix element. Otherwise, it was considered as an element in a carbide.

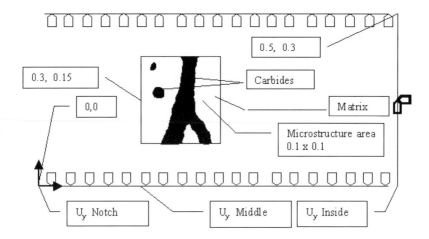

Fig. 3.34 Scheme of the micromodel for the real structure. Coordinates in mm.

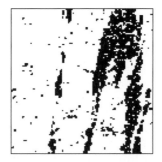

Fig. 3.35 Discretized micrograph of high speed steel HS6-5-3. The black areas present primary carbides, the white area is the "matrix". Region 100x100 mcm.

The criterion of element elimination in the matrix was determined on the basis of available knowledge about the micromechanisms of fracture of steel matrix. Any damage criteria based on the void growth seemed inapplicable in this case due to the mainly brittle macro-behaviour of the matrix. Yet, during SEM-in-situ-experiments, some plastic deformation has been observed at the microlevel (which however is quickly followed by the failure of specimens). Thus, we chose the critical plastic strain as a criterion of the element elimination in the matrix. As follows from SEM-in-situ-experiments described in [17, 18, 23, 24], the critical plastic strain for the matrix of the steels should be very low.

The critical plastic strain value was determined with the use of the numerical experiment technique (this type of modeling is also referred to as the inverse modeling) on the basis of the qualitative information about the crack behaviour in high speed steels and the experimentally (in SEM–in-situ-tests, see [12, 22, 24]) determined carbide failure stress.

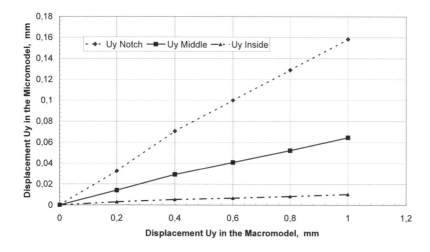

Fig. 3.36 Comparison of the displacements at the boundaries of the macro- and micromodel

It is known [25, 26] that a crack in high speed steels is initiated in carbides, if they are available in the vicinity of the notch tip, grows straightforward in the matrix, and kinks into the carbide rich regions and then follows them to a small part, then jumps to the next carbide band, grows in the carbide band furtherly, and so on. As was noted by Berns et al. [27], "...running crack must follow carbide bands...The width of the crack is restricted to jumps between adjacent carbide bands, but most of the crack surface is produced by cleavage of carbides in one band".

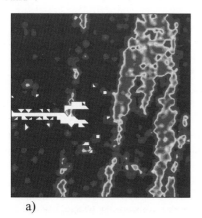
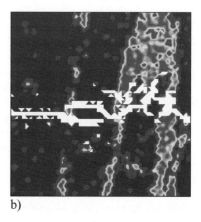

a)   b)

Fig 3.37 Simulated crack growth in the real microstructure of the high speed steel HS6-5-2. a) crack initiation, u = 0.0006 mm, b) crack branching at a carbide, u = 0.0008 mm

We have carried out the simulation of crack initiation with the carbide failure criterion known from [17, 24] (see Table 3.3) and different criteria of crack initiation in the matrix (Rice-Tracey damage criterion, critical stress, different values of the critical plastic strain, etc.), until the above described crack behaviour was obtained in the simulations. As expected, the most appropriate criteria of the crack initiation in the matrix was the critical plastic strain, and the critical level of this value was $\varepsilon_{pl, c} = 0.1$ %. At this level of critical plastic strain, the small crack increment in the matrix is followed by failure of a carbide in the vicinity of the crack tip, as observed in [27], while e.g. for $\varepsilon_{pl, c} = 0.5$ % delayed crack growth occurs.

Figure 3.37 shows the crack growth in the real (band-like) structure of the high speed steel. The real microstructure area contained 5000 finite elements. Generally, the mesomechanical model contained 9888 elements. We used 6-node triangular elements, with full integration.

In Figure 3.37 it can be seen that the crack grows initially in the matrix almost straightforward. The crack began to grow after the displacement reached 0.0006 mm. (In the experiments presented in [17], the crack in the short rod specimens from high speed steels began to grow when the external displacement reached the value 0.0006 mm, too). The calculation of the maximal stress distribution in the first load step (the displacement 0.0002 mm from the notch) shows a higher stress concentration in the vicinity of the crack tip.

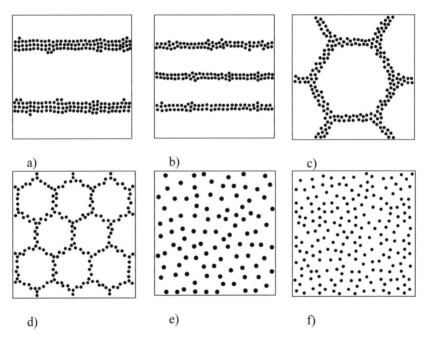

a)   b)   c)

d)   e)   f)

Fig. 3.38 Idealized artificial microstructures of the steel: a) band-like coarse, b) band-like fine, c) net-like coarse, d) net-like fine, e) random coarse and f) random fine microstructures.

The straightforward crack path is stopped due to the carbide bands and the crack begins to grow along the carbide band. Then, the straightforward crack growth in the matrix continued. At a displacement of 0.0008 mm, crack branching caused by the availability of a carbide row in the crack path occurs. Without further increment of load, the full area of the microstructure breaks. So, the direction of crack growth and the structure of the crack are strongly influenced by the carbide rich regions.

Crack Propagation in Idealized Typical Microstructures of Steels

Design of artificial "real" microstructures as a next step in the computational optimization of materials

Using the above model to simulate crack growth in real structures of steels, one can carry out computational experiments to study the effects of the material structure on the fracture behaviour. The next step in the numerical optimization of a material is the simulation of deformation and fracture in artificial quasi-real microstructures.

By testing some typical idealized microstructures of a considered material in such numerical experiments, one can determine the directions of the material optimization and preferable microstructures of materials under given service conditions. Such simulations should be carried out for the same loading conditions and material, as the real structure simulations which proved to reflect adequately the material behaviour.

Among the types of idealized microstructures, one takes usually as a first approximation the random and periodic microstructures. In studying cast and deformed metals, it is advisable to consider also the net-like (typical for the as-cast state) and band-like microstructures (hot formed steels). To investigate the structures of hard alloys and other ceramic materials, one should take into account the degree of clustering of hard (or in some cases, ductile) particles and vary such parameters as the connectivity, degree of clustering [4], etc. as well as the distributions of particle shapes and sizes.

Mesomechanical simulation of crack growth in idealized microstructures

To determine the optimal microstructure of the steel, we considered band-like, net-like and random microstructures. Two types of each microstructure were taken: a fine one with carbide size of 2.5 μm and a coarse one with carbide size of 3.6 μm [23].

The artificial net-like, band-like and fine random microstructures with 200 round particles of a given radius [so that the surface content (in this case, volume content) of the particles is about 10 %] were created with the use of the graphics software XFIG. In the case of the random microstructures, the particles were randomly distributed thereafter. The particles in the band- and net-like structure were distributed in such a manner that the distance between bands or the cell size (respectively) was about 20 times the particle diameter for the coarse structures, and 10 times the particle diameter for the fine microstructures. The coarse random structure possesses only 100 particles, the size of which was so selected that the volume content of carbides is 10%. One should note here that these "limiting case" microstructures are rather typical for the tool steels, but do not exhaust all typical microstructures in other materials, like hard alloys or cermets.

Figure 3.38 gives the considered artificial microstructures. The simulations for these artificial microstructures were carried out in each case with same boundary conditions as for the real structure. The structures in Figure 3.38a und b, 3.38e und f (random and net-like microstructures) are oriented in such a way that crack growth starts from left; on the band-like structures in Figure 3.38c the crack starts up from the lower boundary.

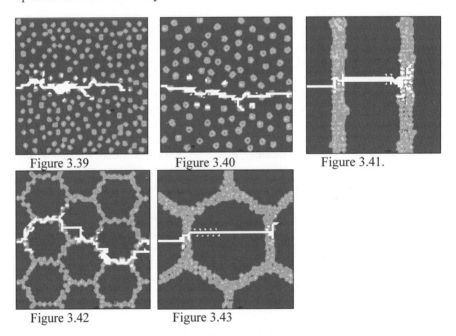

Figure 3.39   Figure 3.40   Figure 3.41.

Figure 3.42   Figure 3.43

Fig. 3.39 Simulated crack growth in the random fine microstructure, u = 0.0006 mm.
Fig. 3.40 Simulated crack growth in the random coarse microstructure, u = 0.0006 mm.
Fig. 3.41 Simulated crack growth in the band-like coarse microstructure, u = 0.0006 mm.
Fig. 3.42 Simulated crack growth in the net-like fine microstructure, u = 0.0008 mm.
Fig. 3.43 Simulated crack growth in the net-like coarse microstructure, u = 0.0006 mm.

Effect of the Microstructure on the Fracture Behavior of the Tool Steels

Crack path in different microstructures: qualitative consideration

Figures 3.39-3.43 show the crack path in the artificial microstructures of the steels. The cracks in these Figures correspond to different applied displacements. The plots represent the distributions of the value KTYP (i.e. the phase distribution). The displacements were chosen in such a way that the cracks in each case almost pass the microstructural area and approach to the embedding.

Let us focus at the crack path in different artificial microstructures. In the fine random structure (Figure 3.39), intensive crack growth was observed at the displacement of ~ 0.0006 mm. Slight deflections of the crack path into the carbides were observed during the crack growth. The carbide arrangement in the coarse random structure (Figure 3.40) causes slight deflection of the crack path from its initial direction as well. After a displacement of ~ 0.0008 mm the crack path is relatively rectilinear.

The crack path in the band-like structure (Figure 3.41) shows markable crack deflection at the carbide bands. In the matrix, the crack grows rectilinearly. The increased carbide density in the bands leads to a very intensive element elimination (i.e., failure of carbides). The force-displacement curve shows a low F_{max} value, which reaches the maximum, however, only at the displacement of ~ 0.0008 mm (Figure 3.44). The carbide deflection can be clearly seen on the second carbide layer. Up to an external displacement of 0.001 mm the crack does not reach the end of the microstructure area.

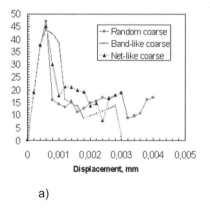

a)

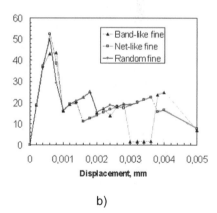

b)

Fig. 3.44 Force-displacement curves for the simulations of crack growth in artificial coarse microstructures

Regarding the force displacement behavior, only a small F_{max} value is achieved, which, however, remains rather long at this level.

In the fine net-like microstructure (Figure 3.42), the crack is instantly directed to the carbide network, and then follows exclusively the carbide network (this mechanism of crack growth was considered theoretically in [28]). It is of interest that the force-displacement curve for this microstructure gives the highest value of the peak force.

As differentiated from the fine net-like microstructure, the crack in the coarse net-like microstructure was initiated in the matrix (Figure 3.43). This influenced the further crack growth sufficiently: the carbide layers are passed by the crack, and lead only to relatively slight crack deflections. The maximal force is only slightly higher than that for the band-like structure (Figure 3.44).

This behavior of the crack in the coarse net-like microstructure is similar to that in the band-like structure (Figures 3.41, 3.42 and 3.43).

Force-displacement curves for different microstructures of steels

Figure 3.44 gives the force-displacement curves for each simulation of the crack growth. Table 3.5 gives some main quantitative characteristics of the crack growth in the different structures. The value F_{max} characterizes the critical load at which the crack begins to propagate. The value G (nominal specific energy of the formation of unit new surface) for each microstructure characterizes the fracture resistance of each of the structures. The physical meaning of the averaged force F_{av} is close to that of G: this value characterizes the fracture resistance of each of the microstructures, yet, not in relation to the unit of nominal new surface, but in the relation to the applied displacement. The difference between these two values (F_{av} and G) is caused by the fact that cracks in each case passed the microstructure area at different applied displacements

The nominal specific energy of the formation of unit new surface for each microstructure was calculated as follows:

$$G = \sum_i (P_i \, \Delta u_i) / L_{RS} \qquad (3.21)$$

where P_i – force for each loading step, Δu_i – displacement increment for each loading step, L_{RS} - linear size of the real microstructure, the summation is carried out for all loading steps until the crack passes the real microstructure.

Almost all force displacement curves achieve the F_{max} values at the displacement of 0.0006 mm (except for the fine band-like structure). The F_{max} is minimal for the band-like structures, and is again sufficiently higher for the fine than for coarse microstructures (especially in the cases of the random and band-like fine microstructures).

For both net-like and band-like microstructures the fracture resistance of the steels is much higher for the fine than for the coarse versions of the structures. Although this effect was not observed for the random microstructure, we assume that it is caused only by the different number of particles in the fine and coarse random microstructures, and that the general tendency remains the same for the random microstructures as well. The fracture energy G is higher for the net-like than for the band-like microstructures.

Table 3.5
Quantitative parameters of fracture behaviour of the artificial structures of steels

Type of the structure		Net-like		Band-like		Random	
		coarse	fine	coarse	fine	coarse	fine
Peak load at the force-displacement curve, N		44.9	52.3	43.8	43.6	47.0	50.2
Nominal specific energy of new surface, J/m^2		436.15	827.0	341.28	676.75	699.18	557.02
Maximal height of the roughness peak, Rmax, μm		13	36	24	14	18	12
Parameters of the crack structure, branching and microcrack density	N_M	1	1	4	4	5	6
	N_{LB}	-	-	3	3	1	1
	N_{MB}	3	3	-	-	-	1
	N_{SB}	-	5	3	2	2	4
Fractal dimension of fracture surface, D		1.285	1.593	1.442	1.40	1.372	1.446

It is of interest that the averaged force F_{av} hardly increases with increasing the peak load. If one considers only the coarse microstructures, one can see that the F_{av} even decreases with increasing the peak load F_{max}. So, increasing the resistance of the materials to the crack initiation does not mean necessarily the increase of the energy consumption in crack growth (i.e. fracture toughness).

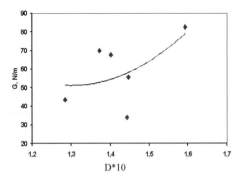

Fig. 3.45 Specific energy of fracture plotted versus the fractal dimension of fracture surface D.

This suggests that an approach which includes both the optimal design of parts (tools) and the optimal design of the tool material should be used for the design of tools from these steels. Namely, in the parts of the cutting tool, in which the tensile stresses are maximal and the crack initiation is therefore most probable, microstructures which ensure maximum F_{max} should be used (first of all, the cutter face, especially in the vicinity of the cutter edge is meant). In the rest of the tool material, a microstructure which ensures the maximum energy consumption in crack propagation should be taken. Therefore, it can be useful to consider the possibility of using gradient microstructures to optimize the tool steels.

Damage Growth and the Structure of Crack Path

Let us look at the form and structure of the crack path in different microstructures. As parameters of the crack appearance, we considered the maximal height of the roughness peak R_{max}, the amount of eliminated elements not connected to the main crack as well as branches of the crack, and the fractal dimension of the crack.

The maximal height of the roughness peak R_{max} was calculated from the crack profile as the distance between highest and lowest points of the crack path measured along the perpendicular to the initial crack direction (horizontal). To characterize the structure of crack path (i.e. the likelihood of branching and microcracking outside the main crack path), we used the following parameters: N_M – the amount of eliminated elements not connected to the main crack, N_{LB}, N_{MB} and N_{SB} – the amount of large (more than 8 μm), medium (between 4 and 8 μm) and small crack branches of the cracks, respectively.

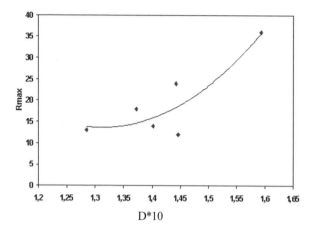

Fig. 3.46 The height of the fracture surface peaks Rmax plotted versus the fractal dimension of the fracture surface D.

The difference in the structure of crack path for different microstructures of steels is rather obvious: whereas a crack grows almost straightforward in the matrix or between carbide layers, it kinks at the carbide layers or even follows the carbide-rich region if this does not require strong deflection of the crack path from its initial direction.

A general tendency is that the maximal height of the fracture surface peak is much higher for the coarse than for the fine microstructure (this was not observed for the net-like microstructure only since the full change of the mechanism of crack growth occured in our simulation – the crack in the coarse net-like microstructure grew like in the band-like microstructures, i.e. it did not follow the carbide network, but just deflected at the carbide layers).

Microcracking at random sites in the microstructure (not in front of the growing crack) was observed in the band-like and random microstructures, but not in the net-like microstructures. Intensive crack branching took place also in the band-like and random microstructures. This can be explained by the comparison with the results by Berns et al. [27], which have shown that when carbides at some distance from the crack tip fail, that leads to the change of the direction of crack growth. Therefore, high intensity of microcracking in random sites of the material causes a high intensity of crack branching.

Fractality of the Fracture Surface

Consider now the fractality of a growing crack in the real materials. The very often observed mechanism of crack growth, when a crack grows by joining microcracks which are formed in front of the crack tip (but not necessarily in the plane of crack propagation) and which are much smaller than the crack [29] causes random variations of the direction of crack growth. These variations determine the fractal dimension D of a fracture surface [30]. The growth of a crack by joining of microcracks is similar to the formation of a fractal cluster from randomly moving particles which may join the cluster [30]. For a fractal cluster which grows by such a mechanism, the fractal dimension of a cluster is given by the formula:

$$i \sim L^D \qquad (3.22)$$

where i – the amount of particles (in our case, the number of unit steps of the crack growth, or eliminated finite elements, L – projected crack length.

This is also the mechanism of the crack growth in our simulations: cracks grow both by joining microcracks (in our model by joining softened or removed finite elements). The removed elements lie in front of the crack tip but not necessarily in the plane of straight mode I crack propagation. The joining of elements, which do not lie in the plane of crack propagation causes random variations of the direction of crack growth, and these variations determine the fractal dimension D of a fracture surface, similarly to the formation of a fractal cluster from randomly moving particles.

To determine the fractal dimension D of the crack path from the above simulations, the number of eliminated elements and the projected length of the crack after each loading step are determined, and the power in the i-L-relationship for each of the structures is calculated. One should note here that the fractal dimensions calculated with the use of this method present just rough estimations. In order to determine the fractal dimension of a crack more exactly, simulations of longer crack growth in much bigger microstructure samples should be carried out.

The fractal dimensions for each microstructure are given in Table 3.5. One can see that the maximal fractal dimension is found in the band-like fine microstructure.

The relationships between the specific energy of fracture, fractal dimension of the fracture surface and the height of peaks on the fracture surface are shown in Figures 3.43 and 3.44. It can be seen that although the results of our numerical experiments (similarly to "normal" experiments) show a great dispersion of results, the tendencies are still rather clear: the specific energy of fracture increases with increasing the fractal dimension of fracture surface. It is worth to note that the appearance of the curve corresponds to the relationship between the surface energy of fracture and the fractal dimension derived in [31]: $G \sim (a/L)^{1-D}$, where a – yardstick length (in [28], it was taken to be equal to an average carbide size), L – unit length (a value of the order of crack length).

The height of the peaks of the fracture surface increases with increasing the fractal dimension as well. It is of interest to compare the last conclusion with the analytical results from [28]. The authors [28] have shown on the basis of the probabilistic analysis of the distribution of peak heights on the fracture surface that the height of the peaks increases with increasing D; yet, the result was obtained only for the net-like microstructures of steels, while our present calculations show this trend for the other microstructures as well.

One should note here that the numerical determination of the fractal dimension of crack is possible only with the use of the above techniques: the element elimination technique (it is evident that prescribing crack path in simulations of the crack growth excludes any possibility to consider the fractality of fracture surface), and the real microstructure simulation (it is also evident that a crack which grows only according to the fracture and continuum mechanics laws, without taking into account "real world" will be not fractal).

Conclusions

Comparing the fracture behaviour of different microstructures of steels, one can draw the following conclusions. The fracture resistance of the steels is much higher for the fine than for coarse version of the same type of microstructures. The resistance of the steels to crack initiation, characterized by the peak load on the force-displacement curve, is lowest for the band-like structures, and is again sufficiently higher for the fine than for coarse microstructures (especially in the cases of the random and band-like fine microstructures).

The roughness of the fracture surface is higher for the coarse than for the fine microstructures. Microcracking at random sites in the microstructure (not in front of the growing crack) as well as the intensive crack branching, were observed in the band-like and random microstructures but hardly in the net-like microstructures.

Generally, the fracture resistance of steels increases in the following order: band-like → random → net-like microstructure.

Furthermore, one may note some interrelations between the geometrical and energy parameters of fracture: the fracture toughness increases with increasing the fractal dimension and height of the roughness profile of the fracture surface. The interrelations between the specific fracture energy, fractal dimension and the roughness height, obtained in our numerical experiments, correspond to the analytical results from [28, 31].

On the basis of the above study, one may speculate about possible directions of the optimization of steels. The following effects which increase the toughness of the steels were observed in the considered structures:

- crack deflection by the carbide layers oriented perpendicularly to the initial crack path (net-like coarse microstructure, band-like microstructures),
- the crack follows the carbide network (net-like fine microstructure),
- and damage formation at random sites of the steels and following crack branching (random microstructures).

One may note that all above effects increase the ratio of the summary area of as-formed surface to the size of failed specimen (projected crack length). Whereas all the modes of steel toughening ensure (or tend to ensure) comparable levels of fracture toughness, the type of microstructure (and therefore the mode of toughening) for given conditions should be chosen on the basis of the economical considerations of the steel production and other requirements like the wear resistance of the tool, etc.

References

[1] Mishnaevsky L. Jr, Dong M., Hoenle S. and Schmauder S. (1999), Computational Mesomechanics of Particle-Reinforced Composites, Comp. Mater. Sci., Vol. 16, No. 1-4, pp. 133-143.

[2] Wulf J. (1995), Neue Finite-Elemente-Methode zur Simulation des Duktilbruchs in Al/SiC. Dissertation. MPI für Metallforschung, Stuttgart.

[3] Mishnaevsky Jr L., Mintchev O. and Schmauder S. (1998), FE-Simulation of Crack Growth Using a Damage Parameter and The Cohesive Zone Concept, In: "ECF 12 - Fracture from Defects", Proc. 12th European Conference on Fracture. Eds. M.W. Brown, E.R. de los Rios and K. J. Miller. London, EMAS, Vol. 2, pp. 1053-1059.

[4] Mishnaevsky Jr L. and Schmauder S., Advanced Finite Element Techniques of Analysis of the Microstructure-Mechanical Properties Relationships in Heterogeneous Materials: A Review, Physical Mesomechanics, Vol. 2, No. 3, pp. 5-20.

[5] Jirasek M. (1998), Finite Elements with Embedded Cracks, LSC Report, Lausanne, EPFL

[6] Rashid Y.R. (1968), Ultimate Strength Analysis of Prestressed Concrete Pressure Vessels. Nucl. Engng. And Design, 7, pp. 334-344.

[7] Tvergaard V. and Hutchinson J.W. (1988), Effect of T-Stress on Mode I Crack Growth Resistance in a Ductile Solid, Int. J. Solids Struct. 31, 6, pp. 823-833.

[8] Xia L., Shih, C. F. and Hutchinson J. W. (1995), A computational approach to ductile crack growth under large scale yielding conditions. J. Mech. Phys. Solids 43, 3, pp. 389-413.

[9] Xia L., and Shih C. F. (1995), Ductile crack growth - II. Void nucleation and geometry effects on macroscopic fracture behavior. J. Mech. Phys. Solids 43, 11, pp. 1953-1981.

[10] Broberg K.B. (1997), The Cell Model of Materials, Computational Mechanics, 19, 7 , pp. 447-452.

[11] Weihe S., Kröplin B., and de Borst R. (1998), Classification of Smeared Crack Models Based on Material and Structural Properties. Int. J. Solids and Structures, 35, 12, pp. 1289-1308.

[12] Weihe S. and Kröplin B. (1995), Fictitiuos Crack Models: A Classification Approach. Proc. FRAMCOS II (2nd Int. Conf. Fracture Mechanics of Concrete and Concrete Structures), Ed. F.H. Wittmann, Aedificatio Publ., Vol. 2, pp. 825-840.

[13] van Vroonhoven J. (1996), Dynamic Crack Propagation in Brittle Materials: Analyses based on Fracture and Damage Mechanics, Eindhoven, Philipps Electronics, p. 195.

[14] Siegmund T., Bernauer G. and Brocks W. (1998), Two Models of Ductile Fracture in Contest: Porous Metal Pasticity and Cohesive Elements, "ECF 12 - Fracture from Defects", Proc. 12th European Conference on Fracture (Sheffield). Eds. M.W. Brown, E.R. de los Rios and K. J. Miller. EMAS, Vol. 2, pp. 981-985.

[15] LARSTRAN, LASSO Ingenieurgesellschaft, Leinfelden-Echterdingen, Germany

[16] Barker L. M. (1981), Short Rod and Short Bar Fracture Toughness Specimen Geometries and Test Methods for Metallic Materials, Fracture Mechanics ASTM STP 743, pp. 456-475.

[17] Mishnaevsky Jr L. and Schmauder S. (1998), 2nd-4th Interim technical reports on the project „Influence of Micromechanical Mechanisms of Strength and Deformation of Tool Steels under Static and Cyclic Load", Creusot Loire, Frankreich

[18] Mishnaevsky Jr L., Lippmann N. and Schmauder S., Micromechanisms of Damage Initiation in Tool Steels: an Analysis with the Use of FE-Simulation of Real Microstructures of Steels and SEM-in-situ-Observations (in preparation)

[19] Trubitz P. (1998), Messung der elastischen Eigenschaften von Karbiden und der Matrix von Stahl X155CrVMo12-1, Prüfbericht, 26.11.1998, TU BA Freiberg

[20] Iturriza I. and Rodriguez J. M., (2000) Ibabe, 5th Interim technical reports on the project „Influence of Micromechanical Mechanisms of Strength and Deformation of Tool Steels under Static and Cyclic Load", Creusot Loire, Frankreich (in preparation)

[21] Lehmann A. (1995), Modellierung des Bruchverhaltens von Schnellarbeitsstählen unter Beachtung bauteilspezifischer Einflüsse, Diplomarbeit, TU Bergakademie Freiberg.

[22] Lippmann N., Lehmann A., Steinkopff Th. and Spies H.-J. (1996), Modelling the Fracture Behaviour of High Speed Steels under Static Loading, Comp. Mat. Sci., 7, pp. 123-130.

[23] Lippmann N. (1995), Beitrag zur Untersuchung des Bruchverhaltens von Werkzeugen aus Schnellarbeitsstählen unter statischer Beanspruchung, Dissertation, Freiberg.

[24] Mishnaevsky Jr L. and Schmauder S. (1999), Pursuance and Analysis of SEM-in-situ-Experiments on the Deformation and Fracture of Cold Work and High Speed Steels, Report for Boehler Edelstahl GmbH, MPA, p. 35.

[25] Broeckmann C. (1994), Bruch karbidreicher Stähle – Experiment und FEM-Simulation unter Berücksichtigung des Gefüges. Dissertation, Ruhr-Universität Bochum.

[26] Gross-Weege A., Weichert D. and Broeckmann C. (1996), Finite element simulation of crack initiation in hard two-phase materials, Comp. Mat. Sci., Vol. 5, pp. 126-142.

[27] Berns H., Broeckmann C., Weichert D. (1997), Fracture of Hot Formed Ledeburitic Chromium Steels, Engineering Fracture Mechanics, 58, 4, pp. 311-325.

[28] Mishnaevsky Jr L. and Schmauder S. (1999), Optimization of Fracture Resistance of Ledeburitic Tool Steels: a Fractal Approach, "Steels and Materials for Power Plants", Ed.: P. Neumann et al., Proceedings of EUROMAT-99 (European Congress on Advanced Materials and Processes, Munich, 1999), Vol. 7, Wiley-VCH Verlag, Weinheim (accepted for publication).

[29] Tetelman A.S. and McEvily Jr A.J. (1967), Fracture of Structural Materials. New York: Wiley.

[30] Mishnaevsky Jr L. (1996), Determination for the Time to Fracture of Solids, Int. J. Fracture, 79, 4, pp. 341-350.

[31] Chelidze T. and Gueguen Y., Evidence of Fractal Fracture, Int. J. Rock Mech. Min. Sci., 1990, Vol. 27, No. 3, pp. 223-225.

3.3.2 FE models of crack propagation tool steels: comparison of techniques and complex microstructures[7]

This paper presents a systematic computational study of the effect of microstructures of metallic materials reinforced with brittle hard particles on their fracture behavior and toughness. Various material microstructures are tested numerically under the same loading conditions. The value of numerical experiments in predicting and improving material performance is demonstrated using the example of high speed steels.

The optimal design of particle-reinforced materials on the basis of computational simulations of their behavior has attracted growing interest of researchers over the last two decades [1-3]. The computational design of materials for industrial needs is possible, if the computational difficulties concerning simulation of complex materials at many scale levels are resolved and corresponding technologies for the creation of the materials are available. Steels are among the group of most investigated materials in the world, and have widely been used industrially in many centuries. Among them, tool steels hold a special position, both due to their wide use in the metal-working industry, and due to the complex requirements on their properties: the microstructure of a tool steel must ensure high hardness and wear-resistance (these properties of tool steels are secured by the availability of hard and brittle primary carbides in the materials), as well as high fracture toughness and lifetime (which are influenced by the properties of the "matrix" of the steels, and the secondary carbides) [4, 5, 6, 7]. On the other hand, microstructures of tool steels can be altered by using different technologies (casting, powder metallurgy, different heat treatment and working), and, therefore, the recommendations to be developed here can be practically realized [5]. That is why we chose tool (high speed) steels as an object for the computational testing of artificially designed microstructures, and analysis of microstructure-fracture resistance relationships. A further reason for the choice of steels as a test object for the computational testing of microstructures is that microcracks initiate in primary carbides in the steels [8-10]; therefore, the initial distribution of primary carbides corresponds to the distribution of potential sites of the crack initiation. Table 3.6 gives a short review of the micromechanical studies of the interrelations between the structure and fracture resistances of tool steels.

---

[7] Reprinted from L. Mishnaevsky Jr., U. Weber, S. Schmauder, "Numerical analysis of the effect of microstructures of particle-reinforced metallic materials on the crack growth and fracture resistance", International Journal of Fracture 125, pp. 33-50 (2004) with kind permission from Springer

Table 3.6
Some research works in the area of the strength and fracture of tool steels

Authors, steel type	Numerical approach, codes	Problems and Main results
Plankensteiner et al. [14, 15]. Steel: Electroslag remelted HSS, with net-like carbide arrangement	PATRAN, ABAQUS. Mesophase Cell Hierarchical Modeling (this approach includes unit cell technique, and transformation field approach or incremental Mori-Tanaka approach).	Overall response of steels as well as mechanisms of local failure in steels with real structures are studied. The effect of progressive carbide cleavage on the stress-strain curve is considered. Stress-strain curves, and distributions of maximum principal inclusion stresses are obtained. It is shown that the carbide grain cleavage is a main fracture mode at microscale.
	HEXGRAIN, ABAQUS. Hexagonal Cell Tiling Concept (unit hexagonal cells containing a number of inclusions).	Stress distribution, and also an effect of initial thermal residual stresses on the parameters of stress distribution are considered. Stress level within carbide particles and in matrix, and the effect of thermal stresses on them are studied.
Broeckmann [16]. Steels: Ledeburitic chromium steel SAE-D3 in as cast condition and hot worked state	CRACKAN. Modeling of real structures. Plasticity of matrix is modeled with J2 flow theory. Carbide particles fail due to cleavage along crystallographic planes.	Influence of carbide particle distribution on the fracture processes in steel is investigated. Stress distribution in real microstructure ahead of main crack, local damage and effect of local triaxiality on cleavage strength of carbides are studied.
Gross-Weege et al. [17]. Steel: as above; HSS manufactured by powder metallurgy and HIP	CRACKAN (like above). Decohesion between carbides and matrix occurs if the stresses normal to the interface reach a critical value. Unit cell model is used to study the interaction between inclusions.	Simulation of crack initiation (also in front of main crack) due to particle cracking and interfacial failure. Particle cracking, damage evolution in front of main crack are studied.
Lippmann et al. [13]. Steel: Electroslag remelted HSS, with 50 % hot reduction	ANSYS. Crack faces are predefined with the use of non-linear spring elements	Crack initiation by carbide failure is simulated. Crack distribution in real structure of steel. It is shown that long and thin carbides fail first and form initial cracks.

	LARSTRAN. Method of multiphase elements (MPE) makes possible to simplify preprocessing. Automatic element elimination technique is used to model crack initiation and growth.	Crack initiation, growth and distribution in real structure of HSS. Stress and crack distribution in real structures is obtained; effect of carbide stringer width of the crack formation is studied.

Numerical methods of the analysis of microstructures in the mechanics of materials

Microstructure modeling at the level of Gaussian points and at finite element level

Information about the microstructure of a material is usually given as a 2D (discretized) micrograph (or, in 3D case, as a series of micrographs) [2, 19]. Figure 3.47 shows schematically the discretization of a real microstructure of high speed steel HS6-5-2 presented in Fig. 3.35 [18].

In order to include the microstructure information in the FE model, there exist in general two methods: first, to impose the complex microstructure on a simple FE mesh without adapting the mesh to the microstructural phase boundaries and second, producing a FE mesh according to the microstructure of the micrograph. The automatic assignment of the digitised microstructure to a simple FE mesh can be done in the framework of multiphase finite elements (MPE) (Figures 3.49-3.49) [1, 2]. The main feature of this method is that the different phase properties are assigned to individual integration (Gaussian) points in the element.

Fig 3.47 Discretized micrograph of high speed steel HS6-5-3. The black areas present primary carbides, the white area is the "matrix". Region 100x100 mcm.

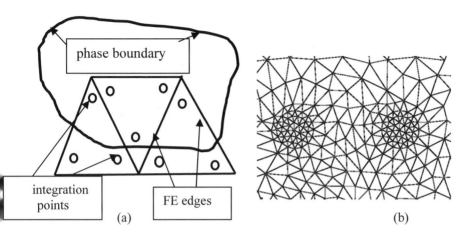

Fig 3.48 Scheme: Multiphase finite elements

Fig 3.49 Scheme: automatic FE mesh generation form microstructures

Contrary to traditional (single-phase) finite elements, a FE-mesh is independent of the phase structure of material in this case, and relatively simple FE-meshes can be used for the simulations of the deformation in a complex microstructure. The possibility of using FE meshes of arbitrary simple structures for the simulation of the behavior of complex materials is the main advantage of the method of multiphase elements.

Therefore, a relatively simple simulation of material behavior in the 3D case becomes possible. One should note however that MPE do not allow one to take into account local effects of interfaces. In some cases this limitation can be useful in order to help to reflect better the gradual transition of the local material properties. In the cases of materials in which neither phase is soluble in any other phases, the interface presents a real boundary and the impossibility taking into account interface effects using MPE can present a serious limitation for the MPE applicability. Another approach is to produce a FE mesh which corresponds to a given microstructure. The main idea of this approach is that a FE mesh is automatically generated in such a way that the boundaries of "surfaces" correspond to the phase boundaries in digitized microstructure micrographs. At the MPA Stuttgart a series of programs was developed, which allow one to generate automatically a FE mesh in PATRAN Pre-Processor which fully corresponds to a given microstructure [4]. Other researchers have developed similar programs. For instance, Iung et al. [20] developed a program which generates FE meshes (to be used by the ABAQUS code) representing the image of a real microstructure. The advantages of this program are that the mesh is generated "in an iterative way by superimposing on the boundaries square grid of growing size", and is refined automatically at the interfaces between the phases.

In our simulations, we focus on the second approach (automatic generation of FE meshes from microstructure micrographs and element elimination). Comparison with the first approach (MPE) is carried out as well.

Finite Element „Softening" and Finite Element Elimination Approaches

Among the main numerical approaches to simulating crack growth, the cohesive zone model [21, 22], smeared crack model [23], representation of cracking by separation of element boundaries [24] and the computational cell methodology [25, 26] should be mentioned. The separation of element boundaries can be done, for instance, if all the elements in the model or the elements along the expected crack path present contact elements. According to Broberg's cell model [27], a material consists of cells (which are defined as the „smallest material unit that contains reasonably sufficient information about crack growth in the material", and which are characterized by their size and cohesion-decohesion relation). If the cell is considered as an element in a FE mesh, such an approach presents a generalization of cohesive models and the computational cell methodology, developed by Xia et al. [25, 26].

In our simulations, crack growth is modeled as an elimination of elements in the FE mesh which represents the body. The evident advantages of this approach are that a body may be discretized into simple finite elements (not special contact or interface elements), and that both damage and crack propagation can be modeled in the framework of one and the same local failure condition. Numerically, such element elimination can be realized in two ways: by element „softening" and by element removal (the last two approaches are often confused) [1, 4, 2]. Element "softening" is done if each element is assumed to be weakened as the local stress ·or damage parameter exceeds a critical level. The Young's modulus of the elements to be softened is set 1...2 orders of magnitude lower than that of the initial material.

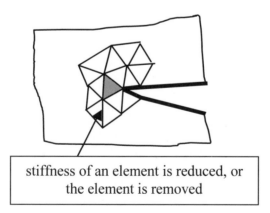

stiffness of an element is reduced, or
the element is removed

Fig 3.50 Scheme: Elimination or softening of a finite element

In some cases, this approach is erroneously called "element elimination" [1]. The possibility to model crack growth in such a way is provided by the subroutines UMAT or USDFLD in the ABAQUS FE code. Another way is to remove the elements from the model, and then to restart the simulation without the eliminated elements, using the RESTART option in ABAQUS. Fig. 3.50 shows schematically the finite element „softening" and finite element elimination approaches.

Problem Statement and FE Model

In order to study the effect of the arrangement of inclusions on the fracture resistance of two-phase materials in a systematic way, different idealized microstructures were designed and crack propagation in these microstructures was simulated numerically. The properties of the constituents of the material (particles, matrix) and the expected mechanism of crack growth were determined from in-situ experiments carried out with simultaneous observation of microprocesses in materials in a scanning electron microscope [8].

FE-Model

A FE model of short rod specimens was developed [28]. The simulations were carried out for 2D plane strain conditions. According to the description of the specimens, the diameter was taken to be 12 mm, height 18 mm, and notch depth 5.32 mm [28].

The model consists of a macromodel and a mesomodel. The macromodel was set up in order to determine the displacement distribution on the boundaries of an area 300 μm x 500 μm in the vicinity of the notch. This area presented then the mesomodel, which included a region with a real steel microstructure. The microstructure of high speed steel was placed in an area 100 μm x 100 μm near the notch in the model.

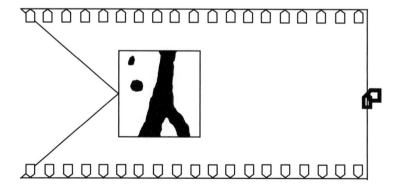

Fig. 3.51 Loading scheme of the mesomodel

The displacement distribution on the boundaries of the mesoscopic model was determined from the macromodel. The scheme of the mesomodel and the position of the real microstructure of the steel in the model are shown in Figure 3.51. The displacement of the points in the vertical direction on the plane of symmetry was set to be zero. The point on the symmetry plane of the specimen, which lies opposite the notch was fixed in the X-direction. Displacement-controlled loading was applied at a point at a distance 1.88 mm from the end of the specimen, according to the real conditions of the loading of short rod specimens. The boundary conditions in the mesomodel (the small area 300 x 500 μm near the notch of the short-rod-specimen) were given as vertical displacements. The total loading displacement was chosen to be 1 mm. Given properties of the components and the steel, and the real microstructure of the steel, crack initiation and growth in the steel are simulated.

Material properties

The elastic and elasto-plastic properties of the carbides and matrix of the steel were determined using different experimental and experimental-numerical methods, including SEM in-situ experiments [8], powder metallurgy production of "matrix alloy" [5], microindentation, etc. The results of the experiments allow us to obtain rather comprehensive information about the properties of the constituents of the steel as presented in Table 3.7.

Table 3.7
Mechanical properties of the primary carbides and matrix in high speed steel

Properties	Carbide	Matrix	Steel	Refs
Young's modulus, GPa	286 (carbides M6C), 351 (MC)	231	240	[11-13]
Poisson's ratio	0.19	0.3	0.3	
Local failure criterion	Maximum normal stress	Plastic strain	-	[9]
Critical level of the local failure criterion	1500 MPa	0.1 %	-	[9]
Fracture toughness, K_{Iv}, MPa m$^{1/2}$	-	49	18.9	[5]
Constitutive law	Elastic, brittle	$\sigma_y = 1500 + 1101 [1-\exp(-\varepsilon_{pl}/0.00369)]$	$\sigma_y = 2200 + 820 [1-\exp(-\varepsilon_{pl}/0.002)]$	[5]

Here: σ_y – von Mises stress, MPa, ε_{pl} – plastic strain.

Types of ideal microstructures

The purpose of this work is to analyze the effect of the microstructure of steels on the fracture resistance by simulating crack propagation in different microstructures. By testing some typical idealized microstructures in these numerical experiments, factors leading to the optimization and preferable microstructures of materials under given service conditions can be determined.

The ideal microstructures were created with the use of the graphics software XFIG. The particles (primary carbides) are supposed to be round, but their distributions in the microstructure region of the mesomodel were heterogeneous and are varied subsequently. The following types of the particles arrangements were considered:

- band-like microstructures (typical for hot formed steels) and continuous net-like (typical for the cast metals) [6, 30],
- random microstructures,
- clustered microstructures: the material consists of regions with high and low density of inclusions; these microstructures were studied in [7]; it was shown in [7] that such microstructures ensure a higher fracture resistance, than random or net-like microstructures (see Fig. 3.52e),
- layered microstructures: a material consists of layers with particles of different sizes (see Fig. 3.52f, g, h).

Simple microstructures (random, net-like and band-like) were considered in [3, 9], and used here only for comparison purposes. In this work, the complex (clustered and layered) microstructures are considered.

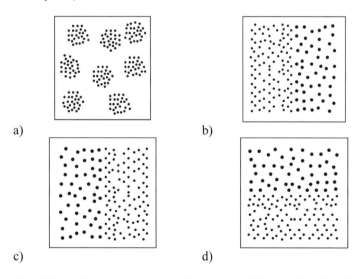

a)   b)

c)   d)

Fig. 3.52 Artificial arrangements of primary carbides considered in this work (along with simple microstructures given in Fig. 3.38): a) clustered, b)-d)- layered (3 different orientations)

Two types of clustered microstructures (fine and coarse) were employed [3]: a fine one with carbide sizes of 2.5 μm and a coarse one with carbide sizes of 3.6 μm.The clustered microstructures contain 200 carbides of diameter 2.5 μm (fine microstructure) and 100 carbides of diameter 3.6 μm (coarse microstructure), grouped in 8 clusters. The surface content (in this case, volume content) of the particles was about 10 % (as for the simple microstructures in [3]). The layered microstructures were supposed to consist of two layers, one with 100 fine particles (2.5 μm) and another one with 50 coarse particles (3.6 μm). The orientation of the layers as related to the initial notch of the specimen was varied such that the expected mode I crack could go first through a coarse layer and then through a fine layer ("coarse -> fine structure", Fig. 3.52g), or conversely ("fine -> coarse structure", Fig. 3.52f), or along the interface between the layers ("coarse/fine structure", Fig. 3.52h).

Simulation of Crack Growth in Different Microstructures

Comparison of different numerical techniques

In our simulation, we use program codes for the automatic generation of FE meshes from the micrographs of microstructures. To model damage and crack propagation, we used the element elimination method with RESTART option.

In order to ensure the compatibility of the results from our previous study (crack growth in simple microstructures) and this work, a comparison of the method of a microstructure-based mesh generation with the multiphase element method/ element softening was carried out.

The coarse random microstructure from [3] was meshed according to both methods: multiphase finite elements [2] and automatic microstructure-dependent mesh generation [4]. Then, two different methods of the simulation of local damage were employed: element softening with relaxation steps and element removal with restarts. Element softening was used in the case of the multiphase elements (since the MPE are used with simple meshes, so the mesh remained intact during the simulation of the crack growth). Element elimination and restarts with new mesh design were employed for the case of the microstructure-based mesh generation.

The crack paths obtained in both cases are shown in Fig. 3.53. Table 3 presents the calculated quantitative parameters of fracture, determined from both simulations. The methods of calculations of the parameters are given below in section 5.1.

Comparing the results, it can be seen that both quantitative parameters of fracture behavior and qualitative crack path simulated by both methods are almost identical. Thus, it may be expected that the results obtained with the use of these methods are compatible.

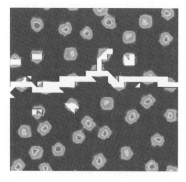 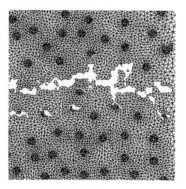

Fig. 3.53 Comparison of the crack path simulated in the artificially designed random coarse microstructure with the use of different numerical approaches: a) MPE, element softening [3], b) automatically generated FE mesh, element elimination

Table 3.8
Quantitative parameters of fracture, determined by two numerical approaches.

	1ˢᵗ **Method:** **MPE**	2ⁿᵈ **Method:** **microstructure-** **dependent mesh design**
Fracture energy G, J/m	699.18	626.77
Fractal dimension of fracture surface	1.372	1.38
Height of roughness peaks, μm	18	18.6

Crack paths in different complex microstructures

Figures 3.54-3.57 show the crack paths in the artificial microstructures of the simulated steels. Generally, fracture occurs as follows: first, several carbides fail and form a "zone of failed carbides" in front of the notch. The "zone" extends and cracks are formed by coalescence of microcracks in the carbides. Intensive microcracking in carbides before a macrocrack is formed has been observed experimentally as well [8].

In the cluster microstructures, the crack grows first in clusters (i.e., it initiates in carbides, and grows from one carbide to another) and then from one cluster to another. Such a mechanism leads to strong deviations of the crack path from its initial direction. In the fine cluster microstructure, two cracks initiated in two different clusters and then propagated by the described mechanism.

In the layered microstructures, crack growth depends strongly on the orientation of the layers: in the "coarse-fine" microstructure, a large crack forms at the initial stage of loading, and propagates first straightforward and then at an angle of about 45 degree with respect to the initial direction. Crack deflection and branching is especially strong in the fine layer.

In the "fine-coarse" microstructure, the "zone of failed carbides" propagates rather far away from the crack tip, but only a small macrocrack forms in that zone (no coalescence of microcracks formed in the carbides). After the microcracks in the large "zone of failed carbides" coalesce, crack with many branches forms. In the "coarse/fine" microstructure (which is in fact symmetric relatively to the notch), the crack grows in the layer of fine carbides, rather than in the layer of coarse carbides: the carbides fail in the layer of fine carbides, and the "zone of failed carbides" extends in the fine layer. When the "zone" becomes large enough, the microcracks coalesce and macrocracks form. However, no such zone and no cracks form in the area of coarse carbides: after the failure of few carbides near the notch, damage growth in the layer of coarse carbides stops. Apparently, the longer distances between primary carbides (potential microcrack initiation sites) in the coarse part prevent the microcracks from joining together in the coarse structures, but not in the layer of fine carbides.

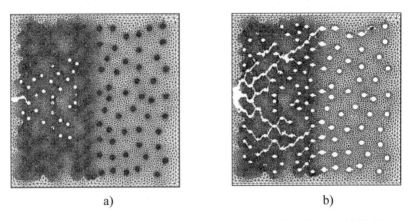

a)   b)

Fig 3.54 Layered microstructure 1 (fine → coarse): a) Step 2, u = 6.00E-03 mm and b) Step 6, u = 7.54E-03 mm.

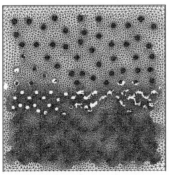

Fig. 3.55 Layered microstructure 2 (coarse/fine): Step 6, u = 7.00E-03 mm

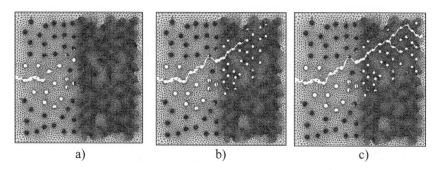

Fig. 3.56 Layered microstructure 1 (coarse → fine): a) Step 8, u = 6.38E-03 mm, b) step 14, u = 6.75E-03 mm, c) step 26, u = 7.13E-03 mm.

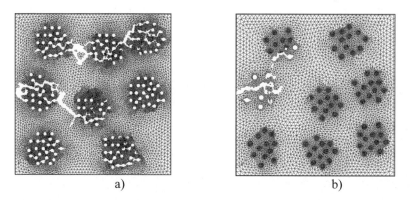

Fig. 3.57 Cluster microstructures: a) Fine, step 6, u = 5.99E-03 mm, b) Coarse, step 6.

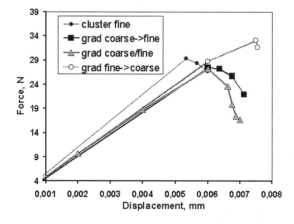

Fig. 3.58 Force-displacement curves for the simulated microstructures.

Crack paths in complex structures were compared with those in simple microstructures obtained in [3]: In the fine net-like microstructure (Figure 3.42), the crack is instantly directed to the carbide network, and then follows exclusively the carbide network. Such a mechanism ensures maximum fracture resistance of the steel (see below). In the band-like (Figure 3.41) and coarse net-like structure, the crack grows rectilinearly in the matrix, and undergoes notable deflections at the carbide bands.

Comparison of different types of microstructures

Fracture resistance of the materials with artificial microstructures

For all simulations performed, the force-displacement curves, energy and geometrical parameters were determined numerically. Fig. 3.58 shows the force-displacement curves for the simulated microstructures.

Table 3.9 provides some main quantitative characteristics of crack growth in the different structures. The value of G (nominal specific energy of the formation of a unit of new surface) for each microstructure characterizes the fracture resistance of each of the structures. This value was calculated as follows[3]: $G = \sum(P_i u_i) / BL_{RS}$, where P_i – force for loading step, u_i – displacement for loading step i, L_{RS} - linear size of the real microstructure in the horizontal direction, the summation is carried out for all i loading steps until the crack passes the real microstructure, B - the thickness of the model.

To characterize crack, we consider the maximal height of the roughness peak R_{max}, the number of eliminated elements not connected to the main crack as well as branches of the crack, and the fractal dimension of the crack. The maximal height of the roughness peak R_{max} was calculated from the crack profile as the distance between highest and lowest points of the crack path measured perpendicular to the initial horizontal crack direction.

The fractal dimension of the fracture surface was calculated using the approach given in the subsection 3.3.1. This will be the subject of future investigations. For comparison purpose, the results for the simple microstructures from [3] are given in the Table 3.9 as well. Fig. 3.59 shows the fractal dimension versus the fracture energy. One can see that the fracture energy increases with increasing fractal dimension of fracture surface. Figure 3.60 shows a comparison of the fracture resistances of the different microstructures. From the figure it can be seen that the heterogeneous microstructures with localized brittle particle distributions (i.e., clustered and layered ones) ensure rather high fracture resistances [34]. This can be explained by the fact that these microstructures lead to strong crack deviations: from one region of high particle density to another one (in a clustered microstructure) or in the layer of fine particles.

Table 3.9
Quantitative parameters of fracture behavior of the artificial structures of steels

	Complex structures				Simple structures (from [3])		
	Layered			Clus-ter fine	Net-like*	Band-like*	Ran-dom*
	"C->F"	"C/F"	"F->C"				
Nominal specific energy of new surface, J/m²	668.9	786.1	733.9	635.7	436.15 / 827.0	341.28 / 676.75	699.18 / 557.02
Maximal height of the roughness peak, R_{max}, μm	44	46	44	35	13 / 36	24 / 14	18 / 12
Fractal dimension of fracture surface, D	1.522	1.556	1.382	1.515	1.285 / 1.593	1.442 / 1.40	1.372 / 1.446

* The values are given for coarse/fine versions of the microstructure, respectively.

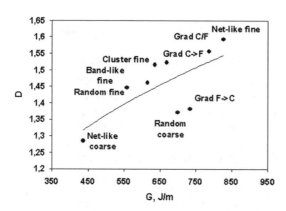

Fig. 3.59 Fractal dimension D plotted versus the calculated fracture energy.

The fracture resistances of the microstructures with spatially localized particle arrangements are always higher than those of simple microstructures, except for the case of the net-like fine microstructure. In the net-like structure, the crack is forced to follow the carbide network, and the crack path becomes much longer than the mode I crack path. This ensures a rather high fracture resistance. However, this toughening mechanism is unstable, since the crack propagates through the carbide network when the network cells are big enough resulting in a very low fracture toughness.

Table 3.10 summarizes the assessments for fracture toughness and mechanisms of toughening for different structures.

Table 3.10
Assessment of Artificial Microstructures

Type	Peculiarities of the mechanism	Mechanism of toughening
Net-like:	Crack follows the carbide network. However, this toughening mechanism is unstable: if the cells are too large, the crack propagates through the cells.	Crack path (determined by the carbide network) is much longer than without the network
Band-like	Crack jumps from one band to another and deflects at the bands.	Crack path deviations at the bands; high toughness of the matrix between bands.
Random	Crack jumps from one carbide to another.	Intensive crack branching and damage formation also apart from the crack path
Layered	A. "coarse → fine": carbides fail in coarse layer, but a crack forms only at high K_I. The fine layers lead to strong deviations of crack direction from the initial path. B. "$\dfrac{\text{coarse}}{\text{fine}}$": the crack preferably grows in the fine layer, rather than in the coarse layers	Crack path deviations in the layer of fine particles; high toughness of the matrix between the particles.
Clus-tered	Crack grows first in clusters (jumps from one carbide to another) and then from one cluster to another.	Crack path deviations from the initial direction due to the jumps to nearest clusters

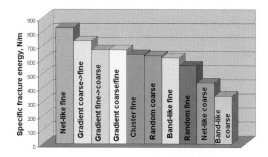

Fig. 3.60 Comparison of fracture resistance of different artificial microstructures

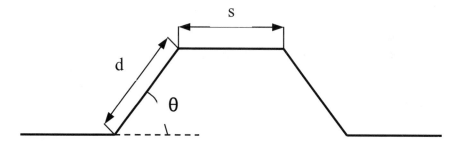

Fig. 3.61 Schematization of the crack deviation according to Suresh

Crack deflection and the fatigue behavior of the crack

Now we estimate the change in the fatigue crack behavior caused by microstructure-induced crack deflections following the model of Suresh [35], assuming that the crack paths in the case of monotonic loading and cyclic loading are similar. According to Suresh [35], the apparent crack propagation rate (i.e., measured along the mode I crack direction) changes due to the crack deviations as:

$$v/v_L \sim m \cos \theta + (1\text{-}m), \qquad (3.23)$$

where v_L is the growth rate of a straight crack under the same loading conditions as the kinked crack, θ is the kink angle, $m = 1/(1+s/d)$, d – distance over which the tilted crack advances along the kink, and s is the distance over which the plane of the growing crack is normal to the far-field tensile axis (see Fig. 3.61).

Table 3.11 gives the value of θ, d, s and m for the cracks simulated above.

Table 3.11
Coefficients of the formula for the reducing fatigue crack growth rate

Type of micro-structures	Kink angle θ	Distance d (tilted)	Distance s (straight)	Coefficient m	Coefficient v/v$_L$
Layered fine-> coarse	~32	46	47	0.515	0.014
Layered coarse/fine	~30	14	5.7	0.710	0.88
Layered coarse->fine	~45	20	14.8	0.574	0.07
Cluster coarse	~ 80*	10.6	27	0.281	0.63
Cluster fine	~ 20*	27	28	0,49	0,036

*Crack kinks from the end of a cluster to the next cluster.

From Table 3.11, it can be seen that the maximum estimated reduction of the crack rate due to the crack deflection is observed in the layered "fine->coarse" and "coarse->fine" microstructures, i.e. on the structures where one of the layers forces the intensive crack kinking and branching. Comparing the values of fractal dimension of the fracture surface (given in Table 3.9) and the values of the relative reduction of crack rate v/v$_L$, it may be seen that the higher values of fractal dimension and roughness of the crack surface correspond to a stronger reduction of the crack growth rate due to deviations of crack path from the mode I direction of crack growth.

Conclusions

On the basis of our numerical investigations, the fracture mechanisms in the materials with different arrangements of brittle round inclusions were clarified.

Clustered and layered heterogeneous microstructures ensure rather high fracture resistances, which are always higher than those of simple microstructures, i.e. band-like, net-like or random ones. This is determined by the fact that these heterogeneous microstructures lead to strong crack deviations: from a region of high particle density to another one (in the clustered microstructure) or into the layer of fine particles. Net-like fine microstructure shows an exception to this rule and forces the crack to follow the carbide network, ensuring the highest fracture resistance, even higher than all the complex microstructures. However, such a mechanism of toughening is unstable: the net-like coarse microstructure (with larger cells) provides very low fracture toughness, since the crack propagates in a straight manner rather than following the carbide network.

The investigations lead to the conclusion that complex (clustered and layered) microstructures possess the potential for improving the fracture toughness of steels. The main mechanisms of the positive toughening effect by complex microstructures are identified: crack path deviations from its initial direction (which increase the crack length without increasing the stress intensity factor K_1), and large areas of the tough matrix between the areas of high inclusion density.

References:

[1] Mishnaevsky Jr L. and Schmauder S. (2001), Continuum Mesomechanical Finite Element Modeling in Materials Development: a State-of-the-Art Review, Applied Mechanics Reviews, Vol. 54, 1, pp. 49-69.

[2] Mishnaevsky Jr L., Dong M., Hoenle S. and Schmauder S. (1999), Computational Mesomechanics of Particle-Reinforced Composites, Comp. Mater. Sci., Vol. 16, No. 1-4, pp. 133-143.

[3] Mishnaevsky Jr L., Lippmann N. and Schmauder S. (2003), Computational Modeling of Crack Propagation in Real Microstructures of Steels and Virtual Testing of Artificially Designed Materials, Int. J. Fracture, Vol. 120, Nr. 4, pp. 581-600.

[4] Mishnaevsky Jr L., Weber U., Lippmann N. and Schmauder S. (2001), Computational Experiment in the Mechanics of Materials, "Computational Modeling of Materials, Minerals, and Metals Processing", Proceeding TMS Conference, Eds. M. Cross, J.W. Evans, and C. Bailey, pp. 673-680.

[5] Le Calvez Ch., Ponsot A., Mishnaevsky Jr L., Iturizza I.. Rodriguez J. M. Ibabe, Lichtenegger G and Schmauder S. (2000), Micromechanical Mechanisms of Strength and Damage of Tool Steels under Static and Cyclic Loading, CLI, Le Creusot, France. Final Reports on ECSC RTD Project.

[6] Berns H., Broeckmann C., Weichert D. (1997), Fracture of Hot Formed Ledeburitic Chromium Steels, Engineering Fracture Mechanics, 58, 4, pp. 311-325.

[7] Berns H, Melander A, Weichert D, Asnafi N, Broeckmann C and Gross-Weege A (1998), A new material for cold forging tool, Comput Mater Sci, 11, pp. 166-180.

[8] Mishnaevsky Jr L., Lippmann N. and Schmauder S. (2003), Micromechanisms and Modelling of Crack Initiation and Growth in Tool Steels: Role of Primary Carbides, Zeitschrift f. Metallkunde (Int. J. Materials Research), 94, 6, pp. 676-681.

[8] Mishnaevsky Jr L., Lippmann N. and Schmauder S. (2001), Computational Design of Multiphase Materials at Mesolevel, ASME International Mechanical Engineering Congress and Exposition, November 11-16, New York, CD-ROM Proceedings, Vol. 2.

[9] Mishnaevsky Jr L., Lippmann N. and Schmauder S. (2001), Experimental-Numerical Analysis of Mechanisms of Damage Initiation in Tool Steels, Proc. 10th International Conference Fracture, 3-7 Dec 2001, Honolulu, USA, CD-ROM.

[10] Lippmann N. (1995), Beitrag zur Untersuchung des Bruchverhaltens von Werkzeugen aus Schnellarbeitsstählen unter statischer Beanspruchung, Dissertation, Freiberg.

[11] Lehmann A. (1995), Modellierung des Bruchverhaltens von Schnellarbeitsstählen unter Beachtung bauteilspezifischer Einflüsse, Diplomarbeit, TU Bergakademie Freiberg.

[12] N. Lippmann, A. Lehmann, Th. Steinkopff and H.-J. Spies (1996), Modelling the Fracture Behaviour of High Speed Steels under Static Loading, Comp. Mat. Sci., 7, pp. 123-130.

[13] Plankensteiner AF, Böhm HJ, Rammerstorfer FG and Buryachenko VA (1996), Hierarchical modelling of high speed steels as layer-structured particulate MMCs, J de Physique IV, 6, pp. 395-402

[14] Plankensteiner AF, Böhm HJ, Rammerstorfer FG and Pettermann HE (1998), Multiscale modelling of highly heterogeneous particulate MMCs, Proc 2nd Europ Conf Mechanics of Materials, Magdeburg, pp. 291-298.

[15] Broeckmann C (1994), Bruch karbidreicher Stähle – Experiment und FEM-Simulation unter Berücksichtigung des Gefüges, Dissertation, Ruhr-Universitaet Bochum.

[16] Gross-Weege A, Weichert D and Broeckmann C (1996), Finite Element simulation of crack initiation in hard two-phase materials, Comput Mat Sci, 5, pp. 126-142.

[17] Mishnaevsky Jr L., Lippmann N. and Schmauder S., Mesomechanical Simulation of Crack Propagation in Real and Quasi-Real Idealized Microstructures of Tool Steels, Proc. 13th Europ. Conf. Fracture "Fracture Mechanics : Applications and Challenges", 6th - 9th September, 2000, San Sebastián, Spain, CD-ROM.

[18] Schmauder S. and Mishnaevsky L. (2001), Damage in Metallic Multiphase Materials: Mechanisms and Mesomechanics, MPA Stuttgart, Preprint, p. 110.

[19] Iung I, Petitgand H, Grange M and Lemaire E (1996), Mechanical behaviour of multiphase materials Numerical simulations and experimental comparisons, Proc IUTAM Symposium on Micromechanics of Plasticity and Damage in Multiphase Materials, Kluwer, pp. 99-106.

[20] Mishnaevsky Jr L., Mintchev O. and Schmauder S. FE-Simulation of Crack Growth Using a Damage Parameter and The Cohesive Zone Concept, In: „ECF 12 - Fracture from Defects", Proc. 12th European Conference Fracture (Sheffield, 1998). Eds. M.W. Brown, E.R. de los Rios and K. J. Miller. London, EMAS, Vol. 2, pp. 1053-1059.

[21] Tvergaard V. and Hutchinson J.W. (1988), Effect of T-Stress on Mode I Crack Growth Resistance in a Ductile Solid, Int. J. Solids Struct. 31, 6, pp. 823-833.

[22] Weihe S., Kröplin B., and de Borst R. (1998), Classification of Smeared Crack Models Based on Material and Structural Properties. Int. J. Solids and Structures, 35, 12, pp. 1289-1308.

[23] Carter, P. B., Wawrzynek J. A, Ingraffea A.R. , (2000), "Automated 3D Crack Growth Simulation" Gallagher Special Issue of Int. J. Num. Meth. Engng. Vol 47, pp. 229-253

[24] Xia L., Shih C. F., and Hutchinson J. W. (1995), A computational approach to ductile crack growth under large scale yielding conditions. J. Mech. Phys. Solids 43, 3, pp. 389-413

[25] Xia L., and Shih C. F (1995), Ductile crack growth II. Void nucleation and geometry effects on macroscopic fracture behavior. J. Mech. Phys. Solids 43, 11, pp. 1953-1981.

[26] Broberg K.B. (1997), The Cell Model of Materials, Computational Mechanics, 19, 7, pp. 447-452

[27] Barker L. M. (1981), Short Rod and Short Bar Fracture Toughness Specimen Geometries and Test Methods for Metallic Materials, Fracture Mechanics ASTM STP 743, , pp. 456-475.

[28] Mishnaevsky Jr L. and Schmauder S. (1999), SEM-in-situ- Experiments on the Deformation and Fracture of Cold Work and High Speed Steels, Report for Boehler Edelstahl GmbH, MPA, p. 35.

[29] Mishnaevsky Jr L. and Schmauder S., Optimization of Fracture Resistance of Ledeburitic Tool Steels: a Fractal Approach "Steels and Materials for Power Plants", Ed.: P. Neumann et al., Proc. EUROMAT-99, Vol. 7, Wiley-VCH Verlag, Weinheim

[30] Mishnaevsky Jr L. (1998), Damage and fracture of heterogeneous materials, Balkema, Rottterdam.

[31] Mishnaevsky Jr L. (1996), Determination for the Time to Fracture of Solids, Int. J. Fracture, 79, 4, pp. 341-350.

[32] Xie H. (1993), Fractals in Rock Mechanics, Balkema, Rottterdam.

[33] Mishnaevsky L. Jr, Shioya T. (2001), Optimization of Materials Microstructures: Information Theory Approach, Journal of the School of Engineering, The University of Tokyo, Vol. 48, pp. 1-13.

[34] Suresh S. (1998), Fatigue of Materials, Cambridge University Press.

[35] Mishnaevsky Jr L. (1997), Methods of the Theory of Complex Systems in Modelling of Fracture: a Brief Review, Eng. Fract. Mech. 56 :, pp. 47-56

3.4 Interface fracture: elastic and plastic fracture energies of metal/ceramic joints[8]

In many technological branches, combinations of materials with different properties are required. One interesting combination consists of metal and ceramic in order to combine ductility, high electrical and thermal conductivity with high strength and chemical inertness [1, 2]. The interface between two materials is often the weakest point of these devices. In the past, the influence of the interface strength on the energy release rate was investigated experimentally [3] as well as numerically [4]. In this subsection, the computational analysis of the effect of the interfacial strength on material properties is carried out using the finite element method. In the model, the interface is treated as ideal, without a transition region. Such ideal interfaces were observed, e.g. in the case of sapphire/niobium [2].

3.4.1 Concept of Modelling

According to earlier experiments [5] four-point bending tests were simulated with a total size of the specimens of (1.6 x 4.0 x 32mm^3) (Fig.3.61). The simulations were performed as two-dimensional under plane stress conditions. This seems to be a good approximation, because the force-deflection curve of plane stress simulations matches nearly exactly with three-dimensional calculations [6]. The four-point bending tests were simulated by a displacement of the upper hard- metal cylinders of the bending device to a maximum of 70 μm.

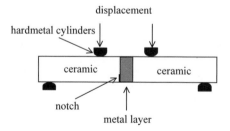

Fig 3.61 Model geometry: Specimen and one half of the hardmetal cylinders of the bending machine

[8] Reprinted from C. Kohnle, O. Mintchev, S. Schmauder, "Elastic and Plastic Fracture Energies of Metal/Ceramic Joints", Computational Materials Science 25, pp. 272-277 (2002) with kind permission from Elsevier

All simulations were done with the commercial code ABAQUS [7] using quad4 elements. The Cohesive Surface Model (CSM) [8,9] is used to model the process of interface fracture. In this model the crack path is prescribed and crack initiation and propagation occurs if a critical normal tension T_n^{crit} is exceeded. This means, that we only model mode I fracture. This is a valid approximation for our model geometry [10]. In the CSM the interface is not modelled by describing an interface zone (e.g. [11]), but the interface is rather modelled as being ideally sharp, which is at least justified for some metal/ceramic combinations as mentioned above. The elastic (el) and plastic (pl) energy release rate occurring during interface fracture is calculated from two specimens with a difference in crack length of $da = a_k - a_l$:

$$G_C^{el} = \frac{E_{a_k}^{el} - E_{a_l}^{el}}{da}$$
$$G_C^{pl} = \frac{E_{a_k}^{pl} - E_{a_l}^{pl}}{da} \tag{3.24}$$

With elastic ($E_{a_k}^{el}$) and plastic ($E_{a_k}^{pl}$) energies:

$$E_{a_k}^{el} = \int_0^t dt \int_V dV \sigma_{ij}^c \dot{\varepsilon}_{ij}^{el}$$
$$E_{a_k}^{pl} = \int_0^t dt \int_V dV \sigma_{ij}^c \dot{\varepsilon}_{ij}^{pl} \tag{3.25}$$

Here σ^e denotes the Cauchy stress tensor, $\dot{\varepsilon}^{el}$ and $\dot{\varepsilon}^{pl}$ the elastic and plastic stain rates, respectively. Both outer (ceramic) parts of the specimen were treated as purely elastic with a Young`s modulus of 390 MPa and a Poisson's ratio of 0.22. The middle (metal) part was modelled with an elastic-plastic constitutive law. The Young's modulus and the Poisson's ratio were the same for all simulations *(E = 104.9 MPa, v = 0.397)*. These elastic data are adjusted to alumina and niobium, respectively [5,6,12]. The plastic behaviour of the stress-strain curve of the metal layer is approximated by a Ramberg-Osgood function [13], which is described in the one-dimensional case by the following:

$$E_c = \sigma + \alpha \left(\frac{\sigma}{\sigma_0} \right)^{n-1} \sigma \tag{3.26}$$

Here, n denotes the hardening exponent, σ the yield offset and σ_0 the yield stress. This material law is nonlinear from the beginning, but for commonly used hardening exponents $n \geq 5$ the divergence from linearity is only slight for stresses below

σ_0. The chosen plasticity theory was the deformation plasticity theory (for details see [9] and references therein), which describes strictly spoken not a plastic material behaviour, but a nonlinear elastic material. This means, that no unloading criterion exists. In order to verify, that the deformation theory is useful in modelling four-point bending specimens, a comparison with the von Mises theory was performed. The obtained results are depicted in Fig. 3.62. The flow curve was implemented as described by eqn. (3.26) with $n = 6$, $\sigma_0 = 180$ MPa and $\alpha = 0.3$. It is obvious, that the results are exactly the same and it is therefore justifiable using the deformation plasticity theory instead of the incremental von Mises theory.

3.4.2 Results

The concept of calculating the energy release rate according to eqns. (3.24) was verified by the following two simulations:

- A correct calculation of the energy release rate according to eqns. (3.24) should use an infinitesimally small value of da, while in the simulation a finite value of da is applied. Therefore it was tested, if the used value of da is small enough to provide results of sufficient accuracy. This was done by performing a second calculation with a difference in crack length of 2 * da. The results are depicted in Fig. 3.63. There is no change in the energy release rates, if we use 2 * da. This means that the value of $da = 0.02$ mm used for all following calculations is applicable.
- The total energy release rate $G_C^{tot} = G_C^{el} + G_C^{pl}$ calculated using eqns. (3.24, 3.25) was compared with a J-integral calculation of the same specimen (Fig. 3.64). The values are found to be identical and the J-integral method does also allow a separation into elastic and plastic contributions.

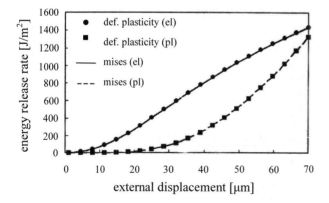

Fig 3.62 Comparison between elastic and plastic parts of the energy release rate versus the external displacement of the hardmetal cylinders for the deformation plasticity theory and the von Mises theory used to model the metal part of the specimen.

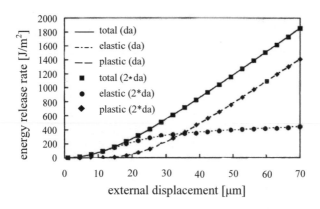

Fig 3.63 Comparison between energy release rate versus the external displacement of the hardmetal cylinders for two values of the difference in crack length used in (1).

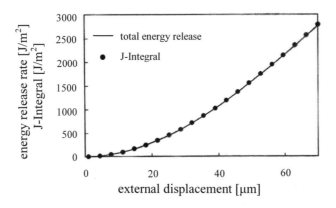

Fig 3.64 Comparison of calculated total energy release rates and J-integral calculations

Taking these results into account, further simulations were performed with varying yield stress σ_0. The corresponding stress-strain functions are depicted in Fig. 3.65. For all following results the Ramberg-Osgood parameters were set to n = 6 and α = 0.3. The thickness of e metal layer was 2mm. For each of these constitutive equations the energy release rate was calculated for different values of the interface strength T_n^{crit}. Due to the stress concentration effects, the maximum normal tensile stress appears at the interface edge of the specimen and at the notch, respectively (compare Fig. 3.61 and 3.69). Fig. 3.66 shows the total energy release rate versus the interface strength for the materials with different yield stresses as shown in Fig. 3.67.

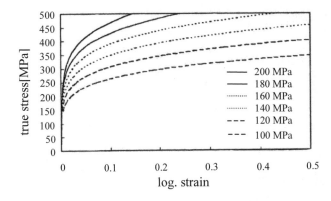

Fig 3.65 Stress-strain curves for different yield stresses σ_0 (E = 104.9MPa, v = 0.397, n = 6, α = 0.3).

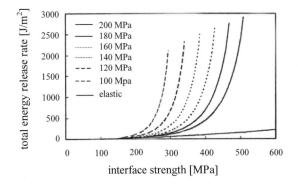

Fig 3.66 Total energy release rate at fracture versus interface strength calculated with the material behaviour of the metal layer as depicted in Fig. 3.67.

For the reason of comparison another simulation with elastically treated niobium was performed. The total critical energy release rate is also shown in Fig. 3.68. The influence of the yield stress on this correlation is obvious. It results from the reduction of the developing normal tension at the interface due to plastic yielding. The specimen with lower yield strength, therefore stores more energy before the critical value T_n^{crit} for interfacial debonding is reached.

The energy release rates were separated in elastic and plastic parts. Fig. 3.69 shows the results for three values of σ_0. As expected, the influence of the yield stress σ_0 on the plastic part of the energy release rate is quite strong. But also in the elastically dominated regime, i.e. for smaller values of the interface strength, where the plastic energy release rate can be neglected, we can find differences between the three $G_C^{el} \sim T_n^{crit}$ -relations. This effect results from small-scale yielding at the crack tip.

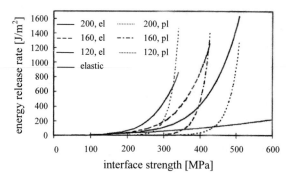

Fig 3.67 Elastic and plastic parts of the energy release rate versus interface strength for three different values of the yield stress σ_0 of the metal layer.

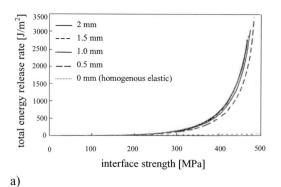

a)

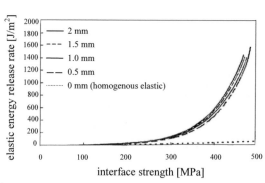

b)

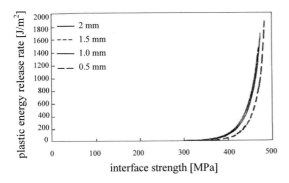

c)

Fig 3.68 a) total, b) elastic and c) plastic energy release rate versus interface strength for thicknesses of the metal layer of 0.5 mm, 1.0 mm, 1.5 mm and 2 mm (0 mm refers to the purely elastic case). The Ramberg-Osgood parameters were $\sigma_0 = 180$ MPa, $n = 6$, $\alpha = 0.3$

Next, the influence of the thickness of the metal layer was studied. Specimens with a metal layer thickness of 0.5mm, 1.0mm, 1.5mm and 2mm were simulated. Fig. 3.68 a) shows the total energy release rate versus the interface strength. We can see that the influence of the thickness nearly vanishes for thicknesses > 1.5mm.

For the reason of comparison a homogenous (elastic) ceramic specimen was simulated as well (Thicknesses between 0 mm and 0.5 mm have not been considered in this study).

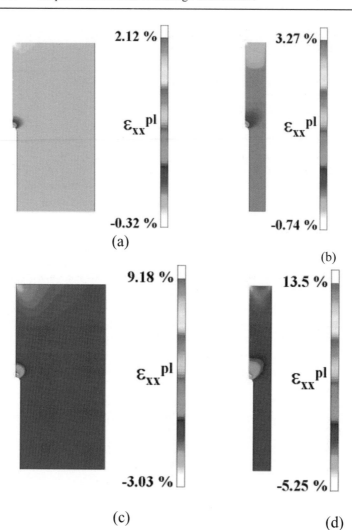

Fig 3.69 Distributions of the ε_{xx}-component of the plastic strain tensor in the metal layer (see Fig. 3.61) for two different external displacements: 29 μm (a, b) and 70 μm (c, d). The thickness of the metal layer is 2 mm (a, c) and 0.5 mm (b, d), respectively (compare with Fig. 3.61).

The behaviour of the corresponding curve in Fig. 3.68 is completely different from the results of the metal/ ceramic joints. Further, if we compare the elastic and plastic parts of the corresponding energy release rates (Fig. 3.68 b) and c)), it can be seen that the influence on the elastic energy release rates also exists for thicknesses > 1.5mm, but the plastic energy release rate is exactly the same for the thicknesses 1.5mm and 2.0mm. This means, that the total energy release rate is dominated by the plastic part.

Now the question arises, why the influence of the thickness on the plastic energy release rate vanishes for thicker metal layers. To answer this question, the developing plastic zone in the metal layer should be analysed for different thicknesses. In Fig. 3.69 the metal layers of two deformation states are depicted (external displacement = 29 μm and 70 μm, respectively) for a layer thicknesses of 2mm and 0.5mm. For the case of the thinner layer it can be seen, that the second interface constrains the development of the plastic zone strongly: even for small deflections of the specimen the plastic zone reaches the second interface, whereas in the metal layer of 2mm thickness the plastic zone is able to develop in an unconstrained manner.

Conclusions

The possibility of separating the energy release rates of interface cracks into elastic and plastic parts is demonstrated. A variation of the yield strength in the Ramberg-Osgood function influenced the results strongly. The study of specimens with varying thicknesses of the metal layer demonstrated, that a constraint of the second interface on the plastic zone exists for smaller thicknesses, but vanishes for thicknesses >1.5mm in the case of the examined metal/ceramic compounds.

References

[1] Doscha H. (1987), Technische Keramik in Produktion und Entwicklung, Metall 41, p. 502.
[2] Elssner. G (1989), Ceramography and metallography of transition zones between ceramics and metals, Prakt. Met. 26, p. 202.
[3] Elssner G., Korn D. Rühle M. (1994), The influence of interface impurities on fracture energy of UHV diffusion bonded metal–ceramic bicrystals, Scripta Metall. Mater. 30, p. 1037.
[4] Hao S., Schwalbe K.H., Cormec A. (2000), Effect of yield strength mis-match on the fracture analysis of welded joints: slip-line solutions for pure bending, Int. J. Solids Struct. 32, p. 5385.
[5] Kohnle C., Mintchev O., Brunner D., Schmauder S. (2000), Fracture of metal/ceramic interfaces, Comput. Mater. Sci. 19, p. 261.
[6] Kohnle C., Mintchev O., Brunner D, Schmauder S (2000), Fracture of metal/ceramic interfaces, in: Proc. Material Week.
[7] ABAQUS Version 5.8, Hibbitt, Karlsson, Sorensen.
[8] Mintchev O., Rammerstorfer F.G. (1996), Microbuckling and delamination effects on the compression behaviour of materials with layered microstructure, in: Proc. 8th. Int. Symp. on Continuum Models on Discrete Systems (CMDS8), p. 250.
[9] Mintchev O., Rohde J., Schmauder S (1998), Mesomechanical simulation of crack propagation through graded ductile zones in hardmetals, Comput. Mater. Sci. 13, p. 81.
[10] O'Dowd N.P., Shih C.F., Stout M.G. (1992), Test geometries for measuring interfacial fracture toughness, Int. J. Solids Struct. 29, p. 571.

[11] Needleman A. (1987), A continuum model for void nucleation by inclusion debonding, J. Appl. Mech. 54, p. 525.
[12] Available from <www.goodfellow.com>.
[13] Ramberg W., Osgood W.R. (1945), Description of stress–strain curves by three parameters, NASA Technical Note No 902.

Chapter 4: Complex, Graded and Interpenetrating Microstructures

In this chapter, different methods of computational modelling of materials with complex microstructures, in particular, interpenetrating and graded materials are considered.

In section 4.1, self-consistent matricity model is presented. This model has been developed to simulate the mechanical behaviour of composites with two randomly distributed phases of interpenetrating microstructures. The model is an extension of the self-consistent model for matrices with randomly distributed inclusions. In addition to the volume fraction of the phases, the matricity model allows a further parameter of the microstructure, the matricity M of each phase, to be included into the simulation of the mechanical behaviour of composites with interpenetrating microstructures. The model is applied to the calculation of stress-strain curves and strain distribution curves of an Fe/18vol. %Ag-composite as well as to stress-strain curves of an Ag/58vol. %Ni-composite and its validity and superiority upon previous models is demonstrated. The matricities of the phases influence the stress-strain behaviour mainly within the bounds between M = 0.3 and M = 0.7. Beyond these bounds, there exists only a minor influence of matricity on the stress-strain behaviour. Good agreement is obtained between experiment and calculation with respect to the composites' mechanical behaviour and the matricity model is thus found to represent well metal matrix composites with interpenetrating microstructures. Further, the mechanical behaviour of different ZrO_2/NiCr 80 20 compositions are analysed and compared with experimental findings. The microwave sintered material is found to possess a slightly dominant ceramic matrix for intermediate volume fractions. Its thermal expansion coefficient deviates from the rule of mixture. The modulus and the stress strain behaviour are simulated with a numerical homogenization procedure and the influence of residual stresses is found to be negligible. The matricity parameter describes the mutual circumvention of the phases and is found to strongly control the stress level of the composite globally as well as locally. Finally, a graded component and a metal/ceramic bimaterial are compared for thermal as well as mechanical loading.

In section 4.2, the methods of explicit micromechanical modelling of graded and interpenetrating phase materials are discussed. Special functionally graded elements are presented, and their application to modeling the deformation and damage behavior of graded hardmetals. The crack-gradient interaction depends on the kind of the gradient and the distance between initial cracks. A considerable effect of Co islands on cracking is observed.

Further, the damage evolution of composites with ductile matrix and hard damageable particles is analysed numerically. It was shown that flow stress and stiffness of composites decrease, and failure strain increases with increasing the degree of gradient of the particle arrangement. The orientations of elongated particles have a strong impact on failure strain and damage growth in the composites reinforced with elongated or plate-like particle: whereas the horizontally aligned particles ensure the highest failure strain, the vertically aligned particles lead to the lowest and the randomly oriented particles to the medium failure strain. The damage growth in the SiC particles in gradient composites begins in the particles, which are located in the transition zone between the zone of high particle density and the particle-free regions.

Finally in this chapter, a method for the reconstruction and generation of 3D microstructures of composites based on the voxel array data is presented. With the use of the program of voxel-based 3D model generation, the deformation and damage evolution in interpenetrating phase composites with isotropic and graded microstructures is numerically simulated. It is shown that the stiffness, peak and yield stresses of a graded interpenetrating phase composite decrease with increasing the sharpness of the transition zone between the region of high volume content of the hard phase and the reinforcement free region. The critical applied strain, at which the intensive damage growth begins, is decreasing with increasing the volume content of the hard phase of the composite.

In order to relate the microstructures of graded hardmetals and their efficiency in milling applications, these materials are investigated experimentally and numerically on several length scales. On the macroscale, milling tests are performed and a numerical model for up-milling is developed. On the mesoscale, the interaction of cracks with the graded surface zones, especially large soft binder islands are studied. On the microscale, the influence of shape and volume of such inclusions are investigated, further on, crack propagation in realistic microstructures is simulated.

4.1. Interpenetrating phase materials: matricity model and its applications

4.1.1 *Matricity model approach*[1]

Coarse isotropic interpenetrating microstructures are often found in powder metal-lurgically fabricated composites in the regime of 25-75% volume fraction, or result from the infiltration of a porous material with a molten metal of a lower melting point. The arrangement of the phases in most of these technical composites is usually random, resulting in an isotropic overall mechanical behaviour of these composites. Besides the volume fraction, such a material requires at least one further parameter to describe the microstructure more closely. In this work the selected further parameter is the "Matricity", which was first introduced by Poech [1] for two-phase steels and later used by Soppa [2] for the characterisation of Ag/Ni composites. This parameter can be measured from a representative micrograph of a microstructure via an image analysing system and also be included in a finite element model which was developed for this kind of microstructure [3-6]. In this work the model is applied to examine the influence of microscopical residual stresses on the macroscopical behaviour of composites and the impact of matricity on these residual stresses.

Matricity

In the following, composites of two phases α and β are considered. Matricity is defined as the fraction of the length of the skeleton lines of one phase S_α, and the length of the skeleton lines of the participating phases.

$$M_\alpha = \frac{S_\alpha}{S_\alpha + S_\beta} \tag{4.1}$$

By definition, the sum of the matricities of all phases equals to one

---

[1] Reprinted from P. Leßle, M. Dong, S. Schmauder, "Self-Consistent Matricity Model to Simulate the Mechanical Behaviour of Interpenetrating Microstructures", Computational Materials Science 15, pp. 455-465 (1999) with kind permission from Elsevier.

$$M_\alpha + M_\beta = 1 \tag{4.2}$$

To obtain the skeleton lines of a certain phase, we must select this phase within an image analysing system and reduce the detected structure to a typically non-connecting line by maintaining the topology. In Fig. 4.1 the matricities have been determined for a ZrO$_2$/NiCr 80 20 cermet which was powder metallurgically fabricated at the University of Dortmund [7]. The structure parameters volume fraction f and matricity M have been determined to be $f_{NiCr} = 0.3$ and $M_{NiCr} = 0.2$.

Matricity model

To take both the parameters into account for the calculation of the mechanical behaviour of the composite, an extension of the embedded cell model [8,9] has been developed.

The embedded cell model has been introduced to simulate the mechanical behaviour of composites with randomly distributed inclusions, where the volume fraction of the inclusions is the main parameter in the model. To take the matricity as second microstructural parameter into account, the self-consistent embedded cell model has been extended by a second self-consistent embedded cell model (Fig. 4.2). In this "matricity model" we are able to define the matricity of the model in the same manner as the matricity is defined in a real microstructure. First the single phases are reduced to skeleton lines.

The lengths of the skeleton lines of the inclusions (Fig. 4.2, left: β; right: α) are zero as the inclusions are spherical and are, therefore, reduced to a point in the process of obtaining the matricity of the phase.

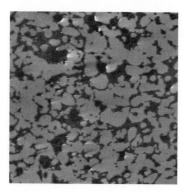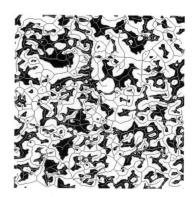

Fig 4.1 Micrograph of a ZrO$_2$ /NiCr 80 20 composite with 30% volume fraction of ZrO$_2$. (Left: greyscale picture, right: binary image with skeleton lines.)

The lengths of the skeleton lines S$_\alpha$ and S$_\beta$ in the matrices are given as the circumference of a circle with a diameter which is calculated as the arithmetic average of the diameter of the embedded cell and the diameter of the inclusion phase (Fig. 4.2, left: S_α; right: S_β).

The diameters of the embedded cells are denominated as W_1 and W_2. The diameters of the inclusion part of the embedded cells depend on the volume fraction of the inclusions and the corresponding factors W_1 or W_2.

$$D_\beta = W_1 \left(\sqrt[3]{f_\beta} \right) \tag{4.3}$$

and, analogous for inclusion α (Fig. 4.2, right) as

$$D_\alpha = W_2 \left(\sqrt[3]{f_\beta} \right) \tag{4.4}$$

Therefore, we derive the skeleton line lengths as

$$S_\alpha = \pi W_1 \frac{\left(\sqrt[3]{f_\beta} + 1 \right)}{2} \tag{4.5}$$

and

$$S_\beta = \pi W_2 \frac{\left(1 + \sqrt[3]{f_\alpha} \right)}{2} \tag{4.6}$$

By taking into account that

$$f_\alpha + f_\beta = 1 \tag{4.7}$$

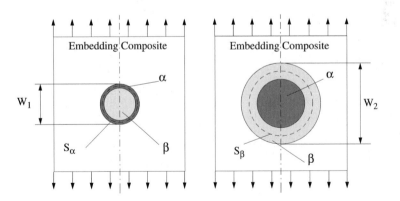

Fig. 4.2 Matricity-model (schematic) with skeleton lines to adjust the measured parameter "Matricity" in the model via the factors W_1 and W_2.

For the 3D case, the diameter of the inclusion b (Fig. 4.2, left) is derived as a function of W_1 and the volume fraction f_β of β in this cell as the matricity M can be calculated as a function of the sizes of the embedded cells and the volume fraction of one of the two phases, as the volume fraction of the phases is held constant in both parts of the matricity model

$$
M_\alpha = \frac{S_\alpha}{S_\alpha + S_\beta} = \frac{W_1 \pi \left(\dfrac{\sqrt[3]{f_\beta} + 1}{2} \right)}{W_2 \pi \left(\dfrac{\sqrt[3]{f_\alpha} + 1}{2} \right) + W_1 \pi \left(\dfrac{\sqrt[3]{f_\beta} + 1}{2} \right)}
\tag{4.8}
$$

or

$$
M_\alpha = \frac{W_1 \left(\sqrt[3]{1 - f_\alpha} + 1 \right)}{W_2 \left(\sqrt[3]{f_\alpha} + 1 \right) + W_1 \left(\sqrt[3]{1 - f_\alpha} + 1 \right)}
\tag{4.9}
$$

Adjusting matricity in the model

As can be seen in Fig. 4.2, the volume fractions of the phases as well as the diameters W_1 and W_2 of the embedded cells are adjustable. To achieve a matricity M in the model, we first realise the measured volume fraction of the phases in the model and then calculate the diameters W_1 and W_2. The corresponding diameters W_1 and W_2 are obtained by rearranging Eq. (4.9)

$$
W_1 = W_2 \left(\sqrt[3]{1 - f_\beta} + 1 \right) \frac{1 - M_\beta}{M_\beta \left(\sqrt[3]{f_\beta} + 1 \right)}, \qquad W_2 = 1,0 \ \text{for} \ M_\beta \geq 0,5
\tag{4.10}
$$

or

$$
W_2 = W_1 \left(\sqrt[3]{1 - f_\alpha} + 1 \right) \frac{1 - M_\alpha}{M_\alpha \left(\sqrt[3]{f_\alpha} + 1 \right)}, \qquad W_1 = 1,0 \ \text{for} \ M_\alpha \geq 0,5
\tag{4.11}
$$

Realisation of the adjustability of matricity by weighting factors

If the geometrical boundary conditions are modelled at a distance of about five times the radius of the embedded cell, they are of almost no influence on the models' mechanical behaviour. If we take care that the boundary conditions keep remote we can model the embedded cell with the surrounding composite in different manners (Fig. 4.3).

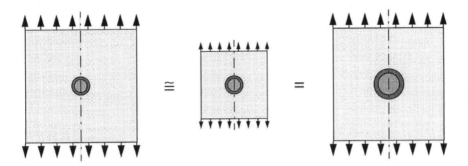

Fig. 4.3 Independence of mechanical behaviour from size variations of embedded and embedding medium (schematic).

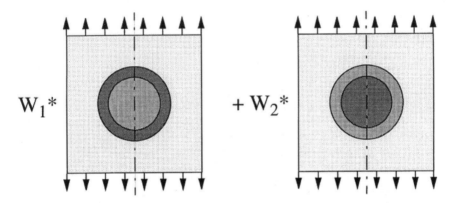

Fig. 4.4 Realisation (schematic) of the adjustability of matricity by weighting factors.

As the remote boundary conditions have almost no influence on the mechanical behaviour of the embedded cell it is assumed that the continuum mechanical stress-strain state in the embedded cell is hardly influenced as well. Taking this into account, a unit cell for a specific volume fraction can be used for each part of the matricity model. Further we can see from the almost independency from remote boundaries, that it is not necessary to model the matricity as a parameter of the FE-mesh but it is possible to introduce the matricity adjusting weighting factors W_1 and W_2 only in the evaluation of the results from the inclusion type geometries. As the results have to be determined by an iterative calculation in about 3-5 iterations, the adjusting weighting factors W_1 and W_2 must be introduced in the evaluation of all iteration steps (Fig. 4.4).

Calculation of stress-strain curves

In principle, stress-strain curves of the two-phase composite are determined from the matricity model in the same iterative manner as it is done for the simple self-consistent embedded cell model. In each increment the components for stress and strain are determined. This is done by a weighted averaging of the stress and strain values over all integration points of both embedded cells. The three-dimensional weighting is done by the "integration point volume" V_{k0} of each corresponding Gaussian integration point, which must be multiplied by W_1^3 and W_2^3, respectively, to account for the matricity effects as described above. The factors W_i are in the power of three as the length of the skeleton lines depend linearly on W_i but the embedded cell volumes depend by the power of three (for the 3D case) on W_i. With these considerations the stress and strain components can be calculated as

$$\sigma_{ij} = \frac{\sum \sigma_{ij}^k V_k}{\sum V_k}, \qquad\qquad j = (x, y, z) \text{ or } (r, z, \Phi) \qquad (4.12)$$

$$\varepsilon_{ij} = \frac{\sum \varepsilon_{ij}^k V_k}{\sum V_k}, \qquad\qquad i, j = (x, y, z) \text{ or } (r, z, \Phi) \qquad (4.13)$$

or more detailed (only for the stress components)

$$\sigma_{ij} = \frac{\left(\sum \sigma_{ij}^k V_{k0} W_1^3\right)_1 + \left(\sum \sigma_{ij}^k V_{k0} W_2^3\right)_2}{\sum V_{k0} W_1^3 + \sum V_{k0} W_2^3} \qquad (4.14)$$

where k is the index of summation and 1 or 2 are the part of matricity model whose embedded cell is weighted by W_1 or W_2.

From the stress components, the von Mises equivalent stress at each strain increment is calculated as (Eqs. (4.15) and (4.16) are only valid for Cartesian coordinates)

$$\sigma_v = \sqrt{\sigma_{xx}^2 + \sigma_{yy}^2 + \sigma_{zz}^2 - \left(\sigma_{xx}\sigma_{yy} + \sigma_{yy}\sigma_{zz} + \sigma_{zz}\sigma_{xx}\right) + 3\left(\sigma_{xy}^2 + \sigma_{yz}^2 + \sigma_{zx}^2\right)} \qquad (4.15)$$

and the equivalent strain is given as

$$\varepsilon_v = \frac{1}{1+\mu}\sqrt{\varepsilon_{xx}^2 + \varepsilon_{yy}^2 + \varepsilon_{zz}^2 - \left(\varepsilon_{xx}\varepsilon_{yy} + \varepsilon_{yy}\varepsilon_{zz} + \varepsilon_{zz}\varepsilon_{xx}\right) + 3\left(\gamma_{xy}^2 + \gamma_{yz}^2 + \gamma_{zx}^2\right)} \qquad (4.16)$$

where μ is the elastic-plastic Poisson's ratio of the composite.

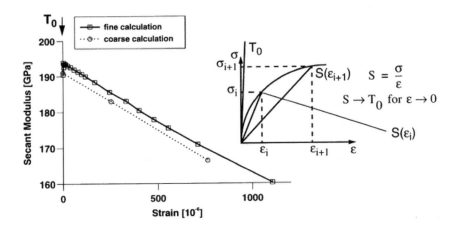

Fig. 4.5 Calculation of tangent modulus T_0 by extrapolation of secant modulus values at zero-strain.

Mechanical constants

Poisson's ratio μ and tangent modulus T_0 at zero strain are calculated from the obtained stress and strain components σ_{ij} and ε_{ij}. Poisson's ratio μ is calculated from the context of elastic constants [10]

$$\mu = \frac{3K - E}{6K} \qquad (4.17)$$

where K is the bulk modulus and E is the Young's modulus. K is defined as [10]

$$K = \frac{\sigma_H}{3\varepsilon_h} = \frac{\left(\sigma_{xx} + \sigma_{yy} + \sigma_{zz}\right)}{3\left(\varepsilon_{xx} + \varepsilon_{yy} + \varepsilon_{zz}\right)} \qquad (4.18)$$

If we consider a material where the elastic modulus E equals the gradient of the stress-strain curve at zero strain, then we can determine the constants K and E from the calculated stress-strain curve. As the moduli change with changing strains we have to extrapolate the bulk modulus K and the tangent modulus T from values near zero-strain to a value at zero-strain (Fig. 4.5).

Yield stress

Here the yield stress is calculated as the stress belonging to the cross point of the stress-strain curve and a straight line through the 0.2% strain point with the gradient that equals to the above calculated tangent modulus. The model does not include any damage parameter or failure criterion. This might lead to unrealistic high yield stresses for composites with dominating linear elastic material behaviour.

Results and discussion

The quality of the matricity model to simulate stress-strain curves and strain distribution frequencies especially for two ductile phases has been demonstrated in [3-6]. Here we compare the matricity model with other models and examine the influence of the matricity parameter on stress-strain curves and on the influence of residual stresses.

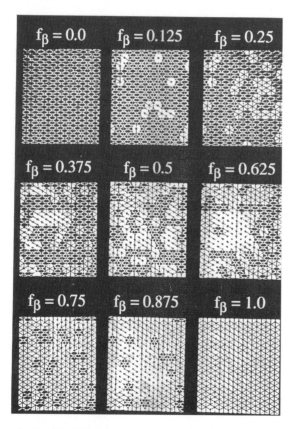

Fig. 4.6 Model microstructures of regular hexagons by Siegmund et al. [11]

Fig. 4.7 Determination of matricity M for the hexagonal model microstructure with $f_\beta = 0.5$ [11]. (Left: original microstructure, right: binary image with skeleton lines.)

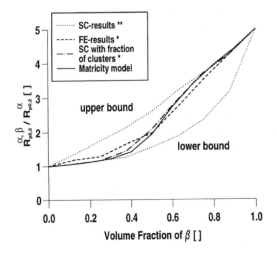

Fig. 4.8 Yield stress of α/β composite. (* = in [11], ** = in [11] calculated from [12].)

Comparison to cluster parameter r

The cluster parameter r_γ ($\gamma = \alpha + \beta$) is defined in [11] as

$$r_\gamma = \frac{N_\gamma}{N_\alpha + N_\beta} \qquad (4.19)$$

where $\gamma = \alpha + \beta$ and N_γ is the number of clusters of phase γ.

Fig. 4.6 shows some model microstructures which have been calculated with distinguishable assumptions. To compare with the numerical yield stress results based on Eq. (4.19) [11] we modelled the microstructures also with the matricity model. To do this the matricity of the microstructures were derived by phase reduction (cf. Matricity) which is shown in Fig. 4.7. The comparison of the matricity model results with the results in [11] shows good agreement with real structure calculations for both models (Fig. 4.8).

The advantage of the matricity model might be that differently sized clusters provide a different contribution to the influence of this cluster to the overall mechanical behaviour, whereas the cluster parameter model assumes that all clusters, independent of the cluster sizes, are equally weighted. An obvious advantage of the matricity model is the good agreement of calculated strain distribution frequencies to experimentally obtained results (cf. [3-6]).

Matricity and stress-strain curves

The influence of the matricity on the overall mechanical behaviour can be shown by comparing the calculated stress-strain curves while keeping the volume fraction f of the phases constant and by varying only the matricity M of the phases (this means to vary W_1 correspondingly for evaluating the results). A parameter study was made for a $ZrO_2/NiCr$ 80 20 composite where the volume fraction f of the phases were kept constant and the matricity M was varied between 0 and 1. This means that the microstructure lies in a range between a pure inclusion microstructure and a pure matrix microstructure for one phase and vice versa for the other phase. In Fig. 4.9, stress-strain curves are shown for two different matricities M_{ZrO2} (= 0.3 and 0.4) with and without residual stresses caused by cooling down the material from processing to environmental temperature. Fig. 4.10 shows the influence of the matricity on the yield stress. The influence of matricity is decisive in the range between M = 0.3 and M = 0.7. Taking residual stresses into account we can recognise that the influence of matricity on the yield stress diminishes. This can be explained with the modelled isotropic hardening of the ductile NiCr 80 20 phase, which is plastically deformed by residual stresses when acting as a matrix. However, no plastification is obtained when this phase acts as a pure inclusion and, therefore, hardly any hardening of the composite will be expected.

Conclusion

The influence of the parameter matricity M besides the parameter volume fraction f on the overall behaviour of two phase composites with coarse interpenetrating microstructures is clearly shown. Especially in cases, when the composite consists of a ductile and a linear-elastic phase, a significant influence of the matricity on stress-strain curves has been found.

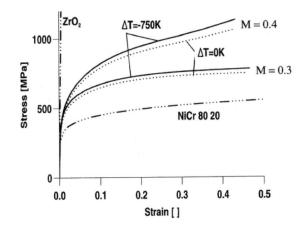

Fig. 4.9 Stress-strain curves of ZrO$_2$/NiCr 80 20 composite with f$_{ZrO2}$ = 0:3 and two different matricities. (E$_{ZrO2}$ = 206 GPa, E$_{NiCr\ 80\ 20}$ = 214 GPa, α$_{ZrO2}$ = 10 x 10^{-6} K^{-1}, α$_{NiCr\ 80\ 20}$ = 14 x 10^{-6} K^{-1})

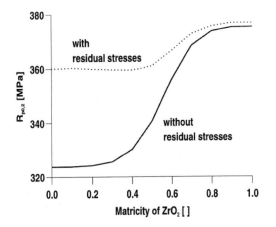

Fig. 4.10 Yield stress of ZrO$_2$/NiCr 80 20 composite with f$_{ZrO2}$ = 0:3 and different matricities M. (E$_{ZrO2}$ = 206 GPa, E$_{NiCr\ 80\ 20}$ = 214 GPa, aZrO2 . 10 x 10^{-6} K^{-1}, α$_{NiCr8020}$ = 14 x 10^{-6} K^{-1}, ΔT = -750 K.)

It is also found that the influence of residual stresses on yield stress depends strongly on the volume fraction and the matricity of the ductile phase. The comparison with model structure calculations shows the high quality of the matricity model. The matricity model must be built once for each volume fraction and the matricity is then considered in the evaluation of the result during the iterative calculation process. The model is, therefore, very simple with respect to FE meshing, and can be quite coarse, as the stress-strain components were evaluated as arithmetical averages over the elements of both embedded cells.

References

[1] Poech M.H., Ruhr D. (1993), Die quantitative Charakterisierung der Gefügeanordnung, Prakt. Metallogr. Sonderband 24, p. 391.

[2] Soppa E. (1995), Experimentelle Untersuchung des Verformungsverhaltens zweiphasiger Werkstoffe, Fortschr.Ber. VDI Reihe 5, 408, VDI-Verlag, Düsseldorf.

[3] P. Leßle P, M. Dong M, E. Soppa E, S. Schmauder S (1997), Selbstkonsistente Matrizitätsmodelle zur Simulation des mechanischen Verhaltens von Verbundwerkstoffen, Verbundwerkstoffe und Werkstoffverbunde, DGM Informationsgesellschaft Verlag, Hrsg.: K. Friedrich, p. 765.

[4] Schmauder S, Dong M, Leßle P (1997), Verbundwerkstoffe mikromechanisch simuliert, Metall 51 (7/8), p. 404.

[5] Schmauder S, Dong M, Leßle P, Simulation von Verbundwerkstoffen mit Teilchen und Fasern, Tagungsband XXIV. FEM Kongress in Baden-Baden, 17.-18 November 1997, p. 445.

[6] Schmauder S, Dong M, Leßle P, Soppa E (1998), Simulation of interpenetrating microstructures by self consistent matricity models, Scripta Mater. 38, p. 1327.

[7] Willert Porada M, Gerdes T, Rödiger K., Kolaska H. (1997), Einsatz von Mikrowellen zum Sintern pulvermetallurgischer Produkte, Metall 51 (1/2), p. 57.

[8] Dong M., Schmauder S. (1996), Modelling of metal matrix composites by a self consistent embedded cell model, Acta Mater. 44, p. 2465.

[9] Dong M., Schmauder S (1996), Transverse mechanical behaviour of fiber reinforced composites ± FE Modelling with embedded cell models, Comp. Mat. Sci. 5, p. 53.

[10] Sautter M (1995), Modellierung des Verformungsverhaltens mehrphasiger Werkstoffe mit der Methode der Finiten Elemente, Fortschr.- Ber. VDI Reihe 5, 398, VDI-Verlag, Düsseldorf, p. 78.

[11] Siegmund T., Werner E., Fischer F.D. (1993), Structure-property relations in duplex materials, Comp. Mat. Sci. 1, p. 234.

[12] Bao G., Hutchinson J.W., McMeeking R.M. (1991), Mech. Mater. 12, p. 85.

4.1.2 Some applications of the matricity model[2]

Functionally graded materials are difficult to simulate because of the lack of material data for different compositions at different locations. The main reason for this situation is the non-linear dependence of elastic-plastic properties on phase compositions of composites. Therefore, a great deal of work has been performed in the recent past to overcome this problem and to estimate the properties of composites.

The following examples are chosen to demonstrate some recent achievements in this respect. In [1] a micromechanical model is applied to study random and discrete microstructures and their interrelations with residual stresses and crystal plasticity effects are taken into account in differently graded FGMs with a layer structure. This model is compared with continuous models and their equivalence is demonstrated with respect to macroscopic behaviour while strong local stress and strain concentrations were found. The analysis of the influence of thermal stresses and failed particles on macroscopic stress-strain curves, based on a constitutive Eshelby type solution is restricted to dilute particle reinforced materials [2]. Calculations of the thermo-elastic response in C/SiC composite systems showed that effective moduli, expansion coefficients and heat conductivities do not require detailed micromechanical analyses, but can be rather derived from homogenization models or in the case of interpenetrating microstructures from self-consistent estimates [3]. Nonlinear effects have not been taken into account in this study. The influence of thermal residual stresses on the coefficient of thermal expansion for metal-matrix, ceramic-matrix and interpenetrating Al/SiC composites taking temperature dependent material properties–especially of Al–into account are given in [4] using unit cell type FE-models. They were compared with upper and lower analytical bounds of composites with homogeneous phase distributions. A FE-analysis of the macroscopic and microscopic elasto-plastic deformation due to thermal and mechanical axial and bending loading of layered Ni/Al_2O_3 composites with graded interfaces based on a single unit cell type model with hexagonal or square packing and with mesomechanical cells of the random arrangement type taking hundreds of hexagonal grains into account has been performed in [5]. FE-models have been also applied in [6] for the same Ni/Al_2O_3 composite with and without graded interfaces for different specimen geometries demonstrating the reduction in maximum residual stresses except for the shear stresses at interface edges by gradation.

However, a major drawback of most literature examples is the lack of experimental comparison for the calculated thermal-mechanical properties. Moreover, these numerical and analytical-numerical models are rather complicated and

[2] Reprinted from S. Schmauder, U. Weber, "Modelling of Functionally Graded Materials by Numerical Homogenization", Arch. Appl. Mech. 71, pp. 182-192 (2001). with kind permission from Elsevier.

typically restricted to two dimensions. In the following, another more promising procedure is described.

Recently, a systematic study has been successfully performed on the strengthening effects of inclusion type 2D and 3D microstructures [7-10]. In this work, thermo-elastic-plastic properties of ZrO_2/NiCr 80 20 composites and FGMs are predicted based on a new numerical homogenization technique [11-17] and results will be compared to experiments.

Composites consisting of phases with strongly different properties have the potential to be applied in new application fields as they comprise otherwise incompatible properties. While the deformation behaviour of inclusion-type of microstructures has been successfully modelled in the past for brittle fiber or particulate reinforced metal matrix composites [7, 8] this was not achieved until recently in the case of interpenetrating microstructures where both phases are connected throughout the material. Such microstructures are typically observed in the composition range of 30÷70% while inclusion type of microstructures are typical for dilute systems with phase volume fractions between 0÷30%. Specifically, functionally graded materials can depict the full composition range in material transitions. As processing techniques are nowadays available to design material transitions from inclusion to interpenetrating type of microstructures, experience in modelling of the full composition range is still lacking. This work is intended to bridge this gap in the case of ZrO_2/NiCr 80 20 composites where the full compositional range is available from a powder-metallurgical route [18], such that comparison in properties and predictions can be made.

Models

Three models are used for the simulation of the elastic properties of ZrO_2/NiCr 80 20 composites with phases $\alpha = ZrO_2$ ($E = 46$ GPa, $\mu = 0.29$, $\alpha = 10.3E-06$) and $\beta = $ NiCr 80 20 ($E = 121$ GPa, $\mu = 0.29$, $\alpha = 17.3E-06$), while the thermo-elastic-plastic behaviour will be analysed numerically. In the case of an inclusion type of microstructure the self-consistent embedded cell model is applied which is described in [7-10, 19].

Interpenetrating microstructures where both phases can show percolation throughout the material are characterized by the above introduced matricity parameter M with values between 0 and 1 describing the mutual material circumscription of the phases in addition to their volume fractions. The matricity model is based on 2D or 3D (axisymmetric) inclusions of a given volume fraction and with circular cross-section in the present context [11-17, 21]. Thus this model shows the same effect as a single model with two included composites α-β and β-α as in the experiment, cf [14,15]. Here all calculations were performed with the 3D (axisymmetric) version of the model.

The model in Fig. 4.41 allows for the additional consideration of thermal residual stresses and can be used to predict the elastic properties, the thermal expansion coefficient and the elastic-plastic stress-strain curves for different phase arrangements as well as to predict phase properties of the phases in the composite.

For comparison reasons the Tuchinskii model is introduced as a second model which allows to predict upper and lower bounds of the elastic modulus E of a composite with interpenetrating microstructures by the following formulae [22],

$$\frac{E}{E_\alpha} = (1-c)^2 + \left(\frac{E_\beta}{E_\alpha}\right)^2 c^2 + \frac{(E_\beta/E_\alpha)c(1-c)}{c+(E_\beta/E_\alpha)(1-c)} \qquad lower\ bound$$

$$\frac{E}{E_\alpha} = \left[\frac{1-c}{(1-c^2)+(E_\beta/E_\alpha)c^2} + \frac{c}{(1-c)^2+(E_\beta/E_\alpha)(2-c)c}\right]^{-1} \qquad upper\ bound$$

(4.20)

where E_i = Young's modulus of phase i, f_i = volume fraction of phase i ($i=\alpha$, β), while $f_\beta = (3 - 2c)c^2$ relation between volume fraction f and geometry parameter c (real solution between 0 and 1)

In a third model by Pompe the calculation of the thermo-elastic constants is also based on the solution of an inclusion problem [23]. Due to the ellipsoidal shape of the inclusions the fields inside the inclusions are homogeneous and can be determined analytically. We assume the special case of spherical inclusions in this model. Interaction between the media can be considered through assumptions about the surrounding material. This is often realized by the effective medium theory (EMA). For the mean stress and strain fields self-consistency is claimed leading to an implicit equation system which allows for the determination of the effective constants. The effective values for Young's modulus, thermal expansion coefficient [14] as well as residual stresses [24] are then determined numerically.

Results and Discussion

Microscopic Results

The matricity character M of ZrO_2 has been measured and found to deviate up to 25 % from a linear 1:1-relationship especially at intermediate volume fractions. This result means that typically ZrO_2 represents rather the matrix than an inclusion. Except when the influence of M was investigated, f=M was, therefore, employed for all values of f (Fig. 4.11).

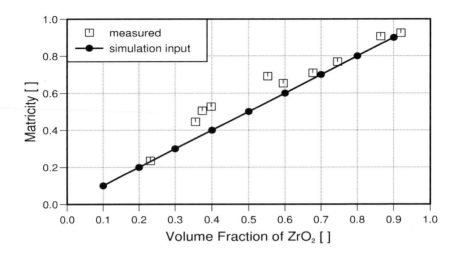

Fig. 4.11 Matricity of ZrO$_2$ vs. volume fraction of ceramic (ZrO$_2$) in ZrO$_2$/NiCr 80 20 composite.

The thermal expansion coefficient obtained using the matricity model and the Pompe model was determined to behave nonlinearly in a similar manner and to decrease with increasing volume fraction of ZrO$_2$ (Fig. 4.12).

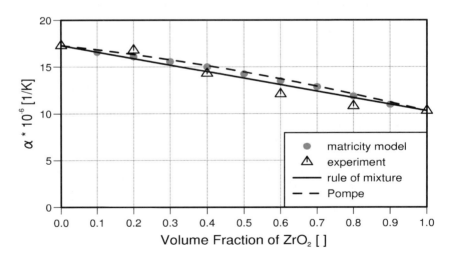

Fig. 4.12 Thermal expansion coefficient vs. volume fraction of ceramic (ZrO2). Comparison between experiment, matricity model, rule of mixture and Pompe model [23].

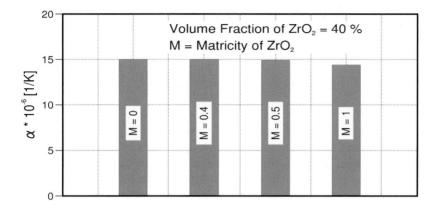

Fig. 4.13 Thermal expansion coefficient as a function of matricity parameter $M = M_{ZrO2}$. Volume fraction of ceramic $f_{ZrO2} = 40$ %.

It is interesting to note that the thermal expansion coefficient is nearly independent on the matricity parameter M (Fig. 4.13).

Furthermore, the elastic modulus was obtained by the Tuchinski model, the Pompe model and the matricity model. Upper and lower bounds of the Tuchinski model as well as the Pompe model were in close agreement to the matricity model. However, the experimental values scattered in a wide range and, therefore, only partly good agreement between experiment and simulation was achieved (Fig. 4.14). Residual stresses are not considered in the models used to calculate the Young's modulus.

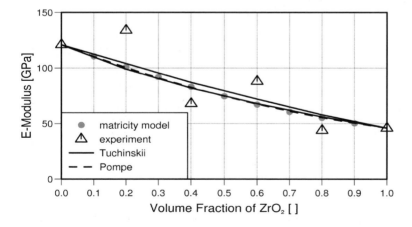

Fig. 4.14 Elastic modulus vs. volume fraction of ceramic (ZrO_2). Comparison between matricity model, Tuchinskii model [22] and Pompe model [23] (no residual stresses).

When residual stresses are taken into account, the stress-strain curve of the composite calculated by the matricity model does frequently not show an initial elastic behaviour. In Fig. 4.15 the influence of volume fraction of ZrO_2 on the stress-strain curves of ZrO_2/NiCr 80 20 composites is studied. Strong variations in the plastic behaviour are found especially for f_{ZrO2} = 30%÷60%, the regime of phase interpenetration. Therefore, this regime of volume fractions is relevant when a variation of the mechanical behaviour of the composite is required without manipulating matricity.

Matricity plays a major role specifically at low ceramic volume fractions (Fig. 4.16) while residual stresses are of less importance at all volume fractions. From this result it is obvious, that the parameter matricity provides a strong potential for designing the mechanical behaviour of the composite. On the other hand residual stresses can be of major importance with respect to failure in the phases.

The ZrO_2-phase depicts typically compressive residual stresses with a sharp peak value while the NiCr phase depicts tensile residual stresses with a broad wide distribution. This is mainly due to the fact that at low volume fractions ZrO_2 is basically present as a small inclusion, thus following Eshelby's constant stress rule for sperical inclusions [25] while NiCr as mainly matrix phase shows a wide variation of stress levels as expected from inhomogeneous strains in local shear bands around the ZrO_2 inclusions [9]. The average stresses (circumferential component) in the phases are shifted to higher values at a higher volume fraction of ZrO_2 (Fig. 4.17a+b) as a result of the increasing influence of the stiffer ceramic.

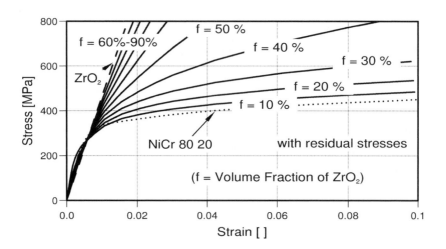

Fig. 4.15 Stress-strain curves for several volume fractions f_{ZrO2} of ceramic. Matricity of cermic phase (ZrO_2) according Fig. 4.11 (with residual stresses).

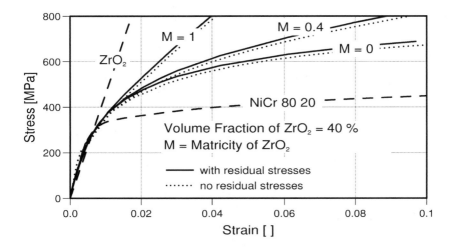

Fig. 4.16 Stress-strain curves for a volume fraction of ceramic $f_{ZrO2} = 40$ %. Variation of Matricity M_{ZrO2} (with and without residual stresses).

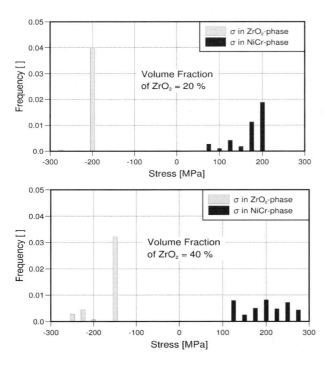

Fig. 4.17 Calculated distribution of circumferential residual stresses obtained with the matricity model in the ceramic (ZrO_2) and metal phase (NiCr) for a volume fraction of ceramic $f_{ZrO2} = 20$ % (a) and $f_{ZrO2} = 40$% (b).

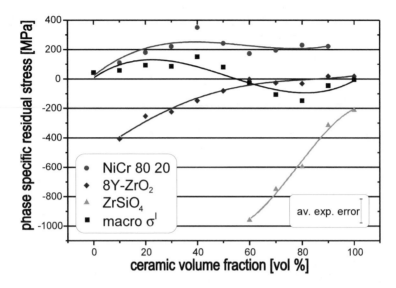

Fig. 4.18 Experimentally measured [24] and numerically (matricity model) analysed residual stresses in a ZrO₂/NiCr 80 20 composite at several phase compositions. (D. Dantz, Ch. Genzel, W. Reimers, HMI Berlin)

It is obvious that the stress distributions depict broadness due to the fact that in one part α surrounds β and v.v in the other part of the model. This is the reason also for the effect that the stresses in the low volume phase ZrO_2 shows two peaks for $f_{ZrO2} = 40$ % in Fig. 4.17b.

Agreements between calculations and experiments for the average stress values in either phase are found to be rather good for both phases (Fig. 4.18) [24]. This fact suggests the effectiveness of the matricity model as a new homogenization procedure. It's superiority in predicting local surface properties have been demonstrated recently for metal/metal composites [15].

Macroscopic Results

The knowledge of the mechanical properties of ZrO_2/NiCr 80 20 composites can be used by applying it to simulate functionally graded materials (FGM). The dependence of the macroscopic behaviour of a graded metal/ceramic composite can be derived by taking the local material behaviour into account. In the present context, locally different microstructural compositions and thus different material properties are considered by a layered model with different material properties in each layer. As a model, a bending specimen is chosen where the transition from the ceramic to the metal phase was realised by four layers (FGM specimen) as well as by a sharp interface (non-graded specimen). FE-mesh, boundary conditions as well as layer subdivision of ungraded and graded specimen are shown in Fig. 4.19. For the FGM specimen, the matricity (and according to section 3.1, therefore, the property) are varied from layer to layer besides the volume fractions. In addition to the measured matricity value M (M~f, Fig. 4.11) three different matricities M = 0, M = 0.5 and M = 1 were assumed in the layers containing 20%÷80% ceramic phase. In the present context, M = M_{ZrO2} , thus M = 1 defines an inclusion phase with ceramic matrix. In a first step, the specimen is cooled down from an assumed stress-free state by 750 K in order to simulate the manufacturing process of the specimen. In Fig. 4.20 the residual stresses parallel to the layers are shown as fringe plots. As expected, the graded specimen (Fig. 4.20a) shows significantly smaller residual stresses compared to the non-graded specimen (Fig. 4.20b). A further reduction of residual stresses parallel to the layers can be expected from even more gradual property transitions in the graded region.

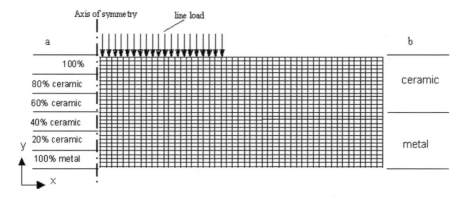

Fig. 4.19 Graded (a) and non-graded (b) bending specimen.

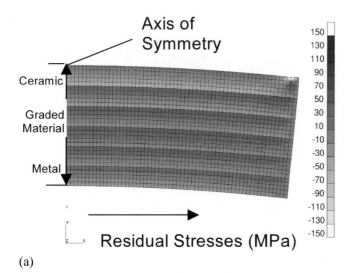

(a)

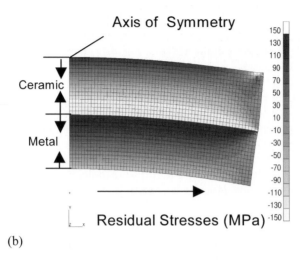

(b)

Fig. 4.20 Distribution of residual stresses in a graded (a) and in a non-graded (b) specimen after cooling down by -750 K. Stresses parallel to the layers.

However, the distribution of stresses perpendicular to the layers inside the specimen are hardly influenced by the composition of the specimen while the non-graded composite shows stronger disturbances of the stresses at the free edge (Fig. 4.21a+b). In a second step, the specimen is heated up to 500 °C and then loaded by an area load of 600 N/mm^2 corresponding to a single force of 24 kN. bending specimen.

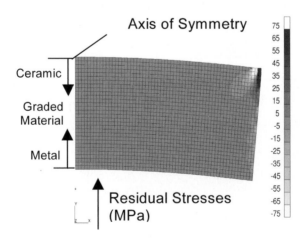

(a)

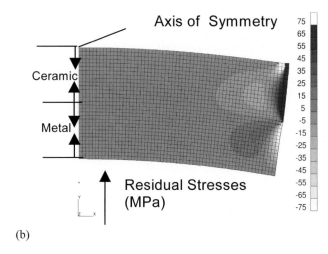

(b)

Fig 4.21 Distribution of residual stresses in a graded (a) and in a non-graded (b) specimen after cooling down by -750 K. Stresses perpendicular to the layers.

By this, the behaviour of the specimen is simulated under thermal-mechanical loading. Fig. 4.22 shows the macroscopic force-displacement behaviour of the graded and ungraded it is found that the macroscopic behaviour of the bending specimen can be significantly influenced by the matricity character of the single layers. The larger M_{ZrO2}–and, therefore, the more metallic phase is circumvented by ceramic phase–the higher the stiffness of the bending specimen. Interestingly, the ungraded metal/ceramic bending specimen shows a similar macroscopic stiffness as compared to the graded specimen with measured matricities, although the stress jumps at the layer interfaces are much smaller in the FGM material. Generally, the macroscopic mechanical behaviour of the bending specimen is strongly influenced by the matricity parameter.

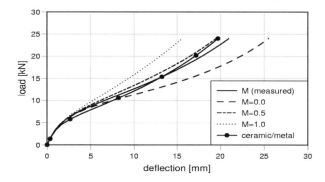

Fig. 4.22 Deflection behaviour of a graded and a non-graded (ceramic/metal) specimen. Influence of Matricity M_{ZrO2}

Conclusions

While the numerical self-consistent embedding cell technique allows for taking into account for microstructures with randomly arranged inclusions, the matricity model is a sophisticated improved homogenization technique which allows for modelling interpenetrating microstructures. This allows for designing microstructures as well as graded composites, while the proof for manufacturing these microstructure is still to be brought.

References

[1]   Dao M., Gu P, Maewal A., Asaro R.J. (1997), A Micromechanical Study of Residual Stresses in Functionally Graded Materials, Acta mater. 45, pp. 3265-3276.
[2]   Noda N., Nakai S., Tsuji T. (1998), Thermal Stresses in Functionally Graded Material of Particle-Reinforced Composite, JSME International Journal A 41, pp. 178-184.
[3]   Reiter T., Dvorak G.J. (1998), Micromechanical Models for Graded Composite Materials: II. Thermomechanical Loading, J. Mech. Phys. Solids 46, pp. 1655-1673.
[4]   Shen Y.L.(1998), Thermal Expansion of Metal-Ceramic Composites: A three-dimensional Analysis, Materials Science and Engineering A252, pp. 269-275.
[5]   Weissenbek E., Pettermann H.E., Suresh S. (1997), Elasto-Plastic Deformation of Compositionally Graded Metal-Ceramic Composites, Acta mater. 45, pp. 3401-3417.
[6]   Williamson R.L., Rabin B.H., Drake J.T. (1993), Finite Element Analysis of Thermal Residual Stresses at Graded Ceramic-Metal Interfaces. Part I. Model Description and Geometrical Effects, J. Appl. Phys. 74, pp. 1310-1320.
[7]   Zahl D.B., Schmauder S., McMeeking R.M. (1994), Transverse Strength of Metal Matrix Composites Reinforced with Strongly Bonded Continuous Fibers in Regular Arrangements, Acta metall. mater. 42, pp. 2983-2997.
[8]   Dong M., Schmauder S. (1996), Modeling of Metal Matrix Composites by a Self-Consistent Embedded Cell Model, Acta metall. mater. 44, pp. 2465-2478.
[9]   Dong M., Schmauder S. (1996), Transverse Mechanical Behaviour of Fiber Reinforced Composites - FE Modelling with Embedded Cell Models, Computational Materials Science 5, pp. 53-66.
[10] Schmauder S., Dong M. (1996), Vorhersage der Festigkeit von Verbundwerkstoffen, Spektrum der Wissenschaft 11, pp. 18-24.
[11] Leßle P., Dong M., Soppa E., Schmauder S. (1997), Selbstkonsistente Matrizitätsmodelle zur Simulation des mechanischen Verhaltens von Verbundwerkstoffen, Vortragstexte der Tagung Verbundwerkstoffe und Werkstoffverbunde, Kaiserslautern, Hrsg.: K. Friedrich, DGM-Informationsgesellschaft Verlag, Oberursel, September 1997, S. 765-770.
[12] Schmauder S., Dong M., Leßle P. (1997), Verbundwerkstoffe mikromechanisch simuliert, Metall 7-8/97, pp. 404-410.
[13] Schmauder S., Dong M., Leßle P. (1997), Simulation von Verbundwerkstoffen mit Teilchen und Fasern, XXIV. FEM-Kongreß in Baden-Baden, 17.-18. November 1997, Tagungsband, Hrsg.: A. Streckhardt, Kongreßorganisation, Ennigerloh, pp. 445-462.
[14] Leßle P., Dong M., Schmauder S. (1999), Self Consistent Matricity Model to Simulate the Mechanical Behaviour of Interpenetrating Microstructures, Computational Materials Science 15, pp. 455-465.
[15] Leßle P., Dong M., Soppa E., Schmauder S. (1998), Simulation of Interpenetrating Microstructures by Self Consistent Matricity Models, Scripta mater. 38, pp. 1327-1332.

[16] Schmauder S., Weber U. (1999) Modelling the Deformation Behaviour of W/Cu Composites by a Self-Consistent Matricity Model", ECM'99, Progress in Experimental and Computational Mechanics in Engineering and Materials Behaviour, 12.-15. September, Urumqi, China, Eds.: D. Zhu, M. Kikuchi, Y. Shen, M. Geni, Northwestern Polytechnical University Press, Xi'an, China, pp. 54-60.

[17] Dong M., Leßle P., Weber U., Schmauder S. (1999), Mesomechanical Modelling of Composites Containing FGM Related Interpenetrating Microstructures Based on Micromechanical Matricity Models, Materials Science Forum Vols. 308-311, "Functionally Graded Materials 1998", pp. 1000-1005.

[18] Jedamzik R., Neubrand A., Rödel J. (1997), Electrochemical Processing and Characterisation of Graded Tungsten/Copper Composites", in Proceedings of the 15th International Plansee Seminar, Vol.1, G. Kneringer, P. Rödhammer und P. Wilhartitz (Hrsg.), Metallwerk Plansee, Reutte, pp. 1-15.

[19] Sautter M. (1995), Modellierung des Verformungsverhaltens mehrphasiger Werkstoffe mit der Methode der Finiten Elemente, Dissertation, Universität Stuttgart.

[20] Poech M.-H., Ruhr D. (1993), Prakt. Met. Sonderbd. 24, pp. 385-391.

[21] Abschlußbericht zum DFG-Forschungsvorhaben "Mikro/Mesomechanische Modellierung des mecha-   nischen Verhaltens von Metall/Keramik-Gradientenwerkstoffen" im SPP Gradientenwerkstoffe under contract Schm 746/12-1 and Schm 746/12-2.

[22] Tuchinskii L.I., Poroshk. (1983) Metall. 7, p. 85.

[23] Kreher W., Pompe W. (1989), Internal Stresses in Heterogeneous Solids, Akademie-Verlag, Berlin.

[24] Dantz D., Genzel Ch., Reimers W., Weber U., Schmauder S. (1998), Analyse von Makro- und Mikroeigenspannungen in Gradientenwerkstoffen (FGM), Vortragstexte der Tagung Verbundwerkstoffe und Werkstoffverbunde, Hamburg, Hrsg.: K. Schulte und K.U. Kainer, DGM-Informationsgesellschaft Verlag, Weinheim, Oktober 1998, S. 704-709.

[25] Mura T. (1987), Micromechanics of Defects in Solids, Second Revised Edition, Kluwer Academic Publishers, Dordrecht, The Netherlands.

4.2 Graded materials: mesoscale modelling

4.2.1 Multilayer model and functionally graded finite elements: application to the graded hardmetals[3]

Tungsten carbide-cobalt hardmetals (WC/Co) are the most important cutting tool materials in modern technology. High hardness and wear resistance on the one hand side as well as high strength and chipping resistance on the other hand make them superior to high speed steels and ceramics [1, 2].

Coating the inserts with thin hard layers (TiN or TiC) by chemical vapor deposition yields a longer edge life in comparison to uncoated inserts. However, residual stresses appear in the tool as a result of cooling after the coating process which may cause initial cracks in the coating [3-7]. A tough surface zone underneath the coating with a high crack resistance prevents crack growth into the tool. Especially, Co enriched functionally graded surface zones provide an improved crack resistance, Fig. 4.23. Therefore, cracks nucleated in the coating are frequently arrested in the ductile binder [8]. Additional Co striations in the gradient close to the surface ensure that cracks arrest before their length becomes critical.

Real structure modelling is much too expensive and time consuming for WC/Co hardmetals. On the other hand, assuming the gradient zone as a homogeneous layer will not reflect the real material behavior. Therefore, a mesomechanical model has been developed and is applied to simulate different graded surface zones in hardmetals.

Mesoscopic model

The linear elastic model presented in this section includes the variation of material properties in the coating, gradient zone and substrate. Isotropic material behavior can be assumed for the coating and the substrate. In the gradient zone the material data are considered as functions of the distance from surface.

This feature can be approximated by a multilayer model where constant material data are assumed for layers of finite thickness

In the case of finite element modelling, the finite element mesh has to be generated with respect to such layers. Each layer represents a material with constant properties. All problems of discontinuity at layer interfaces are introduced artificially by such a model.

---

[3] Reprinted from J. Rohde, S. Schmauder, G. Bao, "Mesoscopic Modelling of Gradient Zones in Hardmetals", Computational Materials Science 7, pp. 63-67 (1996). with kind permission from Elsevier

In this work, a new approach is adopted: functionally graded elements have been developed which allow assigning different material properties to each Gaussian integration point. Thus, the gradient behaviour is approximated in each element which prevents from restrictions with respect to mesh generation. A high accuracy of results can be achieved even with coarse meshes. The mechanical behavior of the gradient is better represented by a functionally gradient model compared to the multilayer model (Fig. 4.24).

Three hardmetal grades with various gradient zones have been investigated [6], such as:

- The grade *yFree* contains a Co enriched zone, free of cubic carbides. The material properties are considered as cubic functions of the distance from the surface.
- The grade *CoStri* contains a Co enriched zone with additional Co striations. Firstly, the thermo-elastic data of this grade are considered as continuous linear functions; in a second model the striations are taken into account. The striations are modelled as surface parallel layers of 1-2 μm thickness with the material data of Co.
- The grade *Conv* is a conventional hardmetal without modified surface zone which is investigated for comparison reasons.

The functions which have been used for the Young`s modulus are shown in Fig. 4.25.

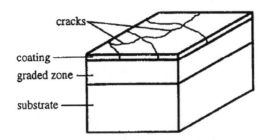

Fig. 4.23 Model of coated hardmetal tool with gradient zone.

(a)

(b)

Fig. 4.24 (a) Multilayer model and (b) functional gradient model (using functionally graded elements).

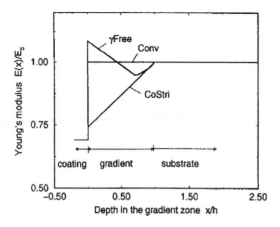

Fig. 4.25 Relative Young's modulus as function of the depth in the gradient zone for the three grades.

Crack / gradient interaction

A mesomechanical model (Fig. 4.26) with equally spaced multiple cracks of length a and the distance 2L is considered. The thickness of the coating is t, the thickness of the gradient zone is designated as h and that of the substrate as H. The variable z describes the depth in the graded zone, the material data are functions of z; Young's modulus $E(z)$ and Poisson's ratio $\upsilon(z)$.

The model is loaded due to a remotely applied uniform strain ε_0, perpendicular to the cracks. The energy release rate G.T is calculated for systematically varied crack lengths between t and t + h.

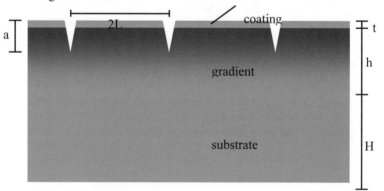

Fig 4.26 Mesomechanical model

The average stress in the untracked model is given as

$$\sigma = \frac{\varepsilon_0}{t+h+H} \int_{-t}^{h+H} \frac{E(z)}{1-v(z)^2} dz \tag{4.21}$$

The energy release rate G, normalized by the square of the average stress σ and the Young's modulus of the substrate E_s as well as the thickness of the gradient zone h results in a dimensionless function of geometric and material data relations. $E(a)$ denotes Young's modulus at a distance z = a from the surface and E_C that of the coating. Fig. 4.27 shows a comparison of ψ for various gradient zones. The differences in energy release rates are directly related to the elastic properties in each grade (cf. Fig. 4.25).

$$\Psi\left(\frac{a-t}{h}, \frac{L}{h}, \frac{t}{h}, \frac{E_C}{E_s}, \frac{E(a)}{E_s}\right) = \frac{GE_s}{\sigma^2 h} \tag{4.22}$$

Therefore, the energy release rate in grade CoStri is much lower than in the conventional grade in the first half of the gradient zone beneath the coating while minor differences between all grades are found for longer cracks. Moreover, the driving force of short cracks (0<(a-t)/h<0.35) is dominated by their proportionality to crack length a.

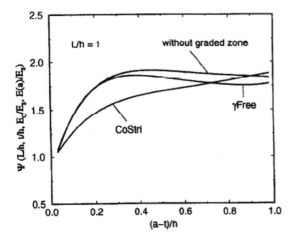

Fig. 4.27 Camparison of different grades

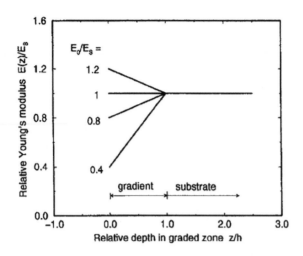

Fig. 4.28 Systematic variation of the Young's modulus in the graded zone.

A parametric study of the dependence of T on *E(z)* is shown in Figs. 4.28 and 4.29. The Young's moduli are considered as linear functions within the graded zone where the values directly at the coating/gradient interface are 20% higher, equal, 20% lower and 60% lower compared to the substrate, Fig. 4.28. It can be seen that the differences in the normalized energy release rates depend non-linearly on the differences of the elastic data, Fig. 4.29.

More importantly, the energy release rate was found to be an increasing function of crack spacing L/h (Fig. 4.30)

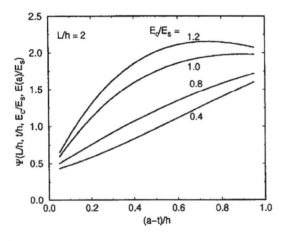

Fig. 4.29 Comparison of different graded zones with linear variation of elastic data.

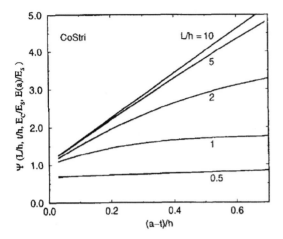

Fig. 4.30 Comparison of different crack spacings.

For large crack spacings (L/h ≥ 10), the crack driving force converges to an asymptotic maximum value, $\psi_{L/h=10}$. In the case of thin coatings only few pre-cracks are expected. Thus, the energy release rate of these surface cracks under mechanical loading is simply described by $\psi_{L/h=10}$. For all grades, $\psi_{L/h=10}$ is a linear function of the crack length, where the abscissa values are identical and the slope depends non-linearly on the variation of the elastic data but nearly linearly on the differences of the Young's moduli just beneath the coating (Fig. 4.31).

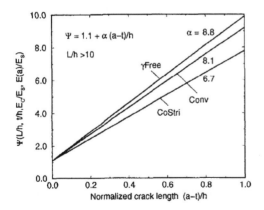

Fig. 4.31 Threshold of the normalized energy release rate for different grades.

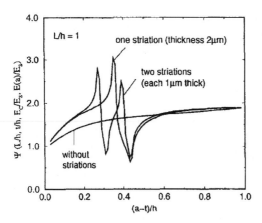

Fig 4.32 Influence of striations

Additional Co striations in the gradient zone underneath the coating may strongly influence the fracture fracture behavior of the materials under consideration. The normalized energy release rate increases dramatically as the crack approaches the striation and decreases rapidly inside the Co, Fig. 4.32. Since the value of T reaches a pronounced minimum inside the striations, cracks will probably arrest inside the Co striations. Moreover, plasticity effects may consume additional fracture energy and thus result in further beneficial effects with respect to cracking of graded hardmetals. The influence of thermal residual stresses from the coating process has been recently shown to effect more initial cracking than crack propagation [9].

Conclusions

A mesomechanical model of a coated hardmetal insert with a tough gradient surface zone was set up which considers material properties as functions of the distance from the surface. For numerical calculations with the finite element method functionally graded elements have been developed to adequately describe the continuously varying material properties in the gradient zone.

Residual stresses resulting from cooling after coating deposition may cause initial cracks in the coating perpendicular to the surface of the material. The crack/gradient interaction has been investigated for different types of gradient zones and different crack spacing in the case of mechanical loading. A non-linear dependence was found between the elastic data and the energy release rate. Crack spacing has an important influence on the crack driving force for dense crack patterns. For widely separated initial cracks the energy release rate reaches a threshold at $L/h \geq 10$. This threshold was shown to be a linear function of the crack length and possesses a predictable slope for a wide range of gradient properties.

The influence of additional Co striations in the gradient zone on the mechanical behavior has been found to be beneficial in our initial elastic calculations.

References

[1] Kolaska H. and Dreyer K. (1992), Hartmetalle und ihr Einsatzfeld, DGM Fortbildungsseminar, Hannover.
[2] Schedler W. (1988), Hartmetall für den Praktiker, VDI Verlag, Düsseldorf.
[3] Koenig W., Gerschwiler K and Fritsch R (1992), Leistung und Verschleiss neueter beschichteter Hartmetalle, in: Beschichten und Verbinden in Pulvermetallurgie und Keramik, ed. H. Kolaska, VDI Verlag, Düsseldorf.
[4] Nordgren A. and Thuvander A. (1992), Comb cracking of TiN, Tic and Al_2O_3,-coated cemented carbide during milling of steel, Report No. IM-2901 Swedish Institute for Metals Research, Stockholm.
[5] Nordgren A (1993), Influence of coating thickness in CVD-TIN coated cemented carbide upon comb cracking during milling of steel, Report No. IM-3034, Swedish Institute for Metals Research, Stockholm.
[6] Nordgren A and Thuvander A. (1990), Residual stresses in CVD coated cemented carbide with structural gradient, Report No. IM-2682, Swedish Institute for Metals Research, Stockholm.
[7] Nordgren A. and Jonsson S. (1994), Residual stress in CVD TiN coatings on unworn and worn cemented carbide and the influence upon crack formation, Report No. IM-3180, Swedish Institute for Metals Research, Stockholm.
[8] Nordgren A. (1990), Influence of tailored microstructural variations in cemented carbide tools upon the. propagation of cracks from surface coatings during machining of steel, Report No. IM-2812, Swedish Institute for Metals Research, Stockholm.
[9] Rohde J. and Schmauder S. (1996), Influence of a functionally gradient surface on cracking in WC/Co hardmetals, Fract. Mech. of Ceram.

4.2.2 Graded multiparticle unit cells: damage analysis of metal matrix composites[4]

The purpose of this part of our work is to investigate the effect of microstructures of functionally graded, SiC particle reinforced Al composites on the strength and damage resistance of the materials using the computational testing of composites with different artificially designed graded microstructures. The gradient composites with aluminum matrix are used, for instance, in electronic packaging industry, for brake rotor assemblies in automobile industry, as armor materials, etc.

In order to study the microstructure-strength relationships of graded Al/SiC composites, a series of numerical mesomechanical experiments was conducted. In the framework of the computational testing of the composites with different (artificially designed) graded microstructures, the tensile stress-strain curves, microcrack density in particles versus applied strain curves, and stress and damage distributions at different stages of loading were determined and compared. It was shown that the flow stress and stiffness of composites decrease and failure strain increases with increasing the gradient degree (i.e., when the particles become more localized in some regions of the material). The orientations of particles have an impact on failure strain and damage growth in the composites reinforced with elongated or plate-like particles: whereas the horizontally aligned particles ensure the highest failure strain, the vertically aligned particles lead to the lowest and the randomly oriented particles to the medium failure strain.

Short literature review: Modeling of gradient composite materials

The problems of the computational analysis of functionally gradient materials and the optimal numerical design of FGMs have attracted a growing interest of scientific community in last decades. Many authors studied the deformation and strength of the gradient materials using the analytical and numerical micromechanical methods [1-3].

One of the classical approaches to the analysis of the strength and stiffness of FGMs is based on the rule-of-mixture. So, Hirano et al. [4] used the rule-of-mixture and the fuzzy set model of the transition from the region of high content of the filler to the matrix to develop an inverse design procedure for the determination of the synthesis method for required properties of FGMs.

Zuiker and Dvorak [5,6] generalized the Mori-Tanaka method of the estimation of overall properties of statistically homogeneous composites to linearly variable overall and local fields. They have shown that the linear and constant field approaches „provide different estimates of overall properties for small representative volumes, but nearly identical estimates for large volumes".

---

[4] Reprinted from L. Mishnaevsky Jr., Functionally gradient metal matrix composites: numerical analysis of the microstructure-strength relationships, Composites Sci. & Technology 66/11-12, pp. 1873-1887 (2006) with kind permission from Elsevier

Buryachenko and Rammerstorfer [7] simulated FGMs as a linear thermoelastic composite medium with elliptical inclusions, arranged in a way that the concentration of the inclusions is a function of the coordinates. They used a generalized "multiparticle effective method", developed by Buryachenko [8], and assumed that the effective field near the inclusion is homogeneous. Considering the joint actions of nonlocal effects, caused by the inhomogeneous inclusion number density and inhomogeneous average applied stress and temperature fields, and taking into account the binary interaction effects of the inclusions, Buryachenko and Rammerstorfer derived a general integral equation for the functional gradient composite, and analyzed the boundary layer and scale effects in this case.

Reiter, Dvorak and Tvergaard [9] developed a micromechanical FE model of graded C/SiC composites consisting of up to thousands inclusions. In the simulations, planar gradient arrangements of hexagonal inclusions with a linear volume gradient, and different transitions between the phases (i.e., microstructures with a distinct threshold between two matrix phases, with the sceletal transition zones, and mixed microstructures) were considered. Further, they presented the FGM as a number of piecewise homogeneous layers, and determined the properties of the layers using the Mori-Tanaka and self-consistent methods. It was shown that the averaging methods can be well used to characterize the graded materials in the framework of the model of piecewise homogeneous layers.

Weissenbek et al. [10] studied the elasto-plastic deformation due to thermal and mechanical loading of layered metal-ceramic Ni-Al_2O_3 composites with compositionally graded interfaces, using the finite element method, as well as analytical models (mean-field approach involving an incremental Mori-Tanaka analysis and the rule-of-mixture approximation). Planar geometries with perfectly periodic arrangements of the constituent phases were considered using the square-packing and hexagonal-packing unit cell formulations for the graded material. Then, unit cells, containing large numbers of randomly placed microstructural units of the two phases were used. It was found that square-packing arrangements provide the best possible bounds for the thermal strains and coefficient of thermal expansion of the graded multilayer, among the different unit cell models examined.

Becker Jr. et al. [13] used the nonlocal brittle fracture model (Ritchie–Knott–Rice (RKR) fracture model), based on the Weibull statistics, to analyze the fracture initiation ("first activated flaw") near a crack in FGMs. The dependencies of the initiation fracture toughness (i.e., the stress-intensity factor that will result in a stated first failure probability) on the phase angle of crack tip as well as on the parameters of the Weibull law, were determined using FEM. Becker Jr. et al. demonstrated numerically and analytically that the gradient in Weibull scaling stress leads to a decrease of initiation fracture toughness, and that „gradients normal to the crack result in a crack growing toward the weaker material". It was shown that the distribution of damage near a crack tip depends strongly on Weibull modulus: for a high Weibull modulus, "failure is dominated by the very near-tip parameters, and effects of gradients are minimized. With low m, distributed damage leading to toughening can be exaggerated in FGMs."

Cannillo et al. [14] used the public domain, image-based FE software OOF and the probabilistic model of brittle fracture to study the crack growth in graded alumina-glass. They analyzed the effect of stochastic placement of the second phase on the hardness and toughness of the material. The variations of the damage parameter versus applied strain curves for different random realizations of the microstructures are determined.

An advanced 3D model of FGMs was developed by Gasik [15,16]. The model, which represents a FGM with "chemical" gradient as an array of subcells (local representative volume elements) and was implemented as own software, allows to calculate elastic and thermal properties of the composite.

On the basis of the above review, one may classify the models of strength and reliability of graded materials as follows. One group of works generalizes and pushes the limit of analytical micromechanical models, developed initially for non-gradient materials (rule-of-mixture, Mori-Tanaka method, multiparticle effective method [8]). Another group is based on the methods of numerical experiments using multiparticle unit cells and FEM. The effects of different transition zones between the phases, the stochastic placement of the second phase, mechanical and fracture properties (as Weibull parameters), residual stresses and strains, as well as applicability limits of different methods and models were studied.

In this work, we use the mesomechanical FE simulations of the deformation and damage evolution in different microstructures of graded composites in order to investigate the microstructure-strength relationships of graded Al/SiC composites, and to develop recommendations for the improvement of the composite properties.

Microstructure design, mesh generation and material properties

Microstructure design and mesh generation

In order to study the effect of the microstructures of materials on the deformation and fracture behavior, the microstructure of the material under consideration should be varied in the required way, so that both the necessary range of structure variation is ensured and most of the interesting cases are considered. This can be done, if microstructures of the considered material are designed artificially, and must not be taken from a real material. The strategy of the numerical testing of the artificial designed microstructures was pursued in this work.

Multiparticle unit cells with many round or elliptical particles, arranged with different gradients, were designed and meshed using two-dimensional version of the program "Meso3D", developed by the author [17,18]. The graded distributions of the particles were generated as follows. The X-coordinates of the particle centers were calculated using the uniform random number generator, whereas the Y-coordinates of the particle centers were calculated as random values distributed by the Gauss law. The mean values of the corresponding normal distribution of the coordinates of particle centers were assumed to be the Y- coordinate of the upper boundary of the box. Fig. 4.33 shows schematically the design of such microstructures. The standard deviations of the probability distribution of the distances between the upper boundary of the cell and the particle center were varied, from very small standard deviation (0.5 mm in the cell of 10 mm height) (highly gradient arrangements) to the deviations comparable with the box size (15mm) (which correspond to the fast uniformly random particle arrangements. The reciprocal of the value of the standard deviation of the distance of the particle centers from the upper boundary of the cell will be called "degree of gradient" further. Therefore, the microstructures with a high "degree of gradient" will have a much localized type of particle arrangement; whereas the low "degree of gradient" means that the particles are arranged almost homogeneously. The type of microstructure will be designated here by its standard deviation: for instance, "grad3" means a graded microstructure with the standard deviation 3 mm.

The ellipsoidal particles with different aspect ratios were oriented randomly, or aligned vertically and horizontally.

The generated microstructures were meshed with the TRIA6 triangular elements. Each model contained approximately 12000 finite elements. The procedure of the microstructure and mesh design is given in more details elsewhere [17-19].

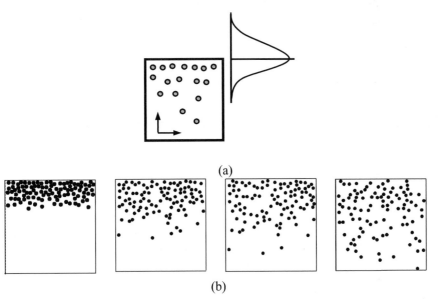

(a)

(b)

Fig 4.33 Schema of the design of artificial gradient microstructures (a) and some examples of the generated microstructures with different degrees of gradient (b)

Finite element model and material properties

Squared unit cell of the sizes 10 x 10 mm, which contained 100 round or elongated SiC, was subject to the uniaxial tensile displacement loading, 2.0 mm. The nodes at the upper surface of the box were connected, and the displacement was applied to only one node. The uniaxial tensile response of each microstructure was computed by the finite element method, using the plain strain model. The FE meshes of the generated microstructures of the composites were generated with the use of the program "Meso3D" [17], and commercial code MSC/PATRAN. The simulations were done with ABAQUS/Standard.

The SiC particles behaved as elastic isotropic damageable solids, characterized by Young modulus $E_P = 485$ GPa, Poisson's ratio 0.165 and the local damage criterion, discussed below. The Al matrix was modeled as isotropic elasto-plastic damageable solid, with Young modulus $E_M = 73$ GPa, and Poisson's ratio 0.345. The experimental stress-strain curve for the Al matrix was taken from [14,18], and approximated by the deformation theory flow relation (Ludwik hardening law): $\sigma_y = \sigma_{yn} + h\varepsilon_{pl}^n$, where σ_y -the actual flow stress, σ_{yn} - the initial yield stress, and ε_{pl}- the accumulated equivalent plastic strain, h and n - hardening coefficient and the hardening exponent. The parameters of the curve for the matrix were as follows: $\sigma_{yn} = 205$ MPa, h = 457 MPa, n = 0.20. The volume content of SiC particles was taken 10 %,

As output parameters of the numerical testing of the microstructures, the effective response of the materials, the microcrack density in particles versus the far-field strain curves and the damage distribution were determined. The far-field applied strain at which many particles fail and the falling branch of the stress-strain curve begins will be called "failure strain" hereafter.

Damage Simulation and Critical Parameters

The micromechanisms of damage evolution in Al/SiC composites under mechanical loading can be described as follows: first, some particles become damaged and fail (in the case of relatively large particles) or debond from the matrix (for smaller particles); after that, cavities and voids nucleate in the matrix (initially, near the broken particle), grow and coalesce, and that leads to the failure of the matrix ligaments between particles, and finally to the formation of a macrocrack in a volume [20-23].

The criteria and conditions of damage and local failure of SiC and Al phases, which were used in our simulations of damage and fracture of the composite, were taken from literature data, mainly, from the research on the damage parameter identification for Al matrix in Al/SiC composites carried out by J. Wulf [22,23], and the investigations of particle failure carried out by Derrien et al. [21].

Wulf [22,23] studied experimentally and simulated numerically damage growth and fracture in real microstructures of Al/SiC composites. By comparing simulated and real crack path, and simulated and experimentally observed force-displacement curves, he determined the correct criterion and the critical values for the local void growth and failure in matrix. According to Wulf [23], finite-element simulations with this damage parameter (called in [25] and hereafter Rice/Tracey damage indicator) produced excellent results for Al/SiC composites: both crack path in a real microstructure of a material and the force-displacement curve were practically identical in the experiments and simulations. According to Fischer et al. [25], who reviewed different damage criteria and carried out the parameter studies, the results of their calculations were "in surprisingly good agreement with experimental observations" as well. The critical value of the damage indicator, verified by Wulf, is $D_{cr} = 0.2$.

A possible alternative to the Rice-Tracey damage indicator for the simulation of crack growth in Al matrix is the approach based on the constitutive equations for porous plasticity developed by Gurson [26] and adapted to practice by Tvergaard. According to Geni and Kikuchi [27], the simulations with the Gurson model give results which are very close to the experimental data as well. Good results can be obtained by using nonlocal version of the Gurson model [28].

In our simulations, the Rice-Tracey damage indicator was used as a parameter of the void growth in the Al matrix.

To model the damage and local failure of SiC particle, the criterion of critical maximum principal stress in the particle material was used. According to [21], the SiC particles in Al/SiC composites become damaged and ultimately fail, when the critical maximum principal stress in a particle exceeds 1500 MPa. This value was used in our simulations as a criterion of damage of SiC particles as well.

The ABAQUS Subroutine User Defined Field (USRFLD), which allows to simulate the local damage growth in both phases of Al/SiC composites as a weakening of finite elements, was developed. In this subroutine, the phase to which a given finite element in the model is assigned, is defined through the field variables of the element. Depending on the field variable, the subroutine calculates either the Rice-Tracey damage indicator (in the matrix) or the maximum principal stress (in particles). If the value of the damage parameter or the principal stress in the element exceeds the corresponding critical level, the field variables of the element is changed, and the stiffness of the elements is reduced. The Young modulus of this element is set to a very low value (50 Pa, i.e., about 0.00001% of the initial value). The critical level of the maximum principal stress can be either a constant value, or a random value with a pre-defined distribution. The numbers of failed elements are printed out in a file, which can be used to visualize the calculated damage distribution.

Finite Element Simulations and Results

Damage evolution in graded composites and the effect of the degree of gradient

The purpose of this part of the investigation was to clarify how the degree of gradient influences the strength and damage evolution in the composites. The deformation and damage evolution of Al/SiC composites with gradient SiC particle arrangements (with different degrees of gradient) were simulated numerically.

As discussed above, the gradient degree of a particle arrangement is determined by the standard deviation of the normal probability distribution of the distances between the Y-coordinates of the particle centers and of the upper boundary of the cell. Since the X-coordinates of particles are generated from a pre-defined random number seed parameter (idum) (which should ensure reproducibility of the simulations), variations of this parameter lead to the generation of new realizations of microstructures with the same gradient. Many graded microstructures with different standard deviations of the distributions of Y-coordinates (which ensured different gradient degrees) and with different random number seed parameter for random X coordinates were generated, meshed and tested. Figure 4.33b shows several examples of the generated microstructures. At this stage of work, only round particles were considered. Fig. 4.34 shows some typical tensile stress-strain curves and the fraction of failed elements in the particles plotted versus the far-field applied strain for the graded particle arrangements with different degrees of gradient.

Table 4.1 gives the critical strains, as well as statistical parameters of the microstructures (averaged distances between nearest-neighbor particles, NND, and the statistical entropy of the nearest neighbor distances). One can see that the gradient degree correlates with the averaged nearest neighbor distances: the lower degrees of gradient lead to the higher average nearest neighbor distances. No correlation between the degree of gradient and the statistical entropy of NND was found.

Figure 4.35a shows the failure strain (critical applied strain) plotted versus the degree of gradient in the composites. Figure 4.35b shows the flow stress of the composite (at the far-field strain u = 0.15) as a function of the gradient degree.

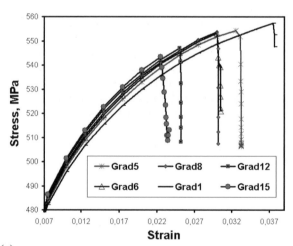

(a)

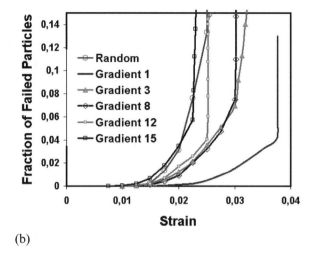

(b)

Fig 4.34 Tensile stress-strain curves (a) and the fraction of failed elements in the particles plotted versus the far-field applied strain (b) for the graded particle arrangements with different degrees of gradient.

It is of interest that the flow stress and stiffness of composites decrease with increasing the gradient degree. Apparently, the more homogeneous is the distribution of hard inclusions in the matrix, the stiffer is the composite. If the particles are localized in one layer in the composite, the regions with low particle density determine the deformation of the material, and that leads to the low stiffness.

One can see from Figure 4.34b that all the microstructures have rather low damage growth rate at the initial stage of damage evolution. At some far-field strain (called here "failure strain"), the intensive (almost vertical) damage growth takes place and the falling branch of the stress-strain curve begins. For all the graded microstructures, the failure strain is higher than for the homogeneous microstructures.

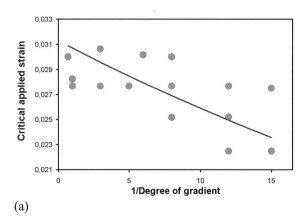

(a)

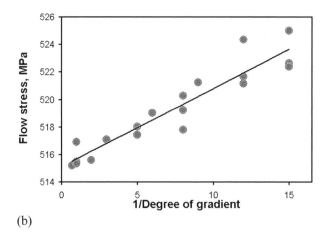

(b)

Fig 4.35 Failure strain (a) and flow stress of the composite (at the far-field strain u=0.15) (b) plotted versus the degree of gradient in the composites.

Failure strain of composites increases with increasing the gradient degree.

Figure 4.36 shows the von Mises stress distribution in a highly gradient (grad3) microstructure. One can see that the stresses are lower in the low part of the microstructure (particle-free region), than in the particle-rich regions. If two particles are placed very closely one to another, the stress level in the particles is much higher than in other particles, especially if these particles are arranged along the gradient (vertical) vector. Then, the stress level is rather high in particles which are located in the transition region between the high particle density and particle-free regions. One could expect that these particles begin to fail at the later stages of loading, and that was observed in the damage simulations indeed. Figure 4.36b shows the damage distribution in the particles and in the matrix (grad3 microstructure, far-field strain 0.29). That the particles begin to fail not in the region of high particle density but rather in the transition region between the particle-rich and particle-free regions, is similar to our observations for the case of clustered particle arrangement: in the case of clustered particle arrangement, the damage begins in the particles which are placed at the outer boundaries of clusters [18]. One can see from Figure 4.36b that the damage in matrix begins near the damaged particles, or between particles which are arranged closely in the direction of gradient vector.

Figure 4.37 shows the mechanism of the damage formation in the composite, observed in our simulations: the void growth begins near the failed particles, and the damaged area expands in the direction to the nearest damaged particle. This mechanism has been observed experimentally as well [20,21].

To verify the numerical results related to the influence of the degree of gradient on the stiffness and failure behavior and obtained in this section, we use the following analytical model. The gradient material is represented as a two-layer material (Figure 4.38).

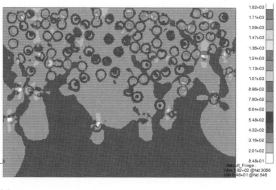

(a)

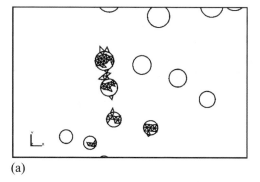

(b)

Fig 4.36 Von Mises stress distribution in a highly gradient (grad3) microstructure (a) and damage distribution in the particles and in the matrix (grad3 microstructure, far-field strain 0.29).

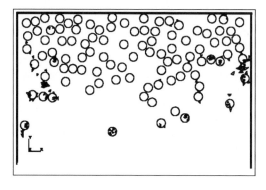

(a)

Fig 4.37 Mechanism of void initiation near a failed particle (a) and of the expansion of the damaged area, observed in the simulations.

The Young modulus of the gradient material is calculated using the Reuss model. The upper layer, which in fact represents the region of the gradient composite with the high particle density, is taken here as a homogeneous material. The thickness of this upper layer is equal to the thickness of the region with the high particle density. The lower layer represents the particle-free regions of the composite. The degree of gradient of microstructures in this model is characterized by two parameters: the thickness and the Young modulus of the upper layer (i.e., of the highly reinforced region of the gradient composite). A highly graded material (like gradient 1, at Figure 4.33) is represented in the framework of this model as a bilayer with thin and hard upper layer, and the lower layer with the properties of the matrix, whereas a material with low gradient degree is considered as a bilayer with a thick upper layer, which properties are rather close to the properties of the matrix. Since the total amount of particles in the cell is assumed to be constant, the volume content of the SiC particles in the upper layer is inversely proportional to the layer thickness. The degree of gradient can be characterized in this model by the ratio of the cell size to the thickness of the upper layer. In [14], the effect of the volume content of SiC particles the on the flow stress and stiffness of Al/SiC composites was analyzed. Approximating the results from [14], one can obtain the following relationship between the Young modulus of the composite and the volume content of SiC particles:

$$E_{up} = a + b * VC, \tag{4.23}$$

where E_{up} – Young modulus of the Al/SiC composite (in this case, of the "upper layer" material) (in MPa), VC – volume content of the SiC particles, a and b – regression coefficients, a= $4.25*10^4$ MPa, b=242.1 MPa. Assuming that the average volume content of SiC particles in the upper layer is 50%, if the thickness of the upper layer is 0.1 (i.e., 10% of the total height of the cell), one obtains the relationship between the thickness of the region of the cell with the high particle density (i.e., of the upper layer) and the volume content of SiC particles in this layer:

$$w_{up} = 0.05/VC \tag{4.24}$$

Substituting these formulas into the Reuss formula for the Young modulus of the bilayer,

$$E = 1/[(w_{up}*1/E_{up}) + (1^2 - w_{up}*1)/E_{matr}], \tag{4.25}$$

where 1 –width of the cell, E_{up} and E_{matr} – Young modulus for the highly reinforced part (upper layer) and the matrix, one can determine the Young modulus of the composite as a function of the degree of gradient (i.e., the ratio $1/w_{up}$).

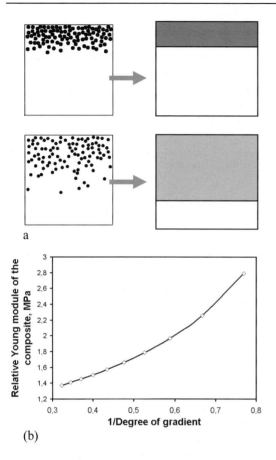

(b)

Fig. 4.38 The model of a gradient material as a bilayer (highly graded and almost homogeneous composites composite) (a) and the normalized Young modulus of the composite plotted

Table 4.1
Critical (failure) strains and statistical parameters of some graded microstructures

	Failure strain	Flow stress (at u = 0.15 mm)	NND	SENND
Grad1	0,038	515,49	0,44	0,83
Grad2	0,025	515,59	0,48	0,90
Grad3	0,028	517,07	0,49	0,36
Grad5	0,033	518,01	0,54	0,54
Grad6	0,030	519,01	0,55	0,54
Grad8	0,030	520,26	0,58	0,75
Grad12	0,025	521,66	0,58	0,43

NND- average nearest-neighbor distances, SENND- statistical entropy of the nearest neighbor distances

Figure 4.38b shows the normalized Young modulus of the composite (E/E_{matr}) as a function of the degree of gradient l/w_{up}. One can see that the stiffness of the composite decreases with increasing the degree of gradient of the composite. Furthermore, it was shown in [14], that the failure strain of a SiC particle reinforced Al composite is inversely proportional to the stiffness of the composite. Taking into account this result and Figure 4.38b, one can conclude that the degree of gradient has the following effect of the failure strain of composites: the failure strain increases when the gradient degree of the composite increases and the particles are highly localized in a layer. This result, obtained with the use of the simple analytical model, confirms our results, obtained in the simulations (Figure 4.35).

Effect of the shape and orientation of the elongated particles on the strength and damage evolution: non-graded composites

In many biomaterials, as nacre, teeth and bones [29-32], one may identify the following type of microstructure at micro- and nanolevel: staggered gradient arrangement of platelets or elongated mineral particles. The materials, which have such microstructures, show rather high damage resistance and strength, and it has been shown that the high performances of the biomaterials can be attributed to this kind of microstructure [30-32]. That is why the effect of the arrangement, shape and orientation of elongated or platelet-like particles is of especial interest for us. At this stage of the work, the effect of the arrangement of elongated particles, their shapes (aspect ratio) and orientations on the effective response and damage behavior of graded and homogeneous composites was studied numerically.

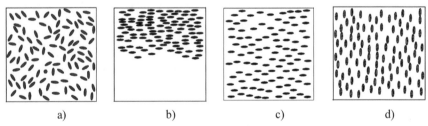

| a) | b) | c) | d) |

Fig. 4.39. Examples of the designed microstructures with elongated particles.

The following microstructures of composites were generated and tested: composites reinforced with elongated particles (aspect ratios 2. and 3.33), aligned horizontally and vertically, and oriented randomly, with graded and homogeneous arrangements. Figure 4.39 shows some examples of the designed microstructures.

First, consider non-graded microstructures with elongated reinforcing particles, with different aspect ratios and orientations of particles. Figure 4.40 shows the stress-strain curve and the damage-strain curves for the non-graded microstructures with different orientations of particles (aligned vertically and horizontally, and randomly oriented). One can see from Figure 4.40, that the failure strain of the composites with elongated particles increases in the following order: vertical aligned < randomly oriented < horizontal aligned particles. The failure strain of the microstructures with round particles is always higher than that for the elongated particles.

Then, let us consider the effect of the aspect ratio of the particles on the strength and failure strain of the composite. Figure 4.41 shows the stress-strain curve and the damage-strain curves for the non-graded microstructures with different aspect ratios of particles (where rr = smaller particle radius divided by bigger particle radius, rr = 0.3, 0.5, 0.7 and 1.). It can be seen that the higher is the aspect ratio of particles, the higher (slightly) are the flow stress and stiffness of the composites. An increase in the value of rr by 0.2 (0.3 -> 0.5, or 0.5 -> 0.7) (i.e., an increase of the aspect ratio by 40...60%) leads to an increase of the flow stress by 1.4 %. The failure strain decreases with increasing the aspect ratio of the particles: when the value of rr increases by 0.2 (0.3 -> 0.5, or 0.5 -> 0.7, what corresponds, again, to the increase of the aspect ratio by 40...60%), the failure strain increases by 13%.

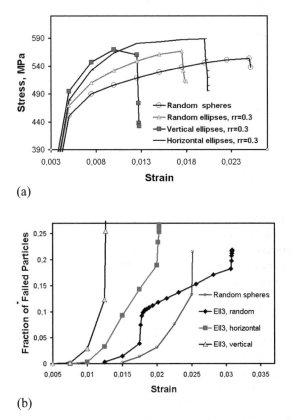

(a)

(b)

Fig. 4.40 Stress-strain curves and the damage-strain curves for the non-graded microstructures with different orientations of particles (aligned vertically and horizontally, and randomly oriented, aspect ratio 3.33).

In order to analyze the mechanisms of deformation and damage evolution in the composites reinforced by elongated or platelet-like particles, one may look at the von Mises stress and damage distributions in a composite with randomly oriented elongated particles (rr=0.3) (Figure 4.42). It can be seen that the damage in matrix begins most often in the places between two particles which are arranged closely along the vertical direction (i.e., along the loading and gradient direction). The void growth in the matrix begins near the sharp ends of the particles. Then, the damaged areas extend and link with other voids, formed near other particles (Fig. 4.42b), rather similar to the mechanism of the damage growth in the composites with round particles (Figure 4.36).

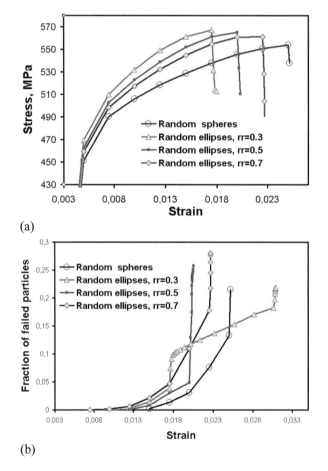

(a)

(b)

Fig. 4.41 Stress-strain curve and the damage-strain curves for the non-graded microstructures with different aspect ratios of particles (rr=0.3, 0.5, 0.7 and 1., rr=smallest particle radius/biggest radius)

Figure 4.43 shows the damage distribution in the microstructures with aligned (vertical and horizontal) elongated particles. One can see that the damage in matrix begins often between closely placed particles, which are arranged one above another in the loading direction, similar to the mechanism in the case of the randomly oriented particles.

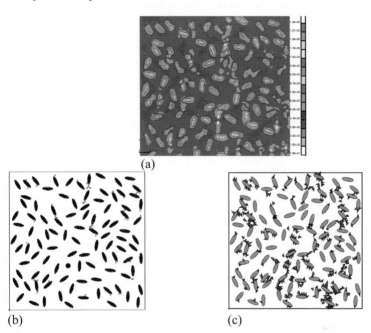

(a)

(b)   (c)

Fig. 4.42 Von Mises stress (u = 0.18 mm) (a) and damage distributions in a composite with randomly oriented elongated particles (rr = 0.3, u = 0.18mm and u = 0.29 mm) (b,c).

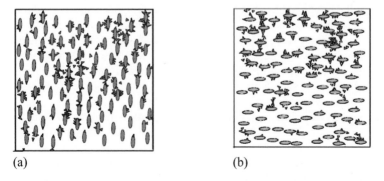

(a)   (b)

Fig. 4.43 Damage distribution in a composite with aligned vertical (a) and horizontal elongated particles (b) (rr = 0.3, u = 0. 0.30 mm and 0.20 mm, respectively)

Effect of the shape and orientation of the elongated particles on the strength and damage evolution: the case of graded composite materials

At this stage of the work, the effect of the graded arrangement of elongated particles on the strength and damage evolution in the composites was considered. Figure 4.44 shows the stress-strain curve and the damage-strain curves for the graded microstructures with different orientations of particles.

The shapes of the stress-strain and damage-strain curves for graded and non-graded composites are similar, but the damage growth rates, and the stiffness and flow stress of the composite are much lower, and the failure strains are sufficiently higher for the graded microstructures than for the homogeneous microstructures. Whereas the damage growth rate, calculated as an increase in the fraction of failed particles divided by the increase in the far-field applied strain, for a homogeneous microstructure (elongated particles, aspect ratio 3.33, randomly oriented) is 19.4, this value for the same, but graded microstructure is equal to 5.3.

One can see that the failure strain for the graded composite increases in the same order, as in the case of homogeneous microstructures: vertical aligned < randomly oriented < horizontal aligned elongated particles. The stiffness of the composite is a little bit higher for aligned (vertical or horizontal) ellipsoids, than for the randomly oriented ellipsoids.

It is of interest that the curves of the fraction of failed elements plotted versus the far-field applied strain for the random orientation of ellipsoidal particles (both graded and homogeneous arrangements) have a shape, which is different from the curves for the microstructures with aligned particles: whereas the fraction of failed particles increases monotonically with increasing applied strain for the case of aligned particle microstructures, the curves of the fraction of failed particles versus strain for the randomly oriented particles have plateaus. After the intensive damage evolution begins and continues for some time, it slows down, and goes on at much slower rate. At some strain level (approximately, two times the strain level of the first intensive damage growth), the intensive damage growth starts again.

This effect can be explained by the following reasoning. Under tensile loading, not only the particles become damaged, but also the some rotation of the randomly oriented particles can take place [33]. As a result, the angles between the particles and the vertical axis can be reduced for many particles. Since the vertically aligned elongated particles show much lower damage growth rate than the horizontally arranged particles, such rearrangement of particles during the plastic deformations can lead to the slowing down the damage rate.

Then, it can be seen from Figure 4.44 that whereas the more localized and highly gradient microstructures have lower stiffness and higher failure strain, than the homogeneous microstructures in all other cases, the first critical strain (i.e., the critical strain, at which the falling branch of the stress-strain curve begins) is the same for both gradient and non-gradient microstructures in the case of the microstructures with the randomly oriented elongated particles. After the damage growth slows down, the damage growth rate is much less for the graded microstructure, than for the homogeneous microstructure. The second critical strain for these microstructures is much lower for the homogeneous, than for the graded version of these microstructures.

This effect (i.e., that the randomly oriented elongated particles may ensure much lower damage growth rate than the aligned particles, see Figure 4.44) corresponds also to the theoretical analysis of the effects of the randomization of materials on the failure strength, carried out by Mishnaevsky Jr. and Shioya [34]. On the basis of the phenomenological model of fracture, Mishnaevsky Jr. and Shioya [34] demonstrated that the randomization of microstructures of multiphase materials (including, among others, random orientation of weaker planes and brittle elements in a tough matrix) can lead to the higher fracture resistance of materials. This numerical result confirms their theoretical conclusions [34].

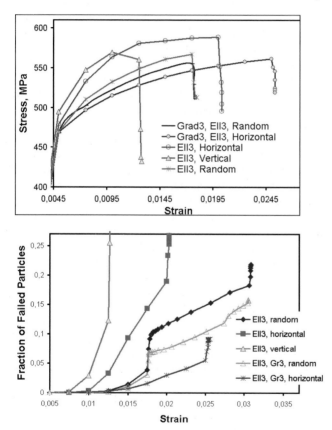

Fig. 4.44 Stress-strain curves (a) and the damage-strain curves (b) for the graded microstructures with particles with aspect ratio 3.333, and different particle orientations

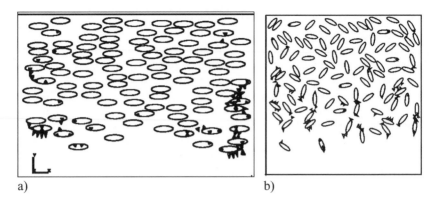

a)   b)

Fig. 4.45 Damage distribution in the matrix in the case of a graded composite reinforced by aligned horizontal (a) and randomly oriented ellipsoids (rr = 0.3) (b).

Now let us consider the mechanism of deformation and damage in the graded composites. Figure 4.45 shows the damage distribution in the matrix in the case of a graded composite reinforced by aligned horizontal and randomly oriented ellipsoids (aspect ratio 3.33). It can be seen that both in the case of the aligned and randomly oriented elongated (and similarly to the case of graded microstructure with the round particles), the density of damaged particles in the area where the high particle density region passes into the particle-free region is rather high, and much higher than in the region of high particle density. Apparently, the particles which are located in the "transition" area begin to fail first. The matrix is damaged not in the region of the high particle density but rather in the area, where the region of high particles density passes into the region of low particle density as well (similar to the mechanism of damage initiation in the graded composites reinforced by graded particles). In the case of the randomly oriented particles, the damage initiation in particles takes rather often if "noses" of two or three elliptical particles are placed close one to another.

One should note that the microstructures with particles aligned along the direction normal to the loading direction, which demonstrated the highest failure strain results in our simulations, are rather similar to the microstructures of many biomaterials [31,32], where the platelets or fibers are arranged with a gradient, and aligned normally to the expected loading direction.

On the basis of the simulations, one may draw the following conclusions. The failure strain of the composites with elongated particles increases in the following order: vertical aligned < randomly oriented < horizontal aligned particles. The higher is the aspect ratio of particles, the higher (slightly) are the flow stress and stiffness of the composites. The failure strain decreases with increasing the aspect ratio of the particles. The particles located in the area where the high particle density region passes into the particle-free region begin to fail first.

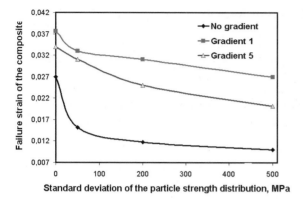

a)

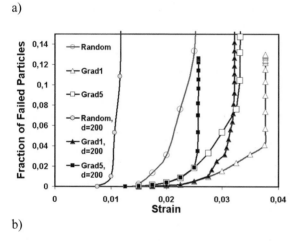

b)

Fig. 4.46 Failure strain of composites plotted versus the standard deviation of the probability distributions of the particle strengths (a) and the damage-strain curves for the graded microstructures with random variations of local strength of particles

The effect of the gradient on the flow stress, stiffness and the failure strain for the microstructures with elongated particles are similar to the effect for the case of round particles: the more localized and highly gradient microstructures have lower stiffness and higher failure strain, than the homogeneous microstructures.

Effect of statistical variations of local strengths of reinforcing particles and the distribution of the particle sizes

Real reinforcing materials have always some statistical variations of the mechanical properties (local strengths, etc.), which have a profound effect on the failure and strength of composites. At this stage of work, we study the effect of the statistical variations of strengths of the particles in the Al/SiC graded composite on the failure behavior of the composite.

The stress-strain and fraction of failed particles versus strain curves were calculated numerically for different microstructures of the composites (random non-gradient, gradient 1, gradient 5) with different degrees of scattering of the strength of particles. The critical maximum stress of SiC particles, which was assumed to be a constant value (1500 MPa) in all above simulations, was a random value here. It was assumed that the critical maximum principal stress is distributed by the Gauss probability law [[1]], with the mean value 1500 MPa (as above) and with the standard deviations 0, 50, 200, 500 and 1000 MPa. In the simulations, the random critical maximum stress was calculated in each element and compared with the current value of maximum principal stress in the element; if the current principal stress exceeded this random critical value, the element was considered to fail, and its stiffness was reduced.

Figure 4.46a shows the failure strain for the different graded microstructures plotted versus the degree of the scattering (standard deviation) of the local strength of particles. Figure 4.46b shows the damage-strain curves for the non-graded and graded microstructures with the different degrees of gradient and different standard deviations of the critical principal stress.

One can see that the failure strain of a composite decreases rapidly when the degree of scattering of local strength of particles increases. However, the negative effect of the scattering of the particle strengths on the failure strain of the composite, observed above, is weakened if the microstructure is graded. The degree of reduction of the failure strain of composite due to the randomness of local strength depends on the microstructure of the composite as well: whereas the increase of the standard deviation of the critical stress from 0 to 500 MPa leads to the 2.7 times reduction of the failure strain in the non-graded microstructure, the same change leads to only 68% and 39% decrease in the failure strain of the graded composites (gradient5 and gradient1, respectively).

Consider now the effect of the variation of the particle sizes on the strength and failure of composites. It has been shown in our previous work that materials with randomly distributed particle sizes have much less failure strain than materials reinforced by particles of the same constant radius [14]. An interesting case of a composite material with both varied sizes of reinforcing particles and the directional gradient is a composite reinforced by particles which radii depend on the position of the particle. Figure 4.47a shows examples of microstructures where the radius of a particle is proportional to the Y-coordinate of the particle. Such microstructures, where the size of particles is proportional to the vertical coordinate will be called further "particle size gradient microstructures".

Two types of microstructures with the graded distribution of sizes of reinforcing particles were considered: "small/big" size gradient microstructure, with small round particles near the upper boundary of the cell and big particles at the lower boundary (called also "south" microstructures, according to the location of big particles in the lower/"southern" part of the cell), and "big/small" size gradient ("north" microstructure). The radius of particles was taken to be proportional to the Y-coordinate of particles, $R \sim L$ ("north" size gradient) or $R \sim (L-Y)$, where L – cell size ("south" size gradient). After the radii of particles were calculated, they were normalized to keep the total volume content of the SiC particles constant.

The numerical testing of these microstructures was carried out for the constant strength of particles, as well as for the case of the random (Gaussian) distribution of the critical stress in particles, with the standard deviations 200, 500 and 1000 MPa. (The average critical principal maximum stress was the same as above, 1500 MPa).

Figure 4.47b shows some typical functions of the fraction of failed elements in the particles plotted versus the far-field applied strain for the "small/big" and "big/small" size gradient microstructures.

One can see that the "small/big" and "big/small" size gradient microstructures have very similar damage growth curves and the same failure strain, when the critical stress of particles does not vary. However, when the statistical variations of the strength of SiC particles are taken into account, the failure strain of the composite is drastically reduced: by 11% for the "small/big" microstructure and by 30% for the "big/small" microstructure. This decline hardly depends on the degree of the variation of the local strength. Figure 4.47c shows the ratio of the failure strains for the "small/big" and "big/small" size gradient microstructures as a function of the standard deviation of the probability distribution of the particle strength. Apparently, since the statistical variations of the strength of particles were included in our continuum mechanical model, the size effects of particles began to play a role in the simulations. Therefore, the greater particles placed near the upper border of the cell in the "big/small" gradient microstructures begin to fail earlier, and that leads to the quicker failure of the composites, whereas this effect does not take place in the "small/big" microstructures.

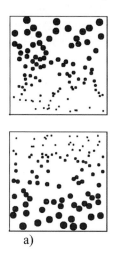

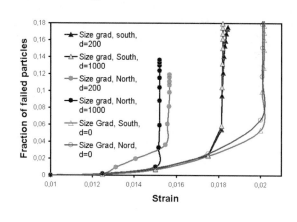

a)   b)

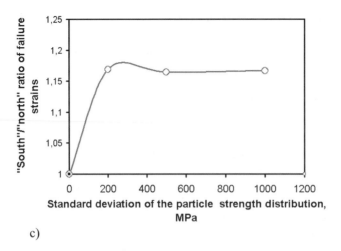

c)

Fig. 4.47 Examples of the "particle size gradient" "small/big"/"big/small" microstructures (a), fraction of failed particles plotted versus applied strain for the microstructures with random variations of the local strength of particles (b) and the ratio of failure strains for the "small/big"/"big/small" microstructures plotted versus the standard deviation of the probability distributions of the particle strengths (c).

On the basis of the simulations, one may draw the following conclusions. The statistical variations of the strength of particles in composites lead to the decrease of the failure strain in the composites. However, this negative effect is weakened, if the microstructures of composites are graded. In the case of the size gradient microstructures (with particles, which strength are varied randomly), the "small/big" microstructures ensure higher failure strains than the "big/small" microstructures.

Hierarchy of microstructural effects

At this stage of work, we would like to identify the microstructural parameters which have strongest influence on the damage resistance of graded composites. To compare the effects of different parameters of microstructures, we determined the ratios of the critical failure strains of materials in extreme points of the variation range of the parameters considered above. Practically, we compared the critical failure strain for composites with the highest localization of particles arrangement (g = 1) and the almost homogeneous particle arrangement (g = 15), with the constant and randomly varied strengths of particles, with the horizontal (normal to the loading vector) and vertical orientation of elongated particles, with round or highly elongated shapes of particles.

Table 4.2 gives the results of this analysis. It can be seen from the table that the random variations of the particle strengths has a biggest effect on the damage resistance of the composites. Further, the orientation of elongated particles, their shapes and the degree of particles localization in the composites play an important role for the damage resistance of the composites as well.

The strong effect of the scattering of local properties (the first factor) on the macroscopic strength of the materials was expected. However, it is of interest that the damage resistance of materials can be increased by 30%...60% only by varying the geometrical microstructural parameters.

Table 4.2.
Hierarchy of parameters of microstructures influencing the damage resistance of the composites

Nr.	Parameters of microstructure	1st Extreme case	2nd Extreme case	Critical failure strain for the 1st case	Critical failure strain for the 2nd case	Ratio
1.	Strength Variations	Constant particles strength	Randomly varied particle strength, deviation 1000 Mpa	0.027	0.01	2.7
2.	Orientation of platelets	Horizontal	Vertical	0.0203	0.01266	1.603
3.	Shape (Aspect Ratio)	1	0.3	0.0251	0.0178	1.41
4.	Gradient degree (g)	1	15	0.31	0.0225	1.33
Combined effects:						
5.	Factors 1+4 (constant strength of particles and graded arrangement)	High gradient (g=3), constant strength of particles	No gradient, random strengths of particles	0,03	0,01	3.
6.	Factors 2+4 (horizontal orientation and graded arrangement of platelets)	High gradient (g=3), horizontal	No gradient, vertical	0.025	0.01266	1.97
7.	Factors 3+4 (Shapes and graded arrangement of particles)	High gradient (g=3), round particles	No gradient, elongated particles	0.03	0.0176	1.7

In order to analyze the possible effects of combining the microstructural factors, we calculated the ratio of the critical strains for the microstructures, which combine both high gradient, constant stress and other positive factors, to some reference cases. It was shown that the changes in the microstructure, which improve the damage resistance of the composites, combined with other positive changes, lead to the multiplying effects. So, composites with both constant strength and the graded arrangement of particles show higher damage resistance, than composites with only constant strength or with only gradually arranged particles (by ~10% in both cases). This holds for other combined effects as well. Thus, one may conclude that the recommendations to the improvement of microstructures of lightweight metal matrix composites, identified in this work, may be combined with other recommendations, and that leads to the multiplication of positive effects.

Conclusions

The effect of the microstructures of graded composite materials on the deformation and damage behavior was studied using the numerical mesomechanical experiments. On the basis of the simulations, the following conclusions may be drawn.

Flow stress and stiffness of composites decrease with increasing the gradient degree, whereas failure strain increases with increasing the gradient degree. The more localized and highly gradient microstructures have lower stiffness and higher failure strain, than the homogeneous microstructures.

The damage evolution in SiC particles begins not in the region of high particle density but rather in the transition region between the particle-rich and particle-free regions.

Failure strain of the composites reinforced with elongated particles increases in the following order: vertical aligned < randomly oriented < horizontal aligned particles. The higher is the aspect ratio of particles, the higher (slightly) are the flow stress and stiffness of the composites. The failure strain decreases with increasing the aspect ratio of the particles.

The statistical variations of the strength of particles in composites lead to the decrease of the failure strain in the composites. However, this negative effect is weakened, if the microstructures of composites are graded.

It can be seen that the graded particle distribution has a very beneficial impact on the damage resistance of the composites: it increases the failure strain, weakens the negative effect of the heterogeneity of particles and slows down the damage growth rate in the particles. These positive effects are the stronger, the higher is the gradient degree.

The availability of the particle-free regions in the composites has a mixed effect on their damage resistance: it reduces the stiffness of the composite, and the damage growth in particles begins at the boundary of the particle-free and particle-rich regions. Yet, as shown in [18,35], the availability of the regions of low particle density have a beneficial effect on the toughness of the composites. One may assume therefore that microstructures, which combine a layer of very high particle density (as in the case of "gradient 1" microstructure), with the rest material, homogeneously reinforced with some low density of particles (which should ensure the required stiffness of the composite), can be the optimal microstructure from the viewpoint of high stiffness, damage resistance and strength.

References:

[1] Suresh S. and Mortensen A. (1998), Fundamentals of Functionally Graded Materials, Institute of Materials, UK.

[2] Mortensen A. and Suresh S. (1997), Functionally Graded Metals and Metal-Ceramic Composites, International Materials Reviews, 40 (6), 239-265, 1996. 42 (3), 85-116.

[3] Miyamoto Y., Kaysser W.A., Rabin B.H., Kawasaki A., Ford R G. (1999), Functionally Graded Materials, Design, Processing and Applications, Kluwer, Dordrecht.

[4] Hirano, T, Teraki J. and Yamada T. (1990), One the Design of Functionally Gradient Materials, Proc. 1st Internat Symposoum, FGM, Sendai, Eds. M. Yamanouchi et al, pp. 5-10.

[5] Zuiker J R and Dvorak G J (1994a), On the effective properties of functionally graded composites-I. Extension of the Mori-Tanaka method to linearly varying fields Compos. Eng. 4 19-35.

[6] Zuiker J R and Dvorak G J (1994b), On the effective properties of composite materials by the linearly field ASME J. Eng. Mater. Technol. 116 428-37.

[7] Buryachenko V., Rammerstorfer F.G. (1998), Micromechanics and Nonlocal Effects in Graded Random Structure Matrix Composites., (Proceedings IUTAM-Symposium on Transformation Problems in Composite and Active Materials (Eds. Bahei-el-Din, Dvorak G.J.), Kluwer Academic Publishers, Dordrecht et al., 197-206.

[8] Buryachenko V.A. (1996):,The overall elastoplastic behavior of multiphase materials with isotropic components., Acta Mechanica 119, 93-117.

[9] Reiter T., Dvorak G.J., Tvergaard V. (1997): Micromechanical Models for Graded Composite Materials; J.Mech.Phys.Sol. 45, 1281-1302.

[10] Weissenbek E., Pettermann H.E., Suresh S. (1997), Numerical Simulation of Plastic Deformation in Compositionally Graded Metal-Ceramic Structures; Acta Mater. 45(8), 3401-3417.

[11] Mishnaevsky Jr L. and Schmauder S. (2001), Continuum mesomechanical finite element modeling in materials development: a State-of-the-Art Review, Applied Mechanics Reviews, Vol. 54, 1, pp. 49-69.

[12] Mishnaevsky Jr L., Dong M., Hoenle S. and Schmauder S. (1999), Computational Mesomechanics of Particle-Reinforced Composites, Comp. Mater. Sci., Vol. 16, No. 1-4, pp. 133-143.

[13] Becker T. L., Jr., Cannon R. M., and Ritchie R. O. (2002) Statistical Fracture Modeling: Crack Path and Fracture Criteria with Application to Homogeneous and Functionally Graded Materials by, Engineering Fracture Mechanics, vol. 69, pp. 1521-1555.

[14] Cannillo V., Manfredini T., Corradi A., Carter W.C. (2002), Numerical Models of the Effect of Heterogeneity on the Behavior of Graded Materials. Key Engineering Materials Vols. 206-213, pp. 2163-2166.

[15] Gasik M.M. (1998), Micromechanical modelling of functionally graded materials, Computational Materials Science 13 (1-3), pp. 42-55.

[16] Gasik M.M. (1995), Principles of Functional Gradient Materials and their Processing by Powder Metallirgy, Acta Polytechnica Scandinavica, Helsinki.

[17] Mishnaevsky Jr L. (2004), Three-dimensional Numerical Testing of Microstructures of Particle Reinforced Composites, Acta Mater, Vol 52/14, pp 4177-4188.

[18] Mishnaevsky Jr L., Derrien K. and Baptiste D. (2004) Effect of Microstructures of Particle Reinforced Composites on the Damage Evolution: Probabilistic and Numerical Analysis, Composites Science and Technology, Vol. 64, No 12, pp. 1805-1818.

[19] Ganguly P. and Poole W.J.(2004), Influence of reinforcement arrangement on the local reinforcement stresses in composite materials, J Mechanics and Physics of Solids, Vol 52, 6, pp. 1355-1377.

[20] Mummery, P. and Derby B., Fracture Behavior, Chapter 14, Fundamentals of Metal-Matrix Composites Edited by: Suresh, S.; Mortensen, A.; Needleman, A., 1993, pp. 251-268 Elsevier.

[21] Derrien K., Baptiste D., Guedra-Degeorges D., Foulquier J. (1999), Multiscale modelling of the damaged plastic behaviour of AlSiCp composites. Int. J. Plasticity, Vol. 15, pp. 667-685.

[22] Wulf J, Schmauder S and Fischmeister HF (1993), Finite element modelling of crack propagation in ductile fracture, Comput Mater Sci, 1, pp. 297-301.

[23] Wulf J (1995), Neue Finite-Elemente-Methode zur Simulation des Duktilbruchs in Al/SiC Dissertation MPI für Metallforschung, Stuttgart.

[24] Rice J.R. and Tracey D.M. (1969), On the Ductile Enlargement of Voids in Triaxial Stress Fields, Int. J. Mech. Phys. Solids 17, pp. 201-217.

[25] Fischer F.D., Kolednik O., Shan G.X. and Rammerstorfer F.G. (1995), A Note On Calibration Of Ductile Failure Damage Indicators Int. J.Fract. 73, pp. 345-357.

[26] Gurson A.L. (1977), Continuum theory of ductile rupture by void nucleation and growth: Part I – Yield criteria and flow rules for porous ductile media, J. Eng. Mat. Tech., pp.2-15.

[27] Geni M and Kikuchi M (1998), Damage analysis of aluminum matrix composite considering non-uniform distribution of SiC particles, Acta mater, 46 (9), 3125-3133.

[28] F. Reusch, B. Svendsen & D. Klingbeil (2003), A non-local extension of Gurson-based ductile damage modeling, Computational Material Science 26, pp. 219-229.

[29] Weiner S., and Wagner H.D. (1998), The material bone: structure-mechanical function relations. Ann. Rev. Mater. Sci. 28, pp. 271-298.

[30] Ji B. and Gao H. (2004), A study of fracture mechanisms in biological nano-composites via the virtual internal bond model, Materials Science & Engineering A, Vol. 366, pp. 96-103.

[31] Jäger, P. Fratzl (2000), Mineralized collagen fibrils - a mechanical model with a staggered arrangement of mineral particles. Biophys. J. 79, pp. 1737-1746.

[32] Tesch, W., Eidelman, N., Roschger, P., Goldenberg, F., Klaushofer, K., Fratzl, P. (2001), Graded microstructure and mechanical properties of human dentin, Calcif. Tissue Int. 69, pp. 147-157.

[33] Agrawal, H.; Gokhale, A.M.; Graham, S.; Horstemeyer, M.F.; Bamman, D.J., (2002), Rotations of brittle particles during plastic deformation of ductile alloys, Materials Sci and Engin: A, Vol. 328, 1-2, pp. 310 – 316.

[34] Mishnaevsky Jr L., Shioya T., (2001) Optimization of Materials Microstructures: Information Theory Approach, Journal of the School of Engineering, The University of Tokyo, Vol. 48, pp. 1-13.

[35] Mishnaevsky Jr L. (1998), Damage and fracture of heterogeneous materials, Balkema Rotterdam, p. 230.

4.2.3 Voxel-based FE mesh generation and damage analysis of composites[5]

In this part of the work, we develop and to test numerical tools for the automatic development of finite element models of complex 3D microstructures of composites, on the basis of voxel array data. The developed cides are applied to model both isotropic and graded interpenetrating phase composites.

This group of materials includes, for instance, biomaterials, tool materials (e.g., WC/Co cemented carbides with high content of WC, in which the WC skeleton ensures high hardness) [1]-[3], sintered Al/SiC composites [4], Ag/Ni composite materials, polymer composites, containing conducting filler particles (e.g., graphite) as well as other dielectric composites. Some graded composite materials have regions with interpercolating phases between the regions of high concentration of each phase 0. Furthermore, the group of materials, for which the analysis in this work can be relevant, includes porous materials and foams, as well as highly damaged ductile materials, where the pores coalesce and form clusters. Some of these materials are widely used industrially (cemented carbides, foams, sintered composites). An improvement of strength and damage resistance of these materials can be of great importance for industry. The voxel-based model generation, as a basis for the analysis of the mechanical behavior of these materials, opens new possibilities for the clarification of microstructure-strength interrelations, and optimization of the properties of the materials.

Short literature review: Incorporation of microstructures of materials into numerical models

Numerical simulations of deformation, damage and fracture of composites present an important tool for the prediction of materials behavior, and the optimization of mechanical properties of materials.

The necessity to incorporate the information about microstructures of materials into the numerical models is one of the challenges of computational mesomechanics of materials 0-[10]. To overcome this problem, different methods and concepts have been used. Let us look at the different methods of incorporating the microstructural information into models of materials.

Automatic microstructure-based mesh generation

One of the most efficient (and widely used) programs for the automatic microstructure-based meshing and microstructural analysis is the C++-based, object-oriented FEM software OOF (="object-oriented finite element analysis"), developed by

---

[5] Reprinted from L. Mishnaevsky Jr., Automatic voxel based generation of 3D microstructural FE models and its application to the damage analysis of composites, Materials Science & Engineering A407/1-2, pp.11-23 (2005) with kind permission of Elsevier

a group of scientists at NIST (USA) [11], [12]. In fact, the software includes several programs: PPM2OOF (which reads image files in the PPM format and creates automatically the FE mesh for OOF on the basis of a microstructure image), OOF solver (which calculates stress and strain distribution in the material, recently also the damage [13]) and OOF2ABAQUS (which converts the geometrical information of the data files created by PPM2OOF or OOF into input files for ABAQUS). Recently, the new version of OOF (OOF2) was made available on the Website of the group [11].

While the PPM2OOF software produces the microstructure-based FE models for the OOF solver, there exist several other programs which generate FE models from microstuctural images directly to the commercial FE programs (e.g., ABAQUS) . So, Tellaeche Reparaz *et al* [14] used their own *Verborde* and *Digit* codes to generate FE models of real structures of duplex steels for ABAQUS from image analysis of micrographs. FE meshes from square elements were automatically associated with corresponding material. Iung *et al* [14] studied the strain heterogeneity in two-phase materials (Ti-alloys, dual-phase steels) on the basis of a FORTRAN program which automatically generates FE meshes (to be used by the ABAQUS code) representing the image of a real microstructure. The mesh is generated "in an iterative way by superimposing on the boundaries square grid of growing size", and is refined automatically at the interfaces between the phases.

Mishnaevsky Jr. et al. [15] simulated the crack propagation in the artificial microstructures of tool steels, using the developed program of automatic mesh generation. The program reads the pgm image files of real or artificial microstructures, and produces a command file for the Pre-Processing FE software MSC/Patran, which generates the microstructural FE model.

Multiparticle unit cells [8], [16]

The unit cells with idealised shapes and/or arrangement of particles are used widely to analyse the microstructure-strength interrelations. This approach of the numerical testing of materials has a longest history and is most widely used. More detailed reviews on this direction of the micromechanics of materials is given elsewhere [6]-[8].

Voronoi cell finite element method

Ghosh and co-workers [21]-[25] developed a very sophisticated and efficient approach to the modeling of deformation and damage initiation in MMCs, called Voronoi cell finite element method (VCFEM). In this method, the FE mesh is created by Dirichlet tessellation of a real microstructure of material. Each polygon, formed by such tessellation (they are called "Voronoi cells") contains one inclusion at most and is used as a finite element. Coupling the VCFEM for mesoscopic analysis and a conventional displacement based FEM for macro-analysis, they developed the "hierarchical multiple scale" model. In the framework of this hierarchical model, the authors used the adaptive schemes and mesh refinement strategies to divide the considered volume into subdomains with periodic and non-periodic microstructures. In the periodic microstructure areas they use the asymptotic homogenization. In non-periodic microstructure subdomains, VCFEM is used. This approach was used to simulate the damage initiation (by particle cracking or splitting) in discontinuously reinforced MMCs.

Surface rendering approach

Shan and Gokhale [38] developed a method to incorporate quantitative description of real complex three-dimensional microstructures into micromechanical models of materials. Using serial sectioning, they generate 3D microstructural image (with the use of the surface rendering approach), which was then embedded into FE model.

Multiphase finite element method [9], [15]

The main idea of this method is that the different phase properties are assigned to individual integration points in the element. Contrary to the traditional (single-phase) finite elements, a FE-mesh in this case is independent of the phase structure of material, and one can use relatively simple FE-meshes in order to simulate the deformation in a complex microstructure. The possibility of using simple meshes for the simulation of the behavior of complex materials (also in 3D case) is the main advantage of the method of multiphase elements. Mishnaevsky Jr. et al. [15] carried out simulations of crack growth in tool steels, using the multiphase and single phase finite elements, and demonstrated that the simulations with both methods yield very close results. Zohdi et al. [17]- [19] carried out large-scale micromechanical simulations of deformation and damage in composites, and determined numerically the optimal shapes of inclusions, using the decomposition of the global domain into a set of computationally smaller, decoupled problems, and statistical genetic algorithms. In so doing, they incorporated the microstructures of materials into the numerical models of the subdomains on the basis of the Gauss point method.

Digital Image-Based (DIB) modeling technique and pixel- and voxel-based mesh generation

The Digital Image-Based (DIB) modeling technique was developed by Hollister and Kikuchi [26] to include the effects of microstructural morphology of bone in the FE simulations in bioengineering. FE models obtained with the use of DIB technique present direct interpretations of micrographs of composite materials. Terada et al. [30] have used the DIB method together with FEM-based asymptotic homogenization method to simulate the overall mechanical behavior of composite as dependent on the geometry of microstructure and properties of components. They have shown that the actual stress-strain curve for the unit cell model obtained with the use of DIB (and reflecting the real structure) is quite different from that obtained in idealized unit cell model (elastic response more compliant, different trend of the strain hardening, etc.).

Garboczi and Day [31] developed an algorithm and a model to incorporate the microstructural information into FE models using the pixel-based approach, and to determine the effective linear elastic properties of random, multi-phase materials. Using the digital image as the finite element mesh should simplify the generation of FE meshes, the authors investigated the effective Poisson's ratio of two-phase random isotropic composites numerically and compared the results with the effective medium theory estimations.

The voxel based automatic mesh generation techniques have been used widely to analyse the mechanical behavior of bones and its dependence on microstructural effects [32]-[35]. The series of numerical investigations with this method, carried out in the group of Keaveny [32], [33] lead to the conclusion that this method allows to analyze the strength of tissues with a great accuracy and reliability. Large scale simulations, which use the voxel-based model generation and special solvers, have been reported in [27], [32]-[35].

Using the pixel/voxel based meshing, Kim and Swan [36], [37] developed and verified a new automated meshing techniques that „start from a hierarchical quadtree (in 2D) or oc-tree (in 3D) mesh of pixel or voxel elements", and then successive element splitting and nodal shifting are carried out in order to create mesh which accurately reflects the microgeometry of the cell. The method was applied to the generation of multielement unit cells, and verified.

In the following, we use the voxel-based approach to generate 3D microstructural models of random and graded microstructures of composites. In order to automate the model generation, a program, which generate automatically the 3D microstructural models of materials, was developed.

Automatic voxel based generation of 3D microstructural FE models of composites

Program for the automatic model generation

An approach and a program for the automatic generation of 3D FE microstructural models of materials with given ideal microstructures has been presented in [8], [6]. The developed program "Meso3D" allows to generate automatically 3D FE microstructural models of volumes of materials, using the exact geometric description of microstructures and the free meshing method [8]. However, this approach works well only for relatively simple geometrical forms of microstructural elements in composite (like spherical or ellipsoidal inclusions). This can be considered as a general drawback of many methods of 3D modeling of microstructures: both shapes of inclusions and their spatial distributions are often oversimplified.

The purpose of this work was to develop and verify a method of automated generation of 3D microstructural models of materials, which is based on the voxel array description of material microstructures, and allows therefore the modeling of arbitrarily complex microstructure of the material.

To automate the generation and meshing of 3D FE models of microstructures, a new program "Voxel2FEM" was developed. The information about the spatial distribution of phases in the representative volume is given as a voxel array. The representative volume of the material is presented as an array of points (voxels), each of them can be either black (0, hard phase) or white (1, soft phase, or matrix) (for a two–phase material). This approach can be simply generalized to the case of a multiphase material as well. As a result of an interactive session, the program "Voxel2FEM" produces a command (session) file for the commercial software MSC/PATRAN, which generates a 3D FE microstructural model of a representative volume of material. The designed microstructures are meshed with brick elements (20-node quadratic brick, C3D20), which are assigned to the phases automatically according to the voxel array data.

The developed program is applicable both to the design and testing of artificial microstructures of materials, and to the reconstruction and analysis of real 3D microstructures. Several built-in subroutines in the program allow reading the microstructure data from an external file (for the case of real microstructures), generation of different phase arrangements, as well as the percolation theory analysis of the microstructures.

Subroutine for generating random microstructures and multiparticle unit cells

The program can read the voxel array data from a text input file, using a built-in subroutine. Alternatively, the program can generate voxel arrays for multiparticle unit cells with different arrangements of spherical particles in a matrix, or random structures (3D random chessboard), as well as graded composite microstructures (s. [61]). The voxel array data for 3D random microstructure models (3D random chessboards) are generated in the program with the use of the random number generator. The voxel arrays for multiparticle unit cells with many spherical particles are generated in a subroutine of the program as well, using the algorithms described in [6], [8].

Subroutine for generating graded composite microstructures with pre-defined gradient

In order to analyze the effect of graded microstructures on the strength and damage in composites, a subroutine for the automated generation of random graded microstructures was included in the program. This subroutine defines the distribution of black voxels as random distributions in X and Z directions, and a

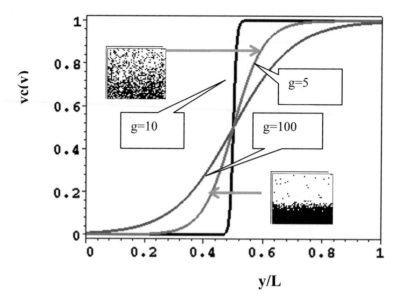

Fig 4.48 Shapes of the curve, which describes the transition between the regions of high content of different phases, for different parameters g (g=5, 10, 100, vc=50%)

graded distribution in Y direction. The graded distribution of black voxels (e.g., grains of hard phase) along the axis Y follows the formula:

$$vc(y)=2*vc0/(1+exp(g-2*g*y/L))), \tag{4.26}$$

Here $vc(y)$ is the probability that a voxel is black at a given point, $vc0$ is the volume content of the black phase, L – length of the cell, g – parameter of the gradient, y – Y-coordinate.

The equation (4.26) allows to vary the smoothness of the gradient interface of the structures (highly localized arrangements of inclusions and a sharp interface versus a smooth interface), keeping the volume content of inclusions constant. If $g < 2..3$, the transition between regions of high content of black or white phases is rather smooth, and if $g > 10$, the transition between the regions is rather sharp. Figure 4.48 gives the shapes of this curve for different g.

Subroutines for the percolation theory analysis of 3D microstructures

When generating the FE models of the representative unit cells, the presence of infinite percolation clusters in the generated microstructure is checked, using the burning algorithm [41]. The subroutine (based on the program, developed by Garboczi et al. [41]) searches all three directions, in the two perpendicular directions for each burn. The subroutine allows either periodic or hard boundary conditions. The information about the presence of percolation clusters for both phases and all the directions, is printed out in the session file. Another subroutine, built-in in the program, carries out the percolation analysis of the generated or reconstructed microstructures with the use of the alternative algorithm of the cluster labeling, suggested by Martin-Herrero and Peon-Fernandez in [42]. This subroutine carries out the labeling of the cluster of voxels, calculates the average and maximum cluster dimensions in all three directions and detects the existence or non-existence of the percolation in all directions. These two subroutines allow to carry out complete percolation analysis of the microstructures, as well as to compare the results obtained with the use of different techniques.

Here, the program developed is used to analyze the effect of the volume content and arrangement of inclusions in composites on the strength and damage evolution.

Routine for damage simulation

In this section, a newly developed ABAQUS subroutine for the damage modeling in Al/SiC composites is presented.

The micromechanisms of damage evolution in most Al/SiC composites (e.g., Alcoa X2080 aluminium alloys, made by powder blending and extruded route [[44]]) under mechanical loading can be described as follows: first, some particles become damaged and fail (in the case of relatively large particles) or debond from the matrix (for smaller particles); after that, cavities and voids nucleate in the matrix (initially, near the broken particle), grow and coalesce, and that leads to the failure of the matrix ligaments between particles, and finally to the formation of a macrocrack in a volume [43]-[46]. These damage mechanisms have been observed, for instance, in Alcoa X2080 aluminium reinforced with different volume fractions of silicon particles. According to Mummery and Derby [43], the interface debonding becomes one of the main damage mechanisms in the case of relatively small particles ($\sim< 10$ μm), but does not play a leading role for the case of bigger particles.

Wulf [46] studied experimentally and simulated numerically damage growth and fracture in real microstructures of Al/SiC composites. By comparing simulated and real crack paths and force-displacement curves, he tested different criteria of local failure and void growth (including the critical equivalent plastic strain, triaxiality factor and the damage indicator [46]). According to Wulf [46], finite-element simulations with the damage parameter, based on the model of a spherical void growth in a plastic material in a general remote stress field with high stress triaxiality, developed by Rice and Tracey [47] produced excellent results for Al/SiC composites: both the crack paths in a real microstructure of a material and the force-displacement curves were practically identical in the experiments and simulations.

A possible alternative to the Rice-Tracey damage indicator for the simulation of crack growth in Al matrix is the approach based on the constitutive equations for porous plasticity developed by Gurson [48] and adapted to practice by Tvergaard. According to Geni and Kikuchi [49], the simulations with the Gurson model give results which are very close to the experimental data as well. Good results can be obtained by using nonlocal version of the Gurson model [50].

In our simulations, the Rice-Tracey damage indicator was used as a parameter of the void growth in the Al matrix. To model the damage and local failure of SiC particle, the criterion of critical maximum principal stress in the particle material was used. According to [44], the SiC particles in Al/SiC composites become damaged and ultimately fail, when the critical maximum principal stress in a particle exceeds 1500 MPa. This value was used in our simulations as a criterion of damage of SiC particles as well.

An ABAQUS Subroutine USDFLD, which calculates the Rice-Tracey damage indicator in the matrix and the maximum principal stress in particles, and allows to visualize the damage (microcrack and void) distribution in the material was developed. The damage in particles was modeled as a local weakening of finite elements in which the damage criterion (maximum principal stress) exceeded a critical value 0. After an element failed, the Young modulus of this element was set to a very low value (50 Pa, i.e., about 0.00001% of the initial value).

Comparison of voxel-based and geometry-based 3D model generation

At this stage of work, the program "Voxel2FEM", which uses the voxel array based method of the reconstruction of 3D microstructures, was tested by comparing its results with the results of the exact geometry-based FE model. Two multiparticle unit cells for identical ideal 3D microstructures were generated using the program "Meso3D" (i.e., exact geometrical shapes plus free meshing) [8] and the program "Voxel2FEM" (voxel-based model generation). The FE analysis of deformation and damage in a composite was carried out, and the results of simulations were compared.

Multiparticle unit cells with 5 spherical particles were considered in both cases. The cells were subject to uniaxial tensile loading. Totally, the geometry-based model contained 7800 elements, and the voxel based model 15625 20-node quadratic brick elements. Each particle contained 370 finite elements in the geometry-based model, and 156 elements in the voxel-based model. The considered material was Al matrix reinforced by SiC particles (volume content 5%). The SiC particles behaved as elastic isotropic damageable solids, characterized by Young's modulus $E_P = 485$ GPa, Poisson's ratio 0.165 and the local damage criterion, discussed below. The Al matrix was modeled as isotropic elasto-plastic damageable solid, with Young's modulus $E_M = 73$ GPa, and Poisson's ratio 0.345. The experimental stress-strain curve for the Al matrix was taken from [40], and approximated by the deformation theory flow relation (Ludwik hardening law): $\sigma_y = \sigma_{yn} + h\varepsilon_{pl}^n$, where σ_y -the actual flow stress, σ_{yn} - the initial yield stress, and ε_{pl}- the accumulated equivalent plastic strain, h and n - hardening coefficient and the hardening exponent.

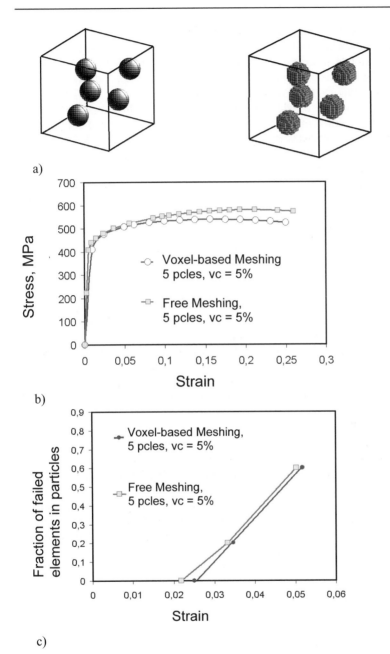

Fig. 4.49 Considered unit cells (a), as well as the stress-strain curves (b) and the fraction of failed elements in the particles plotted versus applied strain (c) obtained numerically

The parameters of the curve for the matrix were as follows: $\sigma_{yn} = 205$ MPa, h = 457 MPa, n = 0.20. The absolute size of this model (10 x 10 x 10 mm) was chosen in such a way so that it corresponds to the models considered in [8]. Apparently, since the considered material models do not have any size dependency, the same results could be obtained using a model with reduced sizes.

Figure 4.49 shows the considered unit cells, as well as the stress-strain curves and the fraction of failed elements in the hard phase versus applied strain curves obtained numerically. 5 realizations of the microstructure were tested. One should note that the damage growth in the ductile phase goes on much more slowly than in the particles: when many (up to 40%) elements in the particles fail, only a few element in ductile phase fail.

In the framework of the element weakening concept, the newly formed surface is represented as removal of volume elements. This should be taken into account when interpreting numerical results: high density of "failed elements" means in the case high microcrack density, and not a formation of a big hole in or crumbling of the material 0.

One can see from Figure 4.49 that the results obtained are rather close: the stress-strain curves differ only by 5%, and the damage-strain curves only by 3..4%. Therefore, the 3D models, generated from the voxel data arrays, give the results which are quite similar to the results of the models generated on the basis of the exact geometrical description of microstructures.

Let us compare our conclusions with the results of the similar works carried out in other groups. Guldberg et al. [34] tested the accuracy of digital image-based finite element models in 3D and 2D cases, and concluded that the "solution at digital model boundaries was characterized by local oscillations, which produced potentially high errors within individual boundary elements". The solution, however, oscillated about the theoretical solution, and was improved by averaging the results over the region of several elements. The observed absolute errors in different simulations were of the order of 1..4 %. Niebur et al. [32] investigated the convergence behavior of finite element models depending on the size of elements used, the element polynomial order, and the complexity of the applied loads. They concluded, that differences in apparent properties at different resolutions were always less than 10 percent when the ratio of mean trabecular thickness to element size was greater than four. Therefore, our conclusions are rather close to the results of other authors.

Numerical simulations and Results

Percolating and near-percolating microstructures of composite

Materials with percolating microstructures have been widely used industrially since long time (an example: cemented carbides with WC skeleton), and attracted a great interest of researchers.

Publications which deal with the analysis of the materials with near percolating and percolating microstructures, can be conditionally divided into several groups: statistical/morphological analysis of microstructures of materials and their effect on the elastic response [51]-[54], percolation theory analysis of critical exponents for mechanical and electrical parameters of composite systems [54], micromechanical analysis of interpenetrating microstructures [55]-[58], analysis of the effect of skeleton, contiguity and connectivity in sintered composites (in particular, cermets, cemented carbides) on their strength [2], [59], etc. Among the main results, obtained in this area, one may list the development of morphological models of random materials and bounds for linear elastic properties of materials with random microstructures [54], determination of interrelations between the critical components for elastic stiffness and electrical conductivity in dielectric composites [54], micromechanical "matricity" model, developed by Schmauder and colleagues [55]-[58] for the analysis of the interpenetrating and graded microstructures of composites, experimental and theoretical analysis of the effect of parameters of skeleton of sintered composites on the deformability and strength [1]-[3],[59], etc. The interest in the modeling of materials with percolating microstructures has increased in last years, as a result of the development of new materials: nanocomposites with nanoscale reinforcement, which forms percolating networks [60], foams and porous materials, etc. Most of the works, which deal with the microstructure-strength relationships of composites with random, percolating and near-percolating microstructures, consider the effect of the material microstructures on the elastic properties. Only a few works deal with plastic behavior or damage of the composites.

In this part of our work, we seek to analyze the effect of random, near-percolating and percolating microstructures on the damage resistance of composites, using the developed program and carrying out numerical mesomechanical experiments. In particular, the effect of the density of hard phase grains, presence of clusters and infinite percolation clusters from the hard phase grains on the deformation, strength and damage in composites should be clarified. In order to solve this problem, a series of 3D FE models of composites with random arrangement of hard phase grains and different volume content of the inclusions (3D "random chessboards") were generated using the program developed, and the commercial code MSC/PATRAN. The Al/SiC composite was taken as a test material, in order to ensure the compatibility and comparability of the results with the results of our previous simulations [8], [6], [61]. Cubic unit cells (of the sizes 10 x 10 x 10 mm) were subject to the uniaxial tensile displacement loading, 2.0 mm.

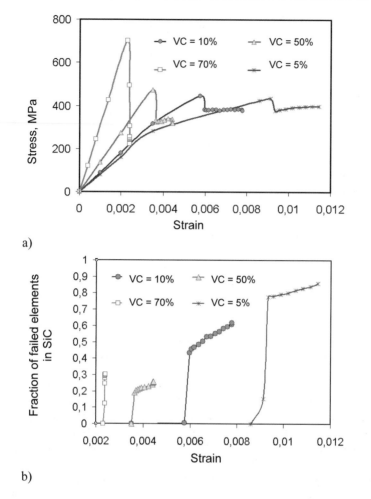

a)

b)

Fig 4.50 Typical tensile stress-strain curves and the fraction of failed elements in the hard phase plotted versus the far-field applied strain for the different volume contents of the hard phase

The nodes at the upper surface of the box were connected, and the displacement was applied to only one node. Tensile stress-strain curves, microcrack density in hard phase as a function of the load and stress, strain and damage distributions were computed by the finite element method. The simulations were done with ABAQUS/Standard. At very high deformation, when numerical problems appeared, the simulations were stopped. Figure 4.50 shows some typical tensile stress-strain curves and the fraction of failed elements in the hard phase plotted versus the far-field applied strain for the different volume contents of the hard phase.

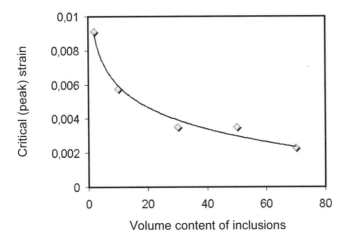

Fig 4.51 Critical applied strain, at which the intensive damage growth in the hard phase begins and goes on, plotted as a function of the volume content of hard phase.

It is seen that the falling branches of the stress-stress curves begin, when the intensive failure of hard phase goes on. After many hard grains fail, the damage growth slows down, and the stiffness of the materials is not reduced further. The damage growth in the ductile phase proceeds much lower than the damage growth in the hard phase. Apparently, the constant stress branches of the curves correspond to the stage of the material behavior, when many particles failed and don't bear any load, while the ductile phase remains almost intact, and only slow damage accumulation in the ductile phase. The next stage of the composite destruction, the void coalescence and crack formation in the ductile phase, could not be simulated due to numerical difficulties.

Figure 4.51 shows the critical applied strain (at which the falling branch of the stress-strain curve and the intensive damage growth in the hard phase begin) plotted as a function of the volume content of hard phase. One can see that the critical strain decreases with increasing the volume content of the hard phase.

It is of interest to correlate the strength, deformation and damage resistance of the composites with the presence of percolation clusters of hard phase grains. When generating the FE models, the percolation analysis for all three directions (X, Y, Z) and for both phases was carried out, and the presence of infinite clusters of the white and black voxels in each direction in the considered representative volume was tested. As expected [62], infinite percolation clusters from grains of the hard phase were not detected at the volume content of the hard phase (vc) < 31%, but were detected (in 1 direction) at vc=32%. Infinite clusters from grains of the hard phase are available in all three directions at vc=70%, but infinite clusters of ductile phase are available only in two directions at this volume content. In the case of the volume content between 32% and 69%, the microstructure is interpenetrating, and both phases form infinite clusters.

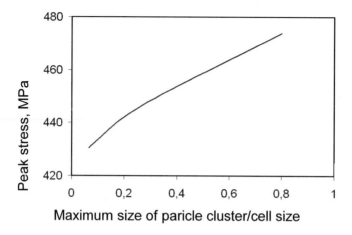

Fig 4.52 Peak stress plotted versus the maximum size of a cluster of the grains of hard phase.

Comparing these data with the results shown in Figure 4.50, one can draw the conclusion that a composite starts to behave as an elastic-brittle material (i.e., the linear stress-strain dependence up to the peak stress and then vertical falling branch of the stress-strain curve), when a percolation cluster of grains is formed (i.e., at VC>32%). When percolation clusters from grains in all three directions are formed, that lead to the strong (nonlinear) increase in the stiffness of the composite: whereas the increase of the volume content of hard phase from 10% to 50% leads to the Young's modulus increase of 50% (90 GPa -> 137 GPa) and the peak stress increase of 6% (447 MPa -> 474 MPa), the increase in the volume content of the hard phase from 50% to 70% leads to the Young's modulus increase of 122% (137Gpa -> 305 GPa) and the peak stress increase of 50% (474 MPa -> 701 MPa).

Figure 4.52 shows the peak stresses of the stress-strain curves plotted versus the maximum size of a cluster of the elements of the hard phase. The linear sizes of all the SiC grain clusters have been calculated for the generated 3D FE models, using the built-in percolation analysis subroutine in the developed program. One can see from Figure 4.52 that the stiffness and the peak stress of a composite increase almost linearly with increasing the linear size of the biggest hard phase cluster up to the percolation threshold. The formation of clusters from the hard grains therefore plays an important role for the stiffness and strength of composites.

Summarizing, one formulate the following conclusions. The increase in the volume content of hard phase leads, as expected, to the proportional increase in the Young's modulus of the composites, and to the strong increase of the peak stress at the stress-strain curves. On the other side, it leads to a decrease in the critical applied strain, at which the falling branch of the stress-strain curve begins. The stiffness and the yield stress of a composite increase almost linearly with increasing the linear size of the biggest hard phase cluster up to the formation of an infinite percolation cluster of hard grains. After an infinite percolation cluster of particles is formed, the material (consisting of the ductile and hard phases) starts to behave as a brittle material (i.e., the linear stress-strain dependence up to the peak stress and then vertical falling branch of the stress-strain curve).

Composites with graded microstructures

Graded composite materials have a great potential for applications in industry, and attract a growing interest of many research groups. Many approaches to the analysis of the strength and damage in graded composites are based on the generalization of homogenization techniques, which have been developed for the non-graded composites. So, the rule-of-mixture was applied to the analysis of gradient materials by Hirano et al. [63]. Zuiker and Dvorak [64] generalized the Mori-Tanaka method of estimation of overall properties of statistically homogeneous composites to the "linearly variable overall and local fields". Buryachenko and Rammerstorfer [65] generalized their "multiparticle effective method" to simulate FGMs, which are considered as linear thermoelastic composites with elliptical inclusions, arranged in a way that the concentration of the inclusions is a function of the coordinates. One should note however that the theoretical models of graded materials, based on the generalization of homogenization methods, can not be directly used to study the damage growth in the composites.

Another direction of the micromechanical analysis of graded materials is the use of multiparticle unit cells and real and quasi-real microstructures. So, Reiter et al. 0 developed a micromechanical FE model of graded C/SiC composites with a linear volume gradient consisting of up to several thousands hexagonal grains. In their simulations, different transitions between the phases (i.e., microstructures with a distinct threshold between two matrix phases, with the skeletal transition zones and mixed microstructures) were considered, using unit cells with large numbers of randomly placed microstructural units of the two phases. Canillo, Carter et al. [13] analyzed the effect of stochastic placement of the second phase on the hardness and toughness of graded alumina-glass using image-based FE software OOF and a probabilistic model of brittle fracture to study the crack growth. Whereas most of these models are 2 dimensional, and only few of them deal with the damage growth in graded composites, the models can serve as a basis of the generalization to the 3D case. An advanced 3D model of FGMs was developed by Gasik [66, 67]. The model, which represents a FGM with "chemical" gradient as an array of subcells (local representative volume elements) and was analyzed using their own software, allows to calculate elastic and thermal properties of the composite. A more detailed review on the micromechanical models of graded composites is given elsewhere [61].

In order to analyze the effect of the graded microstructure of composites on the damage and strength behavior in 3D case, we applied the program described above. A series of 3D FE models of composites with graded random distribution of the hard phase (with different gradient parameter g and different volume contents of the inclusions) were generated and tested.

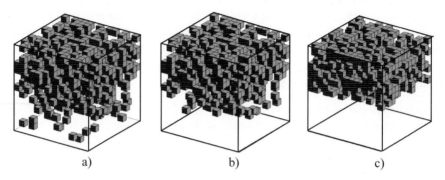

a)   b)   c)

Fig 4.53 Examples of the considered graded microstructures of the material: g = 3, g = 6, g = 100

Figure 4.53 shows some examples of the designed microstructures. The cell sizes, material properties, and damage mechanisms were the same as in the above simulations. Figure 4.54 shows the stress-strain curves of the composites with the volume content of hard phase 10% and 20%, and with varied gradient parameter g (equation (4.26)). One can see from the curves that the critical strain, at which damage growth begins in the materials, does not depend on the parameter of the volume fraction gradient g. Whether the transition between the region of high content of the hard phase to the region of low content of hard phase is sharp or smooth, the critical applied strain remains constant. However, the stiffness of composite and the peak stress of the stress-strain curve increase with increasing the sharpness of the transition between the regions. A reduction of the value *g* from 20 to 1 can lead to the decrease of the peak stress by 6%.

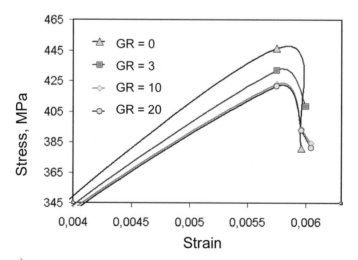

a)

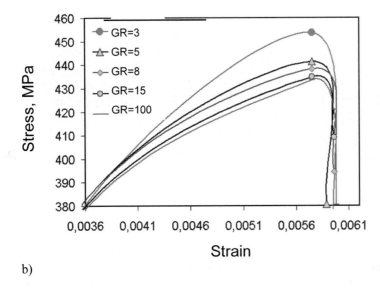

b)

Fig 4.54 Typical tensile stress-strain curves for the different sharpness of the transition zones of the graded composites: a) VC=10%, b) VC=20%

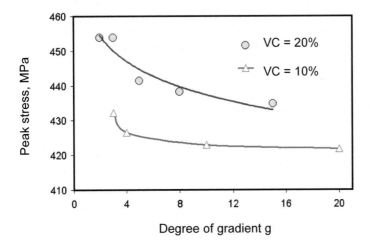

Fig 4.55 Peak of stress of the stress-strain curve plotted versus the sharpness of the transition zone for the graded composites with different volume content of hard phase (10% and 20%).

Figure 4.55 shows the peak stress of the stress-strain curve plotted versus the parameter g of sharpness of the transition between the regions of high and low content of hard phase.

In order to analyze the observed relationship between the peak stress and stiffness of the composite and the sharpness of the transition zone, let us apply the bilayer model of a gradient composite, developed in [61]. In the framework of this model, a gradient material is represented as a two-layer material (Figure 4.56). The upper "layer", which in fact represents the region of the gradient composite with the high content of hard phase (i.e., black or black/white region), is taken here as a homogeneous material. The lower layer represents the SiC-free region of the composite. Thus, one can calculate the Young's modulus of the upper layer (SiC-reinforced region) under uniaxial tensile loading by formula:

$$E_{up} = E_p vc \left(\frac{L}{w_{up}} \right) + E_m \left(1 - vc \frac{L}{w_{up}} \right) \tag{4.27}$$

Here E_{up}, E_p, E_M denote Young's moduli (in normal direction) of the "upper (SiC-reinforced) layer" of the graded composite, of hard and soft phases, L is the cell size, vc is the total volume content of hard phase, w_{up} is the thickness of the region with the high content of hard phase. The ratio w_{up}/L characterizes the function of the volume fraction gradient: if g>20 (sharp transition), $w_{up}/L=0.5$, and if g<5 (smooth transition), $w_{up}/L=0.6..0.9$. Therefore, the sharpness of the transition zone in this model is characterized by two parameters: the thickness and the Young's modulus of the upper layer. Using the Reuss formula for the Young's modulus of a bilayer, one can calculate the total Young's modulus of the gradient material under uniaxial tensile loading:

$$E = \frac{1}{(w_{up}/LE_{up}) + (L - w_{up})/LE_m}, \tag{4.28}$$

Here E_{up} and E_m are the Young's moduli (in normal direction) for the highly reinforced part (upper layer) and the ductile phase under uniaxial tensile loading. Now, one can determine the Young's modulus of the composite in normal direction as a function of the sharpness of the transition zone (i.e., the ratio L/w_{up}). Figure 4.56b shows the Young's modulus of the composite under uniaxial tensile loading, calculated with the use of the simplified model, as a function of the smoothness of the interface between the regions w_{up}. One can see that the increase in the width of the region with the high content of hard phase in the composite (even at a sacrifice of the average volume content of hard phase and stiffness of the region) leads to the proportional increase in the stiffness of the composite. I.e., the stiffness of the composite under uniaxial tensile loading increases with increasing the smoothness of the transition from the highly reinforced region of the composite to the SiC-free region. This result, obtained with the use of the simple analytical model, confirms our numerical results (Figure 4.54).

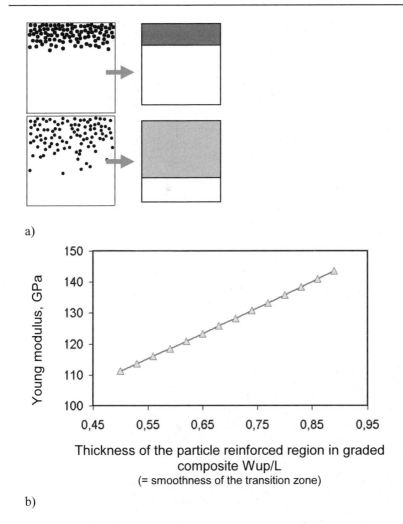

a)

b)

Fig 4.56 A model of a graded composite as a bilayer [[61]], and Young's modulus of a graded composite in normal direction as a function of the width of the particle reinforced region (=smoothness of the transition from the region of high volume content of the hard phase to the reinforcement free region) (b)

Thus, the stiffness of graded composites, considered here and subject to the uniaxial loading, can be improved by making smoother the transition region between the region with the high content of hard phase and the reinforcement-free region.

It is of interest to compare this conclusion with the results from [8], [61]. In [8], it was shown that the graded phase distribution in composites ensures much higher damage resistance but lower tensile stiffness than composites with random and homogeneous phase distribution. Furthermore, the more localized is the phase distribution and the more the gradient degree, the higher is the damage resistance of the composite [61]. Taking into account the result above, one may summarize that Al/SiC graded composites (considered above) with high gradient degree and smooth transition between the region with the high content of SiC phase and reinforcement-free region can ensure both highest damage resistance and relatively high stiffness.

Porous plasticity: open form porosity

3D numerical models of porous material with different porosity and random distributions of pores were generated using the developed program. The relative porosity (volume content of voids) was varied from 10% to 70%. The properties (constitutive law, Young's modulus, Poisson's ratio) of the matrix of the porous material corresponded to the properties of the aluminum in above simulations. The location of each pore was determined using the random number generator (random values in all three directions). Figure 4.57 shows examples of the considered representative volumes of the material. The tensile stress-strain curves for the porous Al with different volumes of porosity are given in Figure 4.58.

When generating the FE models, the percolation analysis for all three directions (X, Y, Z) and for both phases (pores, matrix) was carried out. The probability of the formation of infinite clusters from the pores in each direction in the considered representative volume was calculated, as described above. As expected [62], there were the following critical points in the material: vc=32% (formation of a first infinite percolation cluster of pores) and vc=69 % (infinite clusters of pores form in all three directions, however, infinite clusters of ductile phase are available only in two directions X and Z). In the case of the volume content of pores between 32% and 68%, both infinite percolation clusters of ductile phase and of pores form all three directions. The volume content of pores, at which the percolation of ductile phase is lost (i.e., 69%), corresponds to the formation of a crack along the Z axis (the material was loaded along the Y axis). At this volume content of voids, there is no percolation of ductile phase, and the material does not bear any load.

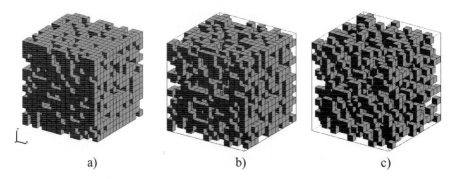

a)   b)   c)

Fig 4.57 Examples of the considered representative volumes of the porous material: porosity 30% (a), 50% (b) and 70% (c).

The yield stress (at the far-field applied strain e=0.03) was plotted versus the porosity (Figure 4.59). As expected, the yield stress of the porous material decreases with increasing the porosity. However, looking at the points of formation of percolation clusters, one can conclude that even formation of a percolation cluster of pores (in one or two directions) in a ductile material does not lead to the stepwise loss of stiffness of the material. Only the lack of percolation of the ductile phase phase corresponds to a stepwise loss of stiffness. Apparently, whereas clusters of connected pores may serve as sites of crack initiation, the formation of a cluster of pores does not necessarily correspond to the formation of a crack in the material.

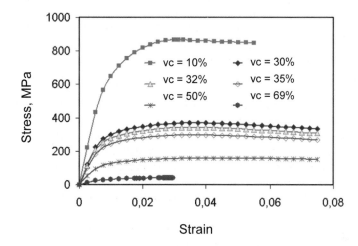

Fig 4.58 Tensile stress-strain curves for the porous Al with different volumes of porosity

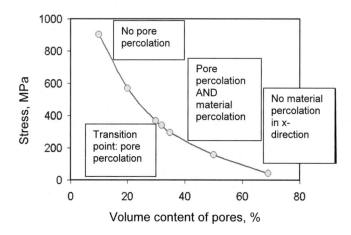

Fig 4.59 Yield stress (at the far-field applied strain e = 0.03 mm) was plotted versus the porosity.

Conclusions

A program for the automatic generation of 3D microstructural models of composite materials on the basis of the microstructure description given in the form of a voxel array has been developed and tested. The program allows to generate 3D models of representative volumes of heterogeneous materials both on the basis of experimental microstructure data (as obtained using the computer tomography or the serial sectioning), or can generate random, percolating, porous microstructures or multiparticle unit cells. The program developed was used to generate a number of 3D FE models of different groups of materials (percolating microstructures, graded composites, porous materials), and to carry out the numerical analysis of the microstructure-strength interrelations for these groups of materials.

It was shown that the stiffness and the yield stress of a composite increase almost linearly with increasing the linear size of the biggest cluster of hard phase grains up to the formation of an infinite percolation cluster of hard phase grains. After an infinite percolation cluster of hard phase elements is formed, the material (consisting of the ductile and hard damageable phases) starts to behave as a brittle material (i.e., the linear stress-strain dependence up to the peak stress and then vertical falling branch of the stress-strain curve).

The effect of the graded microstructure of composites on the damage and strength behavior was studied numerically. It was shown that the stiffness of composite and the peak stress of the stress-strain curve increase with increasing the smoothness of the transition between the region of high volume content of the hard phase to the region of low content of the hard phase. However, the critical strain, at which the damage growth begins in the materials, does not depend on the parameter of the interface smoothness g. Whether the transition between the regions is sharp or smooth, the critical applied strain remains constant.

The yield stress of the porous material decreases with increasing the porosity. However, even the formation of a percolation cluster of voids (in one or two directions) in a ductile material does not lead to the stepwise loss of stiffness of the material. Apparently, whereas clusters of connected pores may serve as sites of crack initiation, the formation of a cluster of pores does not necessarily correspond to the formation of a crack in the material.

References:

[1]   Mishnaevsky Jr L. (1998) Damage and fracture of heterogeneous materials, Balkema, Rotterdam.
[2]   Loshak M. G. (1984) Strength and durability of hard alloys. Naukova Dumka, Kiev.
[3]   Mishnaevsky Jr L. (1995) Appl. Comp. Mat., 1, pp. 317-324.
[4]   Purohit R., Sagar R. (2001) Int. J. Adv. Manufact. Technol., 17, pp. 644 – 648.
[5]   Reiter T., Dvorak G.J., Tvergaard V. (1997) J. Mech. Phys. Sol., 45, pp. 1281-1302.
[6]   Mishnaevsky Jr L. (2004), Derrien, Baptiste, Comp. Sci. & Technol., 64, pp. 1805-1818.
[7]   Mishnaevsky Jr L., Schmauder S. (2001) Appl. Mech. Rev., 54-1, pp. 49-69.
[8]   Mishnaevsky Jr L. (2004), Acta Materialia, 52/14, pp. 4177-4188.

[9]   Mishnaevsky Jr L., Dong M., Hoenle S., Schmauder S. (1999), Comp. Mater. Sci., 16 1-4, pp. 133-143.

[10] Povirk G. L. (1995), Acta Materialia, 43 (8), pp. 3199-3206.

[11] Carter W. C., Langer S., Fuller Jr E., The OOF Manual, OOF: Object-Oriented Finite Element Analysis of Real Material Microstructures, http://www.ctcms.nist.gov/oof/.

[12] García R. E., Reid A. C. E., Langer S. A., Carter W. C. (2004) In: "Continuum Scale Simulation of Engineering Materials: Fundamentals - Microstructures - Process Applications", Eds. Raabe D., Roters F., Barlat F., L-Q. Chen, Wiley, pp. 573-585.

[13] Cannillo V., Manfredini T., Corradi A., Carter W. C. (2002), Key Engineering Materials, 206-213, pp. 2163-2166.

[14] Tellaeche Reparaz M., Martinez-Esnaola J. M., Urcola J. J. (1997), Key Engineering Materials, 127-131, pp. 1215-1222.

[15] Iung I., Petitgand H., Grange M., Lemaire E. (1996), In: Proc. IUTAM Symposium on Micromechanics of Plasticity and Damage in Multiphase Materials, Kluwer, pp. 99-106.

[16] Böhm H. J., Han W. (2001), Modelling Simul. Mater. Sci. Eng., 9 pp. 47-65.

[17] Mishnaevsky Jr L., Weber U., Schmauder S. (2004), Int. J. Fracture, 125, pp. 33-50.

[18] Zohdi T. I., Wriggers P. (2001), Comp. Methods Appl. Mech. and Eng. 190 (22-23), pp. 2803–2823.

[19] Zohdi T. I., Wriggers P., Huet C. (2001), Comp. Methods Appl. Mech. and Eng. 190, 43-44, pp. 5639-5656.

[20] Zohdi T. I., Phil (2003), Trans. Roy. Soc: Math., Phys. and Eng. Sci. 361(1806), pp. 1021–1043.

[21] Moorthy S., Ghosh S. (1998), Comput. Methods Appl. Mech. Eng., 151, pp. 377-400.

[22] Ghosh S., Lee K., Moorthy S. (1995), Int. J. Solids Struct., 32 (1), pp. 27-62.

[23] Lee K., Moorthy S., Ghosh S. (1999), Comput. Methods Appl. Mech. Eng., 172, pp. 175-201.

[24] Li M., Ghosh S., Richmond O. (1999), Acta materialia 47 (12), pp. 3515-3532.

[25] Ghosh S., Moorthy S. (2004), Comput. Mech., 34, pp. 510-531.

[26] Hollister S. J., Kikuchi N. (1994), Biotechnol. Bioeng., 43(7), pp. 586–596.

[27] Hollister S. J., Brennan J. M., Kikuchi N. (1994), J. Biomech., 27, pp. 433-444.

[28] van Rietbergen B., Müller R., Ulrich D., Rüegsegger P., Huiskes R. (1999), J. Biomech., 32, pp. 165-173.

[29] Guldberg R. E., Hollister S. J., Charras G. (1998), J. Biomech. Eng. - Trans ASME, 120, pp. 289-295.

[30] Terada K., Miura T., Kikuchi N. (1997), Comput. Mech., 20, pp. 331–346.

[31] Garboczi E. J., Day A. R. (1995), J. Phys. Mech. Solids, 43, pp. 1349-1362.

[32] Niebur G. L., Yuen J. C., Hsia A. C., Keaveny T. M. (1999), J. Biomech. Eng., 121, pp. 629-635.

[33] Crawford R. P., Rosenberg W. S., Keaveny T. M. (2003), J. Biomech. Eng. 125(4), pp. 434-438.

[34] Guldberg R. E., Hollister S. J., Charras G. T. (1998), J. Biomech. Eng., 120, pp. 289-295.

[35] Ladd A. C. J., Kinney J. H. (1998), J. Biomech. 31, pp. 941-945.

[36] Kim H. J., Swan C. C. (2003), Int. J. Numerical Methods Eng., 56 (7), pp. 977–1006.

[37] Kim H. J., Swan C. C. (2003), Int. J. Numerical Methods Eng., 58, pp. 1683–1711.

[38] Shan Z., Gokhale A. M. (2001), Acta Mater., 49, 11, pp. 2001-2015.

[39] Lippmann N., Steinkopff Th., Schmauder S., Gumbsch P. (1997), Comput. Mater. Sci., 9, pp. 28-35.

[40] Soppa E. (2003), Personal communication.

[41] Garboczi E. J., Bentz D. P., Snyder K. A., Martys N. S., Stutzman P. E., Ferraris C. F., Bullard J. W., Butler K. M., Modeling and measuring the structure and properties of cement-based materials, Part III/2, Electronic monograph, http://ciks.cbt.nist.gov/garbocz/.

[42] Martín-Herrero J., Peón-Fernández J. (2000), J. Phys. A: Math. Gen., 33, pp. 1827-1840.
[43] Mummery P., Derby B., (1993), In: Chapter 14, Fundamentals of Metal-Matrix Composites Eds. S. Suresh, A.Mortensen, A. Needleman, Elsevier, pp. 251-268.
[44] Derrien K., Baptiste D., Guedra-Degeorges D., Foulquier J. (1999), Int. J. Plasticity, 15, pp. 667-685.
[45] Wulf J., Schmauder S., Fischmeister H. F. (1993), Comput. Mater. Sci., 1, pp. 297-301.
[46] Wulf J. (1995), Neue Finite-Elemente-Methode zur Simulation des Duktilbruchs in Al/SiC, Dissertation, MPI für Metallforschung, Stuttgart.
[47] Rice J. R., Tracey D. M. (1969), Int. J. Mech. Phys. Solids, 17, pp. 201-217.
[48] Gurson A. L. (1977), J. Eng. Mat. Tech., pp. 2-15.
[49] Geni M., Kikuchi M. (1998), Acta Mater, 46(9), pp. 3125-3133.
[50] Reusch F., Svendsen B., Klingbeil D. (2003), Comput. Mat. Sci., 26, pp. 219-229.
[51] Torquato S. (2002), Random Heterogeneous Materials: Microstructure and Macroscopic Properties, Springer-Verlag, New York.
[52] Torquato S. (2000), Int. J. Solids Struct., 37, pp. 411-422.
[53] Torquato S., Yeong C. L. Y., Rintoul M. D., Milius D., Aksay I. A. (1999), J. Am. Ceram. Soc., 82, pp. 1263-1268.
[54] Jeulin D. (2002), Lecture Notes in Physics, Springer-Verlag, Vol. 600, Heidelberg, pp. 3-36.
[55] Bergman D. (1978), Phys. Reports, 43, p 377.
[56] Bergman D. (2002), Phys. Rev. E 65, 26, pp. 124-128
[57] Schmauder S. (2002), Annu. Rev. Mater. Res., 32, pp. 437–65.
[58] Lessle P., Dong M., Schmauder S. (1999),Comput. Mat. Sci., 15, pp. 455-465.
[59] Roebuck B., Almond E.A. (1988), Int. Mater. Rev., 33, pp. 90-110.
[60] Buxton G. A., Balazs, A. C. (2004), Molecular Simulation, 30(4), pp. 249 – 257.
[61] Mishnaevsky Jr L., Composites Science and Technology.
[62] Stauffer D., Aharony A. (1992), Introduction to Percolation Theory, Taylor & Francis, London.
[63] Hirano J., Teraki J., Yamada T. (1990), In: Proc. 1st Internat Symposium, FGM, Sendai, , Eds. Yamanouchi M. et al, pp. 5-10.
[64] Zuiker J. R. and Dvorak G. J. (1994), Compos. Eng. 4, pp. 19-35.
[65] Buryachenko V., Rammerstorfer F. G. (1998), In: Proceedings IUTAM-Symposium on Transformation Problems in Composite and Active Materials, Eds. Bahei-el-Din, Dvorak G.J., Kluwer, Dordrecht et al., pp. 197-206.
[66] Gasik M. M. (1998), Comput. Mat. Sci., 13 (1-3), pp. 42-55.
[67] Gasik M. M. (1995), Principles of Functional Gradient Materials and their Processing by Powder Metallirgy, Acta Polytechnica Scandinavica, Helsinki.
[68] Mishnaevsky Jr L. (1997), Eng. Fract. Mech., 56(1), pp. 47-56.

4.3 Material with structure gradient for milling applications: modelling and testing[6]

Milling operations involve severe loading conditions, including tool entry, variable chip load during machining and a distinct tool exit, and consequently, fluctuations in thermal and mechanical loads, resulting in thermally and mechanically induced cracking of the tool [1, 2]. In the case of coated tools, inherent flaws in the CVD coating reduce the fracture strength of the tool [3, 4]. Conventional commercial WC-y-Co hardmetals do not have sufficient combination of load carrying capacity and toughness to resist the load in milling and to stop crack propagation from the coating. A special hardmetal with a graded surface zone of sufficient toughness coated with a thin layer of hard coating could be the solution [5, 6]. In these new hardmetals the layer beneath the coating contains a higher Co binder content to provide a high toughness and sufficient resistance to crack growth. The intention is to stop cracks from the coating within this volume of the substrate. The microstructure changes gradually into the substrate with increasing carbide contents so that the load carrying capacity is increased [7-10].

The aim of this work is to present a method for development of optimised materials for milling operations. This method is based on numerical simulations on the macro-, meso-, micro- and submicroscopic scale. The measured forces, temperature and contact lengths in instrumented milling tests are used to calculate the temperature field as well as the thermal and mechanical stress fields in the tool during a milling cycle. The critical region of the tool, where failure typically occurs, is modelled mesoscopically to simulate the crack-gradient interaction. The failure behaviour of hardmetal depends on the microstructural parameters, especially on the volume fraction and inclusion shape of the binder. This effect is investigated on the microscopic scale. In more detail, the damage behaviour of the binder phase is simulated on sub-microscopic scale to provide sufficient critical values such as damage parameter and fracture toughness for the microscopic and mesoscopic calculations.

Tool Materials

Sintered hardmetals consist of a binder phase, in general cobalt, and carbides, in general WC and as well TiC, TaC and NbC. The latter exist as cubic (Ti,Ta,Nb,W)C, the γ-phase. The microstructural parameters like volume fractions and grain sizes of the phases control the material data and fracture behaviour of hardmetals.

---

[6] Reprinted from S. Schmauder, A. Melander, P.E. McHugh, J. Rohde, S. Hönle, Or. Mintchev, A. Thuvander, H. Thoors, D. Quinn, P. Connolly, "New Tool Materials with a Structural Gradient for Milling Applications", J. Phys. IV France 9, pp. Pr9-147 - Pr9-156 (1999). with kind permission from Elsevier

Three hardmetal grades, CoStri, γFree and Conv, are investigated, the first two variants are graded materials with a tough surface zone on a hard substrate and the third variant is a conventional milling grade. Characteristic microstructures of the graded materials are shown in figure 4.60. The graded zones are 20-30 μm thick for both graded materials. The grade γFree possesses a surface zone free of cubic carbides and with increased Co content. The special merit of this graded zone is the low contiguity.

The two coating variants incorporating TiN/TiC layers with different chemical compositions were performed on the three tool grades. Conventional Chemical Vapour Deposition (CVD) technique was used in one variant, while the other involved plasma assisted CVD (PCVD). The coating thickness was around 5 μm for CoStri and γFree and slightly thinner for grade Conv. The CVD coating shows high residual tensile stresses which lead to cracks in the coating layer. The average distance between these cracks is about 150 - 200 μm. The formation of the brittle η-phase [11] close to the surface was found in grade Conv with conventional CVD coating.

The elastic data of hardmetals mostly depend on the volume fractions of the different phases, [1, 6, 12-15]. For the Young's modulus a linear relation with volume fractions exists. The Young's modulus can be calculated by the formula [6]

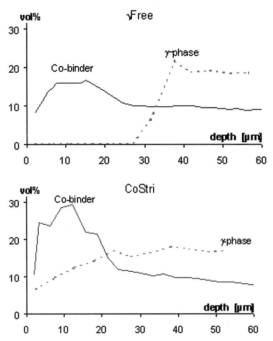

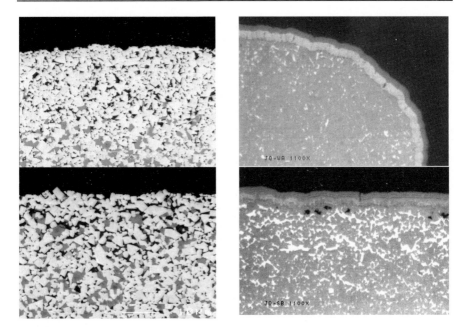

Fig. 4.60 Contents of Cobalt and y-phase, atomic number contrast images and light optic non-etched micrographs on plasma CVD coated yFree (top) and CVD coated CoStri (bottom).

$$E = 133.5 + 93.56(6f_{WC} + 9f_{\gamma Phase} - f_{Co}) \qquad (4.29)$$

where f_i is the volume fraction of phase i.

The fracture toughness is proportional to the square root of the mean free path in the binder [14, 16, 17]. This is important regarding grade CoStri, where locally Co-islands of several μm exists.

Dependencies of yield strength and fracture strength on the temperature were investigated by Bouaouadja et. al. [18]. For hardmetals with low Co content, yielding is visible only for high temperatures, while the temperature dependence of η_y for binder rich hardmetals is remarkable even for small variations in temperature.

Instrumented milling tests

Tool life testing was performed in a cutting data range where the tool life is limited by monotonic fracture or low cycle fracture. Different fracture modes are considered. Furthermore, information of cutting forces, tool/chip contact length and chip deformation are collected for subsequent use in numerical modelling of tool stresses and temperatures. An orthogonal up-milling geometry, implying an increasing chip thickness and cutting force during the actual cut and a hard 90° tool exit was used. Increased feed and work piece hardness reduced the tool life of the inserts substantially. During the tool life studies the milling was stopped every 20 to 50 cycles for visual inspection of the edge. The tool edge was considered fractured for damages greater than 0.4 mm on rake or flank. The edge damages are classified according to appearance and position on the edge, figure 4.61. Great tool life differences were recorded for the six tested variants. The PCVD coated conventional grade performs the best. The gradient structures seem to introduce weaknesses into the material causing them to fracture earlier.

The type of coating used has a bigger effect than the gradient variant on the tool life and the type of chipping that will occur on the edge. CVD coating primarily causes chippings on the flank face while top slice dominates for the PCVD coated tools. There is no clear difference between which fracture types that occurs for grades γFree and CoStri, figure 4.62. The PCVD coated variants performed better than the CVD coated variants for high feed levels where 70% of the latter failed before the first inspection. At lower feed the CVD coated variants improved greatly. The γFree variants also improved, more than the CoStri variants, and at a feed level of 0.115 *mm/rev* and a 50% fracture probability level γFree with CVD coating is the best gradient variant. Both γFree variants give longer tool lives than the CoStri variants.

The coating process also influences the contact length, chip thickness ratio, chip curl, edge radius and friction coefficient during machining, figure 4.63. It may also act as a crack initiation site due to cracks formed in the coating during the coating process. The differences in forces, contact lengths and chip deformation for the coatings hold together well.

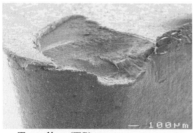
Top slice (TS)

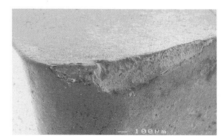
Main edge chipping (MEC)

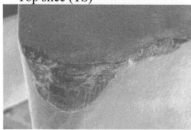
Nose chipping (NC)

Fig. 4.61 Fracture types according to appearance and position on the edge.

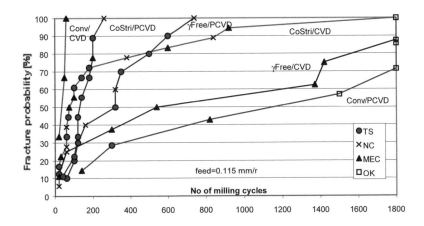

Fig. 4.62 Tool life results at s = 0.115 mm/rev.

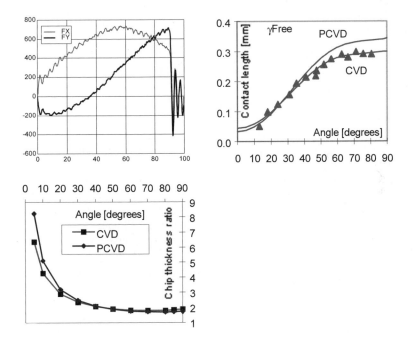

Fig. 4.63 Example of cutting forces, contact length and chip deformation of CVD and PCVD coated γFree.

CVD coating gives a shorter contact length on the tool with chips that are thicker with a smaller radius and a clearer foot formation. The friction coefficient *(F/N)* during cutting is also ~10% higher. The CVD coating is thicker than the PCVD and the edge radius is a little bigger. It is probable that the residual stresses are higher and a number of cracks are present in the coating prior to cutting while no such cracks have been detected in the PCVD coating.

Simulations

Macroscopic simulations

The milling process is simulated numerically with a model based on the Finite Element Method. The model can be used for calculation of macroscopic too stresses due to mechanical and thermal load. The simulations are performed in two separate steps, an initial step to determine the temperature [19] and the subsequent step to calculate the stresses [20, 21]. Process data and tool loads are used as input data. The tool loads are applied as boundary conditions on the tool surfaces and as long as the load distribution can be determined by measurements and simple calculations it is not necessary to simulate the mechanical flow of the work material. The heat transport due to work material flow on the other hand is taken into account. Ten load distributions corresponding to different parts of a milling cycle from tool entry (0°) to tool exit (90°) were entered. Experimental data obtained in the instrumented milling tests are used in all steps except the last one. The tool loads at 89.6° cutter rotation are obtained from a separate numerical simulation of plastic deformation of the chip during the final part of the milling cycle [22]. The loads are ramped down to zero at 90°.

In the temperature calculation heat generation at the primary and secondary deformation zone as well as at the flank surface and heat transport due to work material flow was taken into account. The chip thickness variation is modelled using a mesh corresponding to the maximum chip thickness but only activating those elements which correspond to the present chip thickness.

Both the temperatures and the stresses are cyclic with the largest values close to tool exit. Figure 4.64 shows the calculated temperature distribution in the tool of grade γFree, PCVD coated, just before tool exit. Both the major principal stress and the major shear stress show their highest values at 89.6° cutter rotation angle but at different locations. The level of the largest tensile stresses is about 400 MPa. The largest shear stress is above 3000 MPa and occurs at the position where the highest temperature appears. High shear stresses may cause plastic deformation and possibly shear fracture.

Similar results as those shown above were obtained for γFree tool with the CVD coating. The maximum shear stress level was only slightly higher for the PCVD coated grade. In figure 4.65 the complete cycles of temperature and Tresca stress are shown for both variants at the location where the shear stress has its highest value.

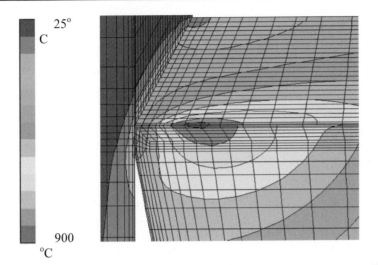

Fig. 4.64 Calculated temperature distribution (OC) in the milling tool, grade γFree, PCVD coating, and the chip at the end of the first cycle (89.6°).

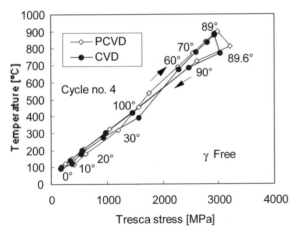

Fig. 4.65 Calculated temperature/Tresca stress cycle for tool grade γFree with CVD and PCVD coating. Evaluated for the point with the maximum temperature.

Mesoscopic simulations

The influence of graded zones on tool failure is investigated by mesoscopically modelling a cut-out of 0.2 x 0.2 mm^2 applying realistic loading conditions from the macroscopic calculation, figure 4.66. The loading conditions are simplified as follows: the temperature distribution is assumed in a radial gradient, maximum temperature and gradient values vary during the milling cycle according to the macroscopic simulations. The mechanical loading is applied by prescribed and suppressed displacements as well as by constraints at the boundaries.Fracture is limited to the defined crack path in the middle of the model. The fracture criterion is based on K_{1c}-theory, whereby different K_{1c} values are chosen for the different materials in the graded zones and temperature dependencies are taken into account.

Different graded zones are studied by comparing the crack resistance [23] of the grades applying Mode I loading. Different coatings are considered and residual stresses appearing from the CVD coating process are taken into account.

Figure 4.67 shows a comparison of grades where the coating was chosen to have a Young's modulus of 400 GPa and a thermal expansion coefficient of 9.3E-6 K^{-1}. After a temperature change of -400 K, initial cracks in the coating appear. Additional mode I loading leads to failure of the model at a global strain of 0.11%. The highest crack resistance was found in grade CoStri while grade Conv and grade γFree show no remarkable differences. In figure 4.68, the results are shown for the same material data, but lower residual stresses; the temperature change in these calculations was only -300 K. All grades show an increased crack resistance, still, grade CoStri is the best grade, but the variation between the grades are less significant. The results shown in figure 4.69 are obtained using a higher Young's modulus of 570 GPa and again a temperature change of -400 K. This combination results in high residual stresses, especially in the coating layer. As a consequence, the crack resistance and the strain of failure decrease and the grades distinguish more remarkable compared to the calculations with lower residual stresses.

Thus, residual stresses are most important for the crack driving force: low residual tensile stresses in the coating result in high crack resistance. The influence of grades on failure is significant in the case of large thermal stresses: a soft graded zone, like grade CoStri may stop cracks from the coating layer.

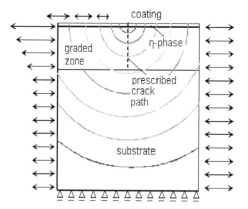

Fig. 4.66 Mesoscopic model (schematic).

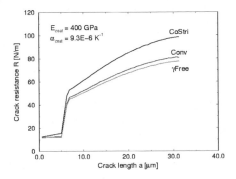

Fig. 4.67 $\Delta T = -400$ K, Mode I, $\varepsilon^{global} = 0.11$ %

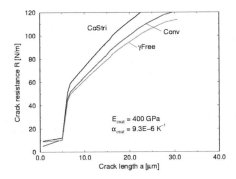

Fig. 4.68 $\Delta T = -300$ K, Mode I. $\varepsilon^{global} = 0.13$ %.

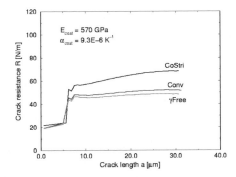

Fig. 4.69 $\Delta T = -400$ K, Mode I. $\varepsilon^{global} = 0.07$ %.

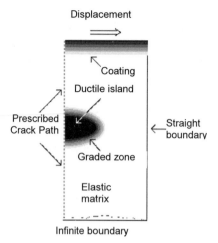

Fig. 4.70 Mesoscpoic model for crack propagation through Co islands.

Especially relative large Co islands can retard crack propagation. The propagation of a crack through a ductile Co-island surrounded by a graded layer, figure 4.60, is therefore simulated as well. The Co-rich islands are elliptic in shape. In the surrounding graded layer, the mechanical properties vary smoothly from the matrix behaviour to the Co island.Deformation theory of plasticity is employed for the inclusion material and the graded interface. The crack propagation is simulated using a cohesive surface model [24]. It allows the evolution of crack initiation from the free surface or from an initial crack, crack growth and crack arrest to be described.

The criterion for decohesion is controlled by the normal traction transmitted through the cohesive surface. Decohesion under mode I loading may occur if the normal traction reaches a critical value, T_n^{crit} the cohesive strength. The variation in properties in the graded zone implies a variation in the cohesive strength. The cohesive strength for the matrix material and parts of the graded zone were determined from the transverse rupture strength of hardmetals [24]. For Co volume fractions more than 40% estimated cohesive strengths for hardmetals vs. volume fraction of Co are shown in figure 4.71. For high binder volume fractions a parametric study was performed to determine the values for which the crack propagates in a stable manner (lower dashed line) or stops due to blunting (upper dashed line) within the island.

Simulations are performed with both criteria. For the lower values of the cohesive strength, the crack propagates in a nearly brittle manner, just a small retardation is observed within the Co island. For values between the two dashed curves the crack propagation is stable. For values higher than the upper dashed line, crack arrest is observed within the Co island. Further on, the presence of a graded coating layer on the top of the free surface was assumed and residual stresses appearing from cooling after coating deposition were taken into account. In all cases, after initiation the crack propagates brittle until the crack tip reaches the graded zone. For this reason, a sharp peak in the force displacement curves is observed, figure 4.72. The residual stresses facilitate crack initiation, resulting in a significant decrease of the peak loading value compared with the case without residual stresses.

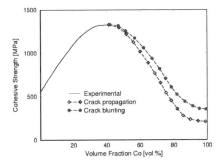 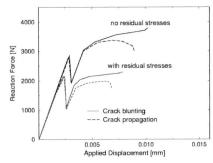

Fig. 4.71 Cohesive strength vs. volume fraction of Co.

Fig. 4.72 Influence of the residual stresses on the crack propagation.

With further loading increase, a retarded crack propagation through the graded zone in the Co island is observed. After some amount of crack propagation a plateau in the force displacement curve is attained. Using the low cohesive strength values, crack propagation is accelerated by residual stresses; the plateau is shorter, showing a smaller amount of applied loading is needed for crack propagation.

Conclusions

Instrumented milling tests were performed and tool life was studied for three different milling grades, where two contain a gradient structure underneath the coating. One of these grades, γFree, has a y-phase depleted zone while in grade CoStri cobalt striations are included. Two coatings were considered, a CVD and a plasma assistant CVD coating. Initial cracks were detected in the CVD coating layer. Clear differences were detected in tool life of the gradient variants γFree and CoStri. The type of coating has strong influence on chipping at the edge. It also effect contact length, chip thickness ration, chip curl, edge radius and friction coefficient.

Numerical simulations of milling have been carried out for the tools to predict tool temperatures and stresses. The largest stresses in up-milling appear just after tool exit due to thermal stresses or just before tool exit due to reversed chip flow. Mesoscopic simulations were set up where the critical region of the tool has been mode lied. The results from the calculations are comparable to the experimental results. The crack resistance curves show the important influence of residual stresses due to the coating procedure. In presence of initial cracks a soft gradient zone may retard crack propagation, in large cobalt islands, crack blunting and arrest may occur.

References

[1] Kolaska H. and Dreyer K (1992), DGM-Fortbildungsseminar Werkstoffgefüge und Zerspanung, Hannover.

[2] Chandrasekaran H (1981), in: Science of Hard Materials, eds. Viswandaham, Rowcliffe and Gurland.

[3] König W, Gerschweiler K and Fritsch R (1992), in: Proc. Symp. Beschichten und Verbinden in Pulvermetallurgie und Keramik, Hagen.

[4] Quinto D.T., Santhanam A.T. and Jindal P.C. (1988), Materials Sci. and Eng. A105/106, pp. 443-452.

[5] Nordgren A. (1991), Report IM-2812, Swedish Institute of Metal Research, Stockholm.

[6] Nordgren, A. and Thuvander A. (1990), Report iM-2628, Swedish Institute of Metal Research, Stockholm

[7] Nordgren A. (1990), Report no IM-2573, Swedish institute of Metal Research, Stockholm

[8] Schwarzkopf M., Exner H.B., Fischmeister H.F. and Schintlmeister W. (1988), Materials Sci. and Eng A105/106, pp. 225-231.

[9] Gustafson P. and Östlund A. (1994), Refract. Met. & Hard Materials 12, pp. 129-136.

[10] Gustafson P. and Akesson L. (1996), Materials Sei. and Eng A209, pp. 192-196.

[11] Skogsmo J. and Norden H. (1992), Refract. Met. & Hard Materials 11, pp. 129-136.

[12] Schedler W. (1988), Hartmetall für den Praktiker, VDI-Verlag Düsseldorf.

[13] Doi H, Fujiwara Y, Miayke K and Osawa Y. (1970), Metall. Trans. 1.

[14] Murray MJ. (1977), Froc. Roy. Soc. London 356, pp. 483-508.

[15] Ravichandran K.S. (1994), Acta. Metall. Mater. 42(1), pp. 143-150.

[16] Sigl L.S. (1986): VDI-Fortschrittsberichte 5, No. 104, VDI-Verlag, Düsseldorf.

[17] Porat R. and J. Malek J. (1988), Materials Sci. and Eng A105/106, pp. 289-292.

[18] Bouaouadja N., Hamidouche M. and Fantozzi G. (1994), Rev.lnt. Hautes Temper. Refraa 29, pp. 115-121

[19] Thuvander A., Nordgren A. and Chandrasekaran H. (1995), Report IM-3244, Swedish Institute for Metal Research.

[20] Chandrasekaran H., Thuvander A. and Wisell H. (1985), Report IM-2194, Swedish Institute for Metal Research, Stockholm.

[21] Thuvander A, and Chandrasekaran H (1986), Report M-2185, Swedish Institute for Metal Research, Stockholm.

[22] Thuvander A and Thoors H (1986), Report IM-2185, Swedish Institute for Metal Research, Stockholm.

[23] Rohde J., Bao G. and Schmauder S. (1996), Comp. Mat. Sci. 7, pp. 63-67.

[24] Mintchev Or., Rohde J. and Schmauder S. (1998), Comp. Mat. Sci. 13, pp. 81-89.

[25] Spiegler R., Schmauder S., and Exner H.E. (1992), Journal of Hard Materials 3, pp. 143-151.

[26] Roebuck B., Ahnond E.A., and Cottenden A.M. (1984), Materials Sci. and Eng. A66, pp. 179-194.

[27] Quinn D.F et al. (1997), Int. J. Mech. Sci. 39, No. 2, pp. 173-183.

[28] O'Regan T.Let al. (1997), Camp. Mechanics 20, pp. 115-121.

[29] Gurson A.L. (1977), Journal of Engineering Materials Technolagy, 99, pp. 2-15.

[30] Rice J. and Tracey D.M (1969), J. Mech. Phys. Solids 17, pp. 201-217.

[31] Hancock J.W. and Mackenzie A.C. (1976), J. Mech. Phys. Solids 24, pp. 147-169.

[32] Wulf J. (1995), VDI-Fortschrittsberichte 18, No. 173, VDI-Verlag, Düsseldorf.

[33] Arndt J., Majedi H. and Dahl W. (1996), J de Physique IV, C6, pp. 23-32.

[34] Hönle S. and Schmauder S. (1998), Comp. Mat. Sci. 13, pp. 56-60.

Chapter 5: Atomistic and Dislocation Modelling

In this Chapter, atomistic and dislocation methods of modelling the material behaviour and damage initiation and growth are discussed.

In section 5.1, an empirical interatomic potential of the embedded atom type is
developed for the Fe-Cu system. The potential for the alloy system was constructed to reproduce known physical parameters of the alloy, such as the heat of
solution of Cu in Fe and the binding energy of a vacancy and a Cu atom in the α-
Fe matrix. The potential also reproduces first-principle calculations of the properties of metastable phases in the system. This atomic interaction model was used in
simulation studies of the interface of small coherent Cu precipitates in α-Fe and of
dislocation core structures.

In section 5.2, atomistic molecular statics simulations are employed in order to
obtain the stress–strain curves of α-Fe single crystals under uniaxial tensile deformation for the whole deformation process. Effects of model sizes, boundary conditions, crystal orientations and displacement increment on the stress–strain curves
arey investigated. Various deformation evidence such as dislocation movement,
dislocation piling up and twinnings are clearly observed. The deformation and
fracture characteristics of α-Fe and their dependence on the boundary conditions
or the stress states are studied.

In section 5.3, the void growth in face centre cubic (fcc) single crystals on the
atomic level (10^{-9} m) is simulated. An atomistic model is used to capture the large
amount of irregularities such as vacancies and void-like vacancy clusters which
exist in even the purest real material. The embedded atom method (EAM) is used
to describe the atomic interaction and so the material behaviour in a realistic manner. The potential energy of the entire system is minimised to find the equilibrium
configuration of any deformed state. Stress-strain curves are plotted for various
initial void configurations and crystal orientations under uniaxial tension. The resulting curves show good qualitative agreement with that expected from experiment but predict stress levels close to the theoretical strength of the material.
Crack propagation and dislocation glide are seen to occur along the theoretically
predicted directions. Comparison is made with crystal plasticity calculations for
similar geometries.

Molecular dynamics (MD) calculations are employed to investigate the interaction between a moving edge dislocation in an α-Fe crystal and a copper precipitate
in section 5.4. In the absence of external stresses, two edge dislocations with the
same slip plane and opposite Burgers vectors within a perfect α-Fe crystal lattice
are investigated. In agreement with Frank's rule, the movement of the dislocations
under mutual attraction is found and attention is focused on the interaction between

one of the dislocations and the Cu precipitate. The critical resolved shear stress of the Fe was calculated and the influence of different sizes of Cu precipitates on the dislocation mobility was studied. The pinning of the dislocation line at the Cu inclusion as derived from the atomistic modelling agrees with previously published continuum theoretical behaviour of pinned dislocations. Therefore, nanosimulation as a way to model precipitation hardening could be established as a useful scientific tool.

Molecular dynamics and micromechanics are combined in section 5.5 to study the low temperature fracture of tungsten. In the simulations a pre-crack was introduced on the (110) planes and cleavage was observed along the (121) planes. Cleavage along (121) planes has also been observed in experiments. Simulations were performed with three sizes of molecular dynamic regions at 77 K, and it is found that the results are independent of the size. Brittle fracture processes are simulated at temperatures between 77 K and 225 K with the combined model. The fracture toughness obtained in the simulations showed clear temperature dependency, although the values showed poor agreement with experimental results. A brittle fracture process at 77 K is discussed considering driving dislocation emissions and cleavage in an atomic scale region of the crack tip. The driving dislocation emissions is saturated after the first dislocation emission, whilst the driving force for cleavage gradually increases with the loading K-field. The increased driving force causes cleavage when it reaches a critical value. The critical values of driving force, which are close to the theoretical strength of the materials, were not influenced by temperature. This indicates that the temperature dependency of fracture toughness is not caused by the temperature dependency of dislocation emissions, but by that of dislocation mobility.

In section 5.6, atomistic computer simulations of the formation of precipitates are used to get a deeper understanding of the mechanical behaviour of Cu-alloyed steels. A model is presented which is able to simulate the 'diffusion' of atoms by vacancy jumps. The underlying Monte Carlo method is presented and a binary system with components A and B is considered. Starting with a random distribution of atoms, the formation and growth of precipitates is simulated at a constant temperature of 600 °C. In a second simulation, an initial temperature of 700 °C is lowered to 400 °C. At 700 °C precipitates of radii between 1.1 and 1.7 nm are formed within seconds. At 400 °C a part of the still dissolved atoms forms smaller precipitates while other atoms increase the size of the larger precipitates. At longer simulation times a significant decrease of the number of small precipitates and an increase of the averaged precipitate radius is found.

In section 5.7, classical molecular dynamics simulations of the interaction of edge dislocations in Ni with chains of spherical Ni_3Al precipitates are performed using EAM potentials. The order hardening is investigated at temperature T = 0K by determining the critical resolved shear stresses (CRSSs) for a superdislocation that is dissociated into four partial dislocations. The CRSS is computed as a function of the radius and the distance of the precipitates. It is found that for precipitates with a diameter smaller than the dissociation width of a perfect edge dislocation in Ni, the CRSS of the trailing dislocation of the superdislocation is a fraction of about 0.4 of the CRSS of the leading dislocation.

5.1 Embedded atom potential for Fe-Cu interactions[1]

Atomistic studies of the structures of dislocation cores, grain boundaries, point defects, and cracks in Cu-containing ferritic steels are of great relevance to the understanding of the irradiation embrittlement of pressure vessels and the alloying behaviour in these materials. A large increase in hardness results both from thermal ageing and from irradiation of steels with very low Cu content, and experimental evidence strongly relates this property change to the occurrence of initially small body-centred cubic (bcc) Cu precipitates (~1 to 2 nm in diameter). Upon further ageing, the precipitates grow with a spherical shape to a mean diameter of about 6 nm, when they become unstable and transform apparently in a martensitic mode to a complex presumably 9R [1] structure. Additional growth leads to the stable face-centred cubic (fcc) phase. Although the mechanisms of precipitation are not well understood, it is likely that vacancies - produced profusely by irradiation - are substantially involved [2].

Atomistic simulations require the development of interatomic potentials for the alloy system. Osetsky and Serra developed a pair potential for the Fe-Cu interactions [2] in terms of the generalized pseudopotential theory, but pair potentials are not suitable to correctly calculate surface energies or fracture properties. Ackland and co-workers recently developed a potential that includes many-body terms and they have calculated point defect properties for dilute solutions of Cu in Fe [3]. The recent work of Osetsky and Serra [2] using pair potentials and molecular dynamics underlines the great importance of the precipitate-matrix interface in the phase stability of small Cu precipitates in Fe.

The objective of the present work is to develop a new interatomic potential that will allow one to perform atomistic simulations of coherent Cu-precipitates in α-iron using the embedded atom method (EAM) technique, and to gain insight into the nature and the energies of the precipitate-matrix interface.

A detailed description of the embedded-atom method can be found in [4]. The total energy E_{tot} of a system of atoms is expressed as

$$E_{tot} = \frac{1}{2} \sum_{i,j(i \neq j)} v^{ij}(r_{ij}) + \sum_i F^i(\bar{\rho}_i) \qquad \bar{\rho}_i = \sum_{j(\neq i)} \rho^j(r_{ij}) \qquad (5.1)$$

where the superscripts i and j indicate atom types, and the subscripts indicate the atoms themselves. The F^i are the embedding functions, ρ_i is the total host electron density of atom i which consists of contributions $\rho^j(r_{ij})$ from atoms j that are at the interatomic distances r_{ij} from atom i, and V^{ij} are the pair potentials between atoms i and j.

[1] Reprinted from M. Ludwig, D. Farkas, D. Pedraza, S. Schmauder, "Embedded Atom Potential for Fe-Cu Interactions and Simulations of Precipitate-Matrix Interfaces", Modelling and Simulation in Materials Science and Engineering 6, pp. 19-28 (1998) with kind permission from Elsevier.

Minimization of the total energy E_{tot} using for example a conjugate gradient scheme for a given initial atomic configuration, which may be one or several Cu atoms or even a Cu precipitate in α-iron, leads to a relaxed stable configuration of the atom ensemble. In the derivation of the potentials we have used a scheme similar to that used in previous work on other alloy systems [5].

5.1.1 Interatomic potentials for the pure components

For α-Fe we used the interatomic pair potential given by Simonelli et al. [6] and for Cu we used the interatomic potential given by Voter [7]. Both potentials fit Roses' equation of state [8] and other element properties including lattice constant, cohesive energy, elastic constants and vacancy formation energy. In the present computation, the first derivative of the embedding function (F'^{Fe} or F'^{Cu}) was set equal to zero for the electron density of the perfect lattice (ρ_0). In such a case, the potentials are said to be in their effective scheme [9]. The interatomic potentials for the pure metals were not originally in the effective pair scheme. Thus, we first transformed them into this scheme. The transformation used to convert to the effective pair scheme is as follows [5]

$$
\begin{aligned}
F_{eff}(\rho) &= F(\rho) - \rho F'(\overline{\rho_0}) \\
V_{eff}(r) &= V(r) + 2\rho(r)F'(\overline{\rho_0})
\end{aligned}
\tag{5.2}
$$

where F is the embedding function (F^{Fe} and F^{Cu}, respectively), V the pair potential function (V^{FeFe} and V^{CuCu}, respectively), ρ the electron density function, $\overline{\rho_0}$ the electron density in the perfect lattice (bcc and fcc structure, respectively) of the pure crystal and r is the interatomic distance. The values for the electron density of the perfect lattice of bcc Fe and fcc Cu were normalized to the same value of 0.34.

Properties of bcc Cu

The Voter Cu potential has been developed based on the fcc phase and in this section we summarize the behaviour of the bcc phase that it predicts. The lattice parameter of the bcc Cu phase is calculated as 0.2880 nm. This parameter is the same as calculated by Osetsky and Serra [10] (0.2885 nm), and in excellent agreement with first-principle calculations [11] that give a parameter of 0.2873 nm, but it is significantly lower than that value of 0.296 nm predicted by the potential used by Ackland et al. [12]. These values can be compared with the lattice parameter of bcc Fe, 0.2867 nm. As a result, in our simulations the Cu precipitates are subject to much lower stresses than predicted by Ackland et al. [3]. The strain of the precipitates in the present work is 0.45%, whereas with the potentials of

Ackland *et al.* it is 3.2%. In our calculations, the energy of the strained lattice is 1 meV/atom higher than that of equilibrium. The difference in energy between the fcc and bcc phases given by the present potential is 46 meV/atom.

5.1.2 Results for the Fe-Cu interaction

The data used in the development of the potentials are as follows.
- The vacancy-Cu atom binding energy in the α-Fe matrix, $E^b_{V\text{-}Cu} = 014\text{eV}$. This energy was obtained using muon spin rotation experiments on dilute Fe alloys [12].
- The energy of solution of one Cu atom in the α-Fe matrix $E_{cu} = 1.23$ eV.

This energy was obtained using the partial molar energy for liquid alloys as 1823 K, $\Delta E_{Cu} = 0.49 eV/atom$, as given in [13]. The energy of one Cu atom in the α-Fe matrix E_{Cu} can be calculated with the partial molar energy ΔE_{Cu} as

$$E_{Cu} = E_{Fe}(bcc) - E_{Cu}(fcc) + \Delta E_{Cu} \tag{5.3}$$

The cohesive energies $E_{Fe.}(bcc)$ and $E_{Cu.}(fcc)$ are 4.28 eV [6] and 3.54 eV [7], respectively. (3) The kinetic binding energy between two Cu atoms in the α-Fe matrix defined by Osetsky and Serra [2] as

$$E^b_k(2) = E^b_{2Cu} - E^b_{V-Cu} \tag{5.4}$$

should not exceed 0.05 eV to allow small Cu clusters in Fe to dissociate thermally. E^b_{2Cu} is the binding energy of two Cu atoms and E^b_{V-Cu} has been defined earlier. This condition is consistent with a mechanism of precipitate growth and dissolution mediated by vacancies.

The mixed-pair interaction potential was obtained by empirically fitting these experimental data to the calculated values obtained by a combination of the effective-pair interactions of Fe and Cu. The general form for the combination used is as follows:

$$V^{FeCu}_{eff}(a+bx) = A[V^{Fe}_{eff}(c+dx) + V^{Cu}_{eff}(e+fx) \tag{5.5}$$

where x takes values from zero to unity. The parameters a, b, c, d, e, f and A were adjusted to give good overall fit to the properties considered. The potential V^{FeCu}_{eff} must be calculated in an interatomic distance interval a, a + bx that must include the interval used in the numerical simulations, and this is the only limitation to the values of a and b. As used in equation (5.4), this procedure entails keeping only the functional form of V^{Fe}_{eff} and V^{Cu}_{eff} similar to the interaction potential that applies between atoms of the same species. Thus, changing c and d (and e and f) from the values that apply to the pair interaction only implies a homogeneous expansion (or contraction) in the interatomic distances. Here, the values of c, d, e and f were chosen to yield no change in the intervals c, $c+d$ and e, $e+d$ used for the pair potentials. The other three parameters of equation (5.5) were systematically varied and their effects on the properties of interest were studied to determine the most suitable set of values to fit the entire system properties mentioned above. Varying A affects mainly the cohesive energy of the lattice, while a and b have a stronger effect upon the lattice parameter of mixed ordered phases.

Table 5.1
Parameters obtained for the mixed interaction potential. Distances are in 10^{-1} nm

a	b	c	d	e	F	A
1.000	4.700	0.992	4.095	1.000	4.961	0.5

Table 5.2
Comparison of adjusted properties with experimental results. Energies are in eV.

	E^b_{V-Cu}	E_{cu}	$E^b_k(2)$	E^b_{2Cu}
Experimental value	0.14	1.23	0.05[a]	0.20
Calculated, present work	0.18	1.23	0.01	0.19
Calculated, Ackland et al. [3]	0.09		0.01	0.1

[a] Assumed value (see [2])

Table 5.3
Results for Ll$_2$ structures

	Lattice parameter (nm)	Cohesive energy (eV)	Formation energy (eV)
Fe$_e$Cu, LMTO	0.3546	3.9917	0.125
Fe$_3$Cu, EAM	0.3729	3.9125	0.183
Cu$_3$Fe. LMTO	0.3555	3.6168	0.102
Cu$_3$Fe, EAM	0.371 44	3.541 75	0.178

Best fit values obtained for the parameters are given in table 5.1. With these parameters we obtained the fitting result given in table 5.2. The calculated values match well with the experimental data.

As a check for the mixed pair potential, the cohesive energies and lattice parameters of two Ll$_2$ structures in the system were calculated and compared with available LMTO calculations [3]. These results are given in table 5.3, showing excellent agreement.

Calculation of interface energies

We used the potentials described above to calculate bcc Cu/α-Fe matrix interface energies. In all the cases, the energy used for the Cu atoms in the lattice referred to as perfect corresponded to the energy of the strained bcc Cu lattice. In this way, the interface energy obtained does not include the effects of the strain on the precipitate. First, we considered a spherical precipitate, 4 nm in diameter, in a simulation involving the relaxation of 16 000 atoms (the block comprised an outer shell of 24 000 fixed atoms). The interface energy in this morphology was obtained by calculating the difference in energy of the block containing the precipitate and a block of the same number of atoms in a perfect lattice condition. The energy obtained for the spherical case is 207 mJ m^{-2}. Figure 5.1 shows a cut of this precipitate through the maximum circle of the sphere in the (100) plane.

We also calculated the surface energy corresponding to a cylindrical interface with its axis along the [001] direction. In this instance, the 'infinite' cylindrical precipitate had a diameter of 4 nm. This surface energy was higher, at 245 mJ m^{-2}. As seen in figure 5.1, this interface is composed mostly of small patches of {100} and {110} interfaces. We, therefore, calculated the values for these two interfaces and found 318 mJ m^{-2} and 121 mJ m^{-2}, respectively. For these calculations, prismatic precipitates having a square cross section with a 4 nm side were considered. All these values are listed in table 5.4.

These results suggest that the preferred interface plane is the most compact plane in the bcc structure. They also show that the interface in the cube plane has a significantly higher energy.

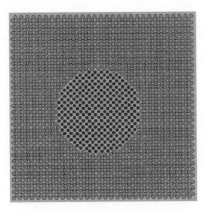

Fig 5.1 Cross section of a spherical bcc-Cu/α-Fe interface showing patches of {110} and {100} interface. Relaxed atomic positions display insignificant deviations from perfect bcc structure.

Table 5.4
Results for interface energies

Morphology	Energy (mJ m^{-2})
Spherical precipitate	207
Cylindrical precipitate along [001]	245
{100} type planes	318
{110} type planes	121

Dislocation cores in Fe-Cu alloys

We studied the dislocation cores in Fe-Cu alloys in two cases. The first case was a ½ <111> screw dislocation in an Fe matrix containing a random distribution of Cu atoms. In pure Fe, this dislocation is known to have a non-planar structure with a core spreading in three different {110} planes [14], according to a core structure simulation study that used the same Fe potential as we use in the present work. In order to study the possible effects of the Cu atoms on the core structure of the dislocation, a relatively large concentration of Cu atoms was included (16%). Figure 5.2 shows the dislocation core using the differential displacement map technique where the length of the arrows indicates the extent of the deviation of the interatomic distances from the perfect lattice positions. Similar studies were conducted previously for the effect of substitutional Cr atoms on the same dislocation [15]. The results show that the effects of Cu and Cr are similar.

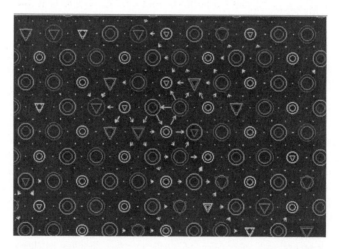

Fig 5.2 Projection on plane {111} of a (111) screw dislocation core structure in an alloy containing a random distribution of Cu substitutional atoms. Differential displacement map with arrows showing extent of atomic displacements. Circles represent iron atoms and triangles represent Cu atoms.

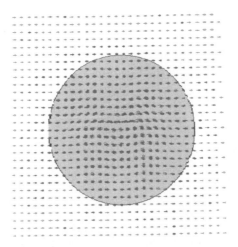

Fig 5.3 A mixed-type dislocation running across a bcc Cu precipitate and α-Fe matrix. Dislocation line normal to view on the {001} plane.

The dislocation core maintains the basic three-plane structure but there are deviations that can be observed at a few interatomic distances from the dislocation centre. The core appears to have a strong attractive interaction with the Cu atoms. This preference suggests that Cu atoms would tend to segregate to the dislocation core.

The second simulation involved a dislocation running through a spherical precipitate 4 nm in diameter. The calculation block included 16 000 free atoms. In this case, the dislocation is of mixed type, with a Burgers vector ½ <111> and the dislocation line along a cube direction. The results of the simulation are shown in figure 5.3. Since this dislocation has a large edge component, it is shown in a direct plot of atomic coordinates. The figure shows the projection of the atomic coordinates of all atoms of the block on a plane perpendicular to the dislocation line. The simulation was performed using periodic boundary conditions along the dislocation line, with a periodicity of about 30 lattice parameters so that the dislocation was free to bow out of the precipitate if this resulted in a lower energy configuration. However, as can be seen in figure 5.3, no bowing occurred and the dislocation inside the precipitate does not display a structure significantly different from the structure seen in the Fe matrix outside the precipitate.

Instability of the bcc precipitates

Initiation of the transformation to an fcc structure (or to another more complex phase) is expected to occur as the size of the precipitate increases. Although the 4 nm diameter spherical precipitate simulated above developed compressive strains, the Cu atoms did not exhibit any deviations from perfect bcc lattice positions.

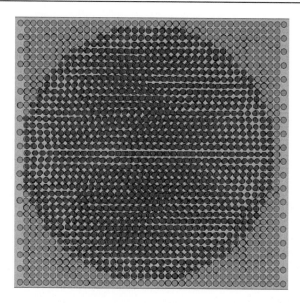

Fig 5.4 Projection on plane {001} of the relaxed positions of a 7.5 nm diameter spherical precipitate.

We then changed the size of the precipitate in order to study its stability against a phase transformation. As seen in figure 5.1, in the projection of relaxed atomic coordinates on a cube plane for a spherical precipitate 4 nm in diameter, the atomic coordinates for the precipitate show that the relaxed atomic positions do not deviate significantly from the perfect bcc lattice positions. This result is another proof that the bcc structure in the Cu precipitate is fully stabilized by the presence of the interface. A simulation performed with a similar precipitate containing a vacancy concentration of 3% did not show any change either. For a precipitate that is 6 nm in diameter substantial distortions from the bcc structure start appearing in the centre of the precipitate. These deviations are particularly important in one out of every six compact planes. This suggests that the incipient transformation might be accompanied by a shear and produces a daughter phase bearing a definite orientation relationship with the parent bcc phase. The precipitate size at which this instability is observed agrees very well with the precipitate size where transformation of the bcc Cu phase is observed experimentally [1]. In this simulation, no vacancies were incorporated, therefore implying that the transition is not vacancy-assisted but is a result of phase instability alone. Further changes occur for a 7.5 nm precipitate, as illustrated in figure 5.4, where a slab of the sphere can be viewed in the (001) plane.

Summary and discussion

A many-body potential for the Fe-Cu system was developed in the framework of the embedded atom approach using a very simple model. The potential was based upon the same-atom interaction potentials developed by Voter [7] for pure Cu and by Simonelli et al. [6] for α-Fe. The procedure that was followed for the mixed-pair interaction potential used three properties of Cu in Fe, viz, the vacancy-solute binding energy, the heat of solution into liquid Fe and the binding energy of a Cu dimer in the ferritic matrix. Although the latter does not come from experimental measurement, it is a very reasonable assumption based on a very likely vacancy mediated precipitate nucleation and growth mechanism, strongly suggested by the two known precipitation conditions, i.e. irradiation with energetic particles and thermal ageing.

The small Cu precipitates have been well characterized as having a bcc structure. Our calculation of the lattice parameter and energy of this lattice using Voter's potential yielded 0.288 nm for the former, the same value obtained by Osetsky and Serra [2] using a pair potential approach. Results of LMTO calculations conducted by Ackland *et al.* [3] allowed for a comparison with two hypothetical intermetallic phases in the L12 structure, Fe$_3$Cu and Cu$_3$Fe, with excellent agreement, rendering additional reliability to the potential developed here. The cohesive energy of bcc Cu yielded 3.494 eV, very close to the value of 3.496 eV obtained by Ackland *et al.* The cohesive energy favours the fcc over the bcc phase by 46 meV, this difference being larger than Osetsky and Serra's reported value of 37 meV by 24% [2]. The results obtained here are thus seen to be in fair agreement with those obtained by other authors. The small lattice misfit with α-Fe requires a relatively low strain energy to immerse the spherical precipitate with a coherent interface into the α-Fe matrix.

Simulations of some low-index coherent interfaces between the bcc Fe matrix and the bcc Cu phase performed here yielded the fact that the more compact planes are favoured. The magnitude of the interface energy for the spherical precipitate is of the same order as that of a coherent fcc/bcc interface in pure Fe as calculated for the $(1\bar{2}1)\,\|\,(3\bar{1}2)$ coherent interface, 178 mJ m^{-2}. Comparison with experimental observations, however, is not straight-forward since the spherical precipitates have been observed in ferritic steels where the presence of other alloying elements may stabilize a spherical interface rather than a faceted one [1].

Our calculation of the of an edge dislocation in a Cu-containing Fe matrix showed a tendency of the Cu atoms to segregate to the core, as could be expected from the strong repulsive interaction between Cu and Fe. A similar calculation for a mixed dislocation running through the precipitate yielded no difference in the dislocation core structure between the two regions, thus indicating that such a dislocation has the same features as in pure α-Fe.

Our studies of phase stability showed onset of a phase transformation in the precipitate at the size of 6 nm. It is worth underlining that no vacancy presence was required for such instability. Our results are at variance with those of Osetsky and Serra who performed molecular dynamics simulations of the phase instability of spherical bcc Cu precipitate. In their studies a 6% vacancy concentration was required to induce structural changes, adding a diffusional contribution to the transformation mechanism [10]. In our calculations, such a contribution does not appear as necessary at the above-mentioned size. Adding vacancies at a 3% level in a smaller precipitate did not prompt any transformation in our static calculations. The relaxation of a 7.5 nm precipitate shows additional displacements as compared to the 6 nm precipitate, clearly revealing the size effect upon the instability of the bcc phase.

In summary, the results obtained in this work indicate that the potential developed here is reliable to conduct further simulations on both point defect and extended defect behaviour in the Fe-Cu system.

References

[1]   Othens P J, Jenkins M L and Smith G D W (1994), Phil. Mag. A 70, p. 1.
[2]   Osetsky Y N and Serra A (1996), Phil. Mag. A 73, p. 249.
[3]   Ackland G J, Bacon D J, Calder A F and Harry T (1997), Phil. Mag. 75, p. 713.
[4]   Daw M S and Baskes M I (1984) Phys. Rev. B 29, p. 6443.
[5]   Farkas D (1994) Modelling Simulation Mater. Sci. Eng. 2, p. 975.
[6]   Simonelli G, Pasionot R and Savino E J (1993), Mater Res. Soc. Proc. 291, p. 567.
[7]   Voter A F (1993), Los Alamos Unclassified Technical Report 93-3901 Los Alamos National Laboratory.
[8]   Rose J H, Smith J R, Guinea F and Ferrante J (1984), Phys. Rev. B 29, p. 2963.
[9]   Johnson R A (1990), Implication of the EAM format Many atom interaction in solids (Springer Proceedings in Physics 48) (Berlin: Springer) p 85.
[10] Osetsky Y N and Serra A (1997), Phil. Mag. 75, p. 1097.
[11] Kraft T, Marcus P M, Methfessel M and Scheffler M T (1993): Phys. Rev. B 48, p. 5886.
[12] Brauer G and Popp K (1987), Phys. Status Solidi a 102, p. 79.
[13] Hultgren R, Desai P D, Hawkins D T, Gleiser M and Kelley K K (1973), Selected Values of the ThermodynamicProperties of Binary Alloys (Metals Park, OH: ASM).
[14] Farkas D and Rodriguez P L (1994), Scripta Met. 30, p. 921.
[15] Farkas D, Schon C G, deLima M S F and Goldenstein H (1996), Acta Met. 44, p. 409; Chen J K, Farkas D and Reynolds W T Jr (1994), Solid–State Phase Transformations ed W C Johnson et al (Warrendale, PA: The Minerals, Metals and Materials Society) p. 1097.

5.2 Atomistic simulations of deformation and fracture of α-Fe[2]

With molecular dynamic techniques, a number of atomistic simulations on fracture processes of quasi-static cracks [1–6], dynamic cracks [7–9], and the effects of temperature and crack orientations on fracture properties [10–12] have been performed for α-Fe single crystals in the recent past. Among their most important results there was evidence for the nucleation and emission of dislocations at crack tips and mechanisms of crack propagation under cracktip stress field. However, such studies of damage mechanisms have been performed only for one crystal orientation and only starting from an already existing macroscopic crack. Unfortunately, in all these studies only one crystal orientation was studied and attention paid typically to only one particular atomic configurational phenomenon. In addition, almost all of these studies started with an existing crack but very few results of calculations starting from a crack-free sample have ever been published (e.g. [3]). In order to provide more general insights into such phenomena, the aim of the present work has been to simulate the whole process of (i) elastic and (ii) plastic deformations, (iii) defect formation and developing and (iv) cracking under (v) different stress states and under (vi) different crystal orientations for a single crystal.

A comprehensive understanding of single-crystal deformation responses is one of the prerequisites for developing accurate micromechanical models for the description of deformation and fracture of polycrystals. In the present work, the deformation characteristics of α-Fe single crystals are investigated using a modern interatomic potential approach and with special emphasis on different stress states and on five typical different crystal orientations. The stress–strain curves and atomic configurations during the deformation process are presented.

The corresponding deformation mechanisms and their competition under different stress states can be detected thus providing a detailed overview on various damage mechanisms occuring for different crystal orientations.

5.2.1 Model and method

The specimen shown in figure 5.5 represents the main simulation cell in this work, where x, y, z are the global coordinates and $(li; mi; ni)$ $(i = 1;2;3)$ are the Miller indices of the crystal directions. The whole simulation cell consists of two parts. One part is referred to as the active zone in which the atoms move according to the interatomic potentials; another part is referred to as the boundary zone where the positions of the atoms are given by the prescribed boundary conditions. A periodic boundary condition is imposed along the z-direction to simulate plane strain

[2] Reprinted from S.Y. Hu, M. Ludwig, P. Kizler, S. Schmauder "Atomistic Simulations of Deformation and Fracture of α-Fe", Modelling and Simulation in Materials Science and Engineering 6, pp. 567-586 (1998). with kind permission from Elsevier

conditions. In the *x*-direction, two kinds of boundary conditions are considered, which either surface stress-free or periodic conditions are. The size of the active zone is indicated by *H*, *D* and *B*. The parameters relative to the specimens considered in this work are listed in table 5.5 in which a_0 is the lattice parameter. Five different kinds of models with different crystal orientations have been set up.

No is an abbreviation for an orientation in a crystal and was introduced in order to link the table to texts and figures later on. p and f denote periodic and free boundary conditions in the *x*-direction, respectively.

Figure 5.6 shows the *x–y*-planes of these specimens, with indices of the axes labelling the number of the crystal orientation. Both standard experimental works as well as several of the previous mentioned theoretical works have pointed out that the main cleavage planes of body-centred-cubic (bcc)-Fe are (100) and (110) and the main slip systems are {110}<111> and {112}<111>. For an overview on the role of particular planes in different crystals with respect to deformation, see, for example, [13]. In order to check and show the deformation mechanisms clearly, orientations 1 and 2 were chosen to check the cleavage planes and orientations 3–5 were chosen to check the slip systems. Nevertheless, it should be kept in mind that the orientation of the slip plane is a function of the orientation of the tensile axil and its activation depends also on the temperature and strain rate.

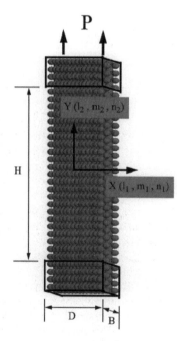

Fig 5.5 Model of an α-Fe single crystal under uniaxial tensile load.

For different model sizes, the number of the atoms in the active zone varies from 400 to 2400 and the total number of atoms in the models are about 6600 to 30 800. In order to simulate uniaxial tensile loading in the *y*-direction, the lower end of the specimen is fixed, and a constant displacement increment Δu_y is applied to all atoms in the upper end block.

Table 5.5

H/a_0	D/a_0	B/a_0	(l_1,m_1,n_1)			(l_2, m_2, n_2)			(l_3, m_3, n_3)			N_0	x-dir.	$\Delta u_y/H$	Figure
30	10	1	1	0	0	0	1	0	0	0	1	1	p	0.008	-
40	15	1	1	0	0	0	1	0	0	0	1	1	p	0.008	-
45	20	1	1	0	0	0	1	0	0	0	1	1	p	0.008	3
30	20	1	1	0	0	0	1	0	0	0	1	1	p	0.008	3,5,8
20	20	1	1	0	0	0	1	0	0	0	1	1	p	0.008	3
26	20	1	1	-1	0	1	1	0	0	0	1	2	p	0.008	5,10
26	20	$\sqrt{2}$	1	0	1	0	1	0	-1	0	1	3	p	0.008	5
26	20	$\sqrt{2}$	1	0	0	0	1	-1	0	1	1	4	p	0.008	5,13
26	20	$\sqrt{6}$	1	-1	0	1	1	0	-1	-1	2	5	p	0.008	5,15
20	20	1	1	0	0	0	1	0	0	0	1	1	f	0.008	4
40	20	1	1	0	0	0	1	0	0	0	1	1	f	0.008	4,6,9
50	20	1	1	0	0	0	1	0	0	0	1	1	f	0.008	4
60	20	1	1	0	0	0	1	0	0	0	1	1	f	0.008	4
20	10	1	1	0	0	0	1	0	0	0	1	1	f	0.008	-
30	10	1	1	0	0	0	1	0	0	0	1	1	f	0.008	-
36	20	1	1	-1	0	1	1	0	0	0	1	2	f	0.008	6,10
36	20	$\sqrt{2}$	1	0	1	0	1	0	-1	0	1	3	f	0.008	6,11
36	20	$\sqrt{2}$	1	0	0	0	1	-1	0	1	1	4	f	0.008	6,12
36	20	$\sqrt{6}$	1	-1	0	1	1	1	-1	-1	2	5	f	0.008	6,14
20	20	1	1	0	0	0	1	0	0	0	1	1	p	0.001	7
36	20	$\sqrt{2}$	1	0	1	0	1	0	-1	0	1	3	f	0.001	7
36	20	$\sqrt{2}$	1	0	0	0	1	-1	0	1	1	4	f	0.001	7

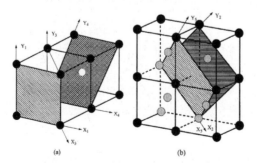

(a)   (b)

Fig 5.6 The X_i–Y_i ($i = 1, 2, ..., 5$) plane of five crystal orientations in α-Fe.

Then the atoms in the active zone are relaxed to an equilibrium configuration using a molecular statics method. In a next step, a displacement increment Δu_y is applied again and the process is repeated. In the present work, program FEAt developed by Kohlhoff *et al.* is employed. This program has already been applied successfully for the atomistic simulation of crack growth and nucleation of dislocations at the crack tip in [4–5], therefore it was an interesting challenge to use the same program to perform calculations starting from undamaged structures.

The interatomic potential of α-Fe used in the present work is the embedded-atommethod interaction potential (Simonelli *et al.* [14]). Equilibrium configurations are obtained by a static relaxation method, in other words, the temperature is chosen as $T = 0$ K. The descriptions about the relaxation process and the boundary conditions are omitted, because they are given in detail in [3–5]. For our typical model size, the computer time required to obtain a stress–strain curve as shown in figures 5.7 and 5.8 is about 5–10 h on a DEC Alphastation 600 5/266.

5.2.2 Results: stress-strain curves and fracture patterns

In the following, results for stress–strain curves are provided in section 3.1, and deformation and fracture patterns are presented in section 3.2.

Stress–strain curves

The strain ε_y can be calculated from the displacement increment applied in each step. When trying to obtain a stress–strain curve, the crucial point is how to calculate the stress σ_y. In the present simulation, the atoms in the upper and lower end blocks are assumed to remain in their perfect lattice positions during deformation. Therefore, there will be a zero-stress region in the middle of the end blocks, if the block size in the y-direction is large enough. A zero-stress region means that the force applied to every atom is zero. The forces acting on the other atoms in the end blocks, however, do not vanish because of the effect of the end surface or the atom position change in the active zone. In an equilibrium configuration, the resulting force F_y on all atoms between the zero-stress middle region of the end block and the active zone stems from the interaction between the end block and the active zone.

Therefore, the stress σ_y can be calculated as F_y/S, where S is the initial cross section of the specimen in the y direction. For all simulations of the present work, the size of the end blocks in the y direction is not less than $5a_0$ to ensure the existence of a zero-stress region in the middle of the end block.

In the following sections 3.1.1, 3.1.2 and 3.1.3, effects of model sizes, crystal orientations, boundary conditions and displacement increments on stress–strain curves are discussed, respectively.

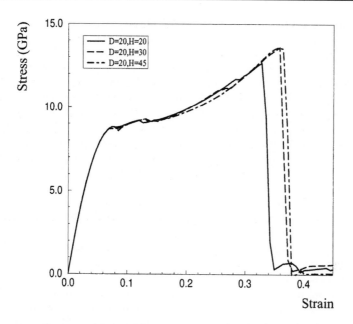

Fig 5.7 Stress–strain curves for crystal orientation No 1, periodic boundary conditions and different model sizes.

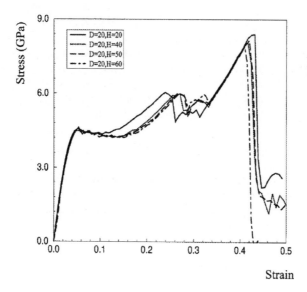

Fig 5.8 Stress–strain curves for crystal orientation No 1, free boundary conditions and different model sizes.

Stress–strain curves and effect of model sizes

The stress–strain curves for the whole deformation process for crystal orientation No 1 including elastic deformation, plastic deformation and fracture were obtained, see figures 5.7 and 5.8. The almost linear onset of the curves up to a strain of approximately 0.05 corresponds to the elastic crystal response whereas for higher strains plastic deformation due to irreversible damage takes place. When strain exceeds a certain level, damage in the structures grows until rupture occurs around a strain of 0.4. In the case of free boundary conditions which is less realistic for bulk materials than for thin metal films, the model has more freedom to allow slipping of planes and other damage mechanisms than in the case of periodic boundary conditions. Therefore, for periodic boundary conditions, the shape of the curves is very similar to experimental stress–strain curves of real bulk materials whereas for the free boundary condition case the crystal becomes softer after initialization of damage. The small drops in the plastic part of the diagram stem from slippings between lattice planes (see section 3.2.1). More details will be discussed in later sections.

In order to examine the effect of model sizes, the specimens with crystal orientation No 1 but different sizes (cf table 5.5) were studied. It was found that the model size had not much effect on the stress–strain curves, especially not on the elastic deformation and early plastic deformation stage. Although some differences existed in the later plastic deformation and final fracture process, the curve profiles corresponding to the same boundary condition with different computational cell sizes were similar. Therefore, it can be concluded that the model size affects only the threshold stress values at which the associated deformation mechanism activates or inactivates, but not the deformation mechanism for a given model. In addition, it was also found that the stress–strain curves tend to an asymptotic one with the increase of model sizes. Their characteristics can be seen from the stress–strain curves in figures 5.7 and 5.8, which were obtained under periodic and free boundary conditions, respectively. In figures 5.7 and 5.8 the stress and strain appear to be a linear relation during the onset of the deformation stage. These linear relations can be approximately described as

$$\sigma_y = 146 \ \varepsilon_y \quad \text{under stress-free boundary conditions in the x-direction} \qquad (5.6)$$

$$\sigma_y = 221 \ \varepsilon_y \quad \text{under periodic boundary conditions in the x- direction} \qquad (5.7)$$

Using the linear elastic constitutive relation of a single crystal with crystal orientation No 1

$$\begin{aligned}
\sigma_x &= c_{11}\varepsilon_x + c_{12}(\varepsilon_y + \varepsilon_z) \\
\sigma_y &= c_{11}\varepsilon_y + c_{12}(\varepsilon_x + \varepsilon_z) \\
\sigma_z &= c_{11}\varepsilon_z + c_{12}(\varepsilon_x + \varepsilon_y)
\end{aligned} \qquad (5.8)$$

one can obtain

$$\sigma_y = \frac{c_{11}^2 - c_{12}^2}{c_{11}} \varepsilon_y \text{ under stress-free boundary conditions in the x-direction} \quad (5.9)$$

$$\sigma_y = c_{11}\varepsilon_y \text{ under periodic boundary conditions in the x-direction} \quad (5.10)$$

with the following assumptions: $\sigma_x = \varepsilon_z = 0$ in the case of stress-free boundary conditions in the x-direction and periodic boundary conditions in the z-direction; $\varepsilon_x = \varepsilon_z = 0$ in the case of periodic boundary conditions in both x- and z-directions, respectively. Insertion of $c_{11} = 242$ GPa and $c_{12} = 146$ GPa [14] into equations (5.9) and (5.10) gives

$$\sigma_y = 154\varepsilon_y \text{ under stress-free boundary conditions in the x-direction} \quad (5.11)$$

$$\sigma_y = 242\varepsilon_y \text{ under periodic boundary conditions in the x-direction} \quad (5.12)$$

Comparing (1), (2) and (6), (7), it can be seen that our simulation results for stress–strain curves are in agreement with theoretical analyses in the elastic deformation stage. The yield stress σ_s can be obtained from the stress–strain curves as well. For comparison, the resolved shear stresses τ_s on the (110) plane are given. They are about 1.76 GPa and 2.25 GPa for the α-Fe single crystal with orientation No 1 under periodic and stress-free boundary conditions, respectively. These values are close to the theoretically estimated value $\tau_s = 2.6$ GPa [15], but much larger than the experimental value $\tau_s = 0.01$–0.1 GPa [16–18]. The reason is that our model assumes a perfect single crystal while the single crystal used in experiments might contain many dislocations and defects which strongly affect the yield stress. For the same reason, the experimental fracture strain of 0.1 for Fe single crystals at $T = 47$ K [19] is smaller than most of the theoretical values, which again means that a real material withstands less stress than an idealized one. Nevertheless, the above comparison concerning the stress–strain relation of elastic deformation stage and yield stress demonstrates that our simulation results are reliable.

Effect of crystal orientations and boundary conditions

In the present work, we set up five specimens with different crystal orientations, see table 5.5. Figure 5.6 shows the x–y-planes of these specimens (No 1; 2; 3; 4; 5). The stress–strain curves for different crystal orientations with periodic and free boundary conditions are shown in figures 5.9 and 5.10 respectively. It is easy to see that yield stresses and fracture strains depend strongly on crystal orientations. As shown in figure 5.10, for example, the yield stress of the crystal with crystal orientation No 5 is about 22 GPa against about 4 GPa for No 1; the fracture strain of No 3 is more than 50%, and about 16% for No 2 in figure 5.9.

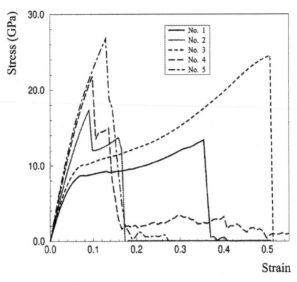

Fig 5.9 Stress–strain curves for different crystal orientations and periodic boundary conditions, $D = 20$, $H = 26$.

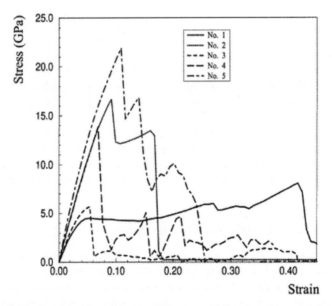

Fig 5.10 Stress–strain curves for different crystal orientations and free boundary conditions, $D = 20$, $H = 26$

Furthermore, differences on the profiles of curves imply that different crystal orientations correspond to different deformation mechanisms. As a matter of fact, different boundary conditions mean different stress states.

Periodic boundary conditions in the x-direction, for instance, represent a state where $\varepsilon_x = 0$, while free boundary conditions in the x-direction represent $\sigma_2 = 0$. Comparing figures 5.9 and 5.10, obvious differences in the stress–strain curves under two different boundary conditions can be observed for a given crystal orientation. That means, different stress states may activate different deformation mechanisms, which will be discussed in section 3.2 by analysing the atomic configurations and the stress components on slip planes.

Effect of displacement increment Δu_y

During the simulation process, the tensile deformation is performed by a well-defined displacement of the atoms in the upper end block of the specimen.

$$\Delta\varepsilon = \frac{\Delta u_y}{H} = \text{constant} \tag{5.13}$$

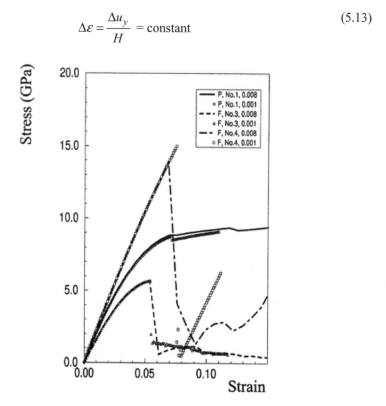

Fig 5.11 Stress–strain curves for different crystal orientations, different boundary conditions, and displacement increments.

The displacement increment per step remains a constant for a given model, i.e. where H is the initial length of the active zone in the y direction. In order to examine the effect of the displacement increment Δu_y, we designed two values for this constant. They are $\Delta\varepsilon = 0.008$ and $\Delta\varepsilon = 0.001$, respectively. The corresponding results are shown in figure 5.11.

The stress–strain curve of the crystal with crystal orientation No 4 shows that Δu_y affects the deformation after yielding. This is due to the randomness of sites of defect nucleation and following dislocation movement.

Deformation and fracture mechanisms

From the analysis of stress–strain curves, it is known that different deformation and fracture mechanisms exist for different boundary conditions and crystal orientations. Now, we analyse the deformation mechanisms by detecting the atomic configurations at different deformation stages.

Crystal orientation No 1 (x:4(1 0 0), y:(0 1 0), z:(0 0 1))

Observing the evolution of atomic configuration, slipping between (110) planes was detected taking place in the crystal with orientation No 1 under periodic boundary conditions. These slippings resulted in the discontinuous stress drops in stress–strain curves of figure 5.7. Owing to the constraint imposed by the periodic boundary condition, however, the dislocation could not slip out of the crystal. Thus, two stacking faults were formed accompanied by partial dislocations at their ends, which can be seen in figure 5.12(a). With further increases of the displacement, microcracks initiated at the locations of the partial dislocations and cleavage fracture occurred on (110) planes (see figure 5.12(b)).

Under free boundary conditions in the x-direction, one may expect to observe slips of the dislocations on the (110) plane, because of the weakening constraint. However, the atomic configuration of figure 5.13 shows that the main deformation under free boundary conditions actually resulted from the expansion in the h010i direction together with the cross-contraction in the <100>direction, but not from the dislocation movement. For strain $= 0.14$, the cross sectional area has decreased by 10%. Note that this area as used in stress calculation was the initial cross section area. Considering the real cross section area, the real stress after the cross-contraction at strain $= 0.04$ should be larger by about 10% than in figure 5,8. Therefore, the real stress actually increases with increase of strain. The contraction of the cross section explains why the crystal does not become softer. But in the case of free boundary in the x-direction, why can dislocations on (110) not be activated? This seems to be unreasonable. An analysis of the stress components on the (110) planes explains this feature.

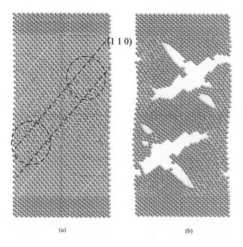

Fig 5.12 Atomic configurations for orientation No 1 with periodic boundary conditions: (a) strain = 0.368; (b) strain = 0.376.

Considering relative boundary conditions and the linear elastic constitutive relations, the resolved stress components on (110) can be expressed as

$$\tau_{(110)} = 0.1973\sigma_y \qquad \sigma_{(110)} = 0.8026\sigma_y$$

$$(5.14)$$

for x- and z- periodic boundary conditions

$$\tau_{(110)} = 0.5\sigma_y \qquad \sigma_{(110)} = 0.5\sigma_y$$

$$(5.15)$$

for x-free and z- periodic boundary conditions

Substituting the yield stresses obtained from figures 5.7 and 5.8 into equations (5.13) and (5.14), respectively, the stress components on (110) are $\tau_{(100)} = 1.76$ GPa and (110) are $\sigma_{(100)} = 7.04$ GPa in the case of periodic boundary conditions, and (110) are $\tau_{(100)} = 2.25$ GPa and (110) are $\sigma_{(100)} = 2.25$ GPa in the case of free boundary conditions.

It can be seen that the shear stress under free boundary conditions is slightly larger than under periodic boundary conditions. However, the normal stress under free boundary conditions is much smaller than that under periodic boundary conditions. Larger normal tensile stress on the (110) plane leads to larger distances between the (110) planes, which decreases the resistance of dislocation slip on (110) planes. This explains why slip on (110) took place under periodic but not under free boundary conditions. Nevertheless, at approximately 30% strain in figure 5.8a pronounced step appears along the stress–strain curve.

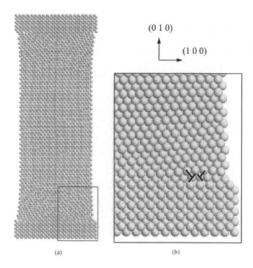

(0 1 0)

(1 0 0)

(a)

(b)

Fig 5.13 Atomic configurations for orientation No 1 with free boundary conditions at strain = 0.28.

The corresponding atomic configuration indicates that this step was caused by the dislocation movement on the (110) planes at the location of the stress concentration (see figure 5.13). Similar behaviour was also observed in the specimen with a square void, i.e. dislocations on (110) planes emitted simply from the corners of the square void.

From these results, it may be concluded that dislocation slip not only depends on shear stress but also on normal stress of the slip plane, because the normal stress will increase or decrease the distance between slip planes, thus changing the resistance against dislocation slip.

Crystal orientation No 2 (x:(1−1 0), y:(1 1 0), z:(0 0 1))

The (110) plane is the closest packed plane in α-Fe. A comparison of the surface energies in table 5.6 calculated with three different potentials also shows that the surface energy s of plane (110) is lowest. Therefore, one can expect that the (110) plane is favourable for cleavage.

Table 5.6
Surface energies of different crystal planes.

Plane	γ_s.(J m^{-2}) in this work	γ_s.(J m^{-2}) in [7]	γ_s.(J m^{-2}) in [11]
(100)	1.554	1.973	1.306
(110)	1.374	1.874	1.206
(111)	1.700	2.296	—

The atomic configurations in figures 5.14(a) and (b) corresponding to the periodic boundary conditions (BC) and stress-free BC respectively demonstrate the fact that cleavage fracture occurred exactly on plane (110). Several other atomic configurations at different deformation stages show that the distance between (110) planes increases with increasing load first in a limited region, then to a larger region. In figures 5.9 and 5.40, there are sharp drops on the stress–strain curves of the crystal with orientation No 2. These drops can be interpreted as the results of sudden decreases of the number of atoms inside the region r_{cut}. The follow-up stress–strain curves with smaller slopes also indicate this process. Furthermore, it is very interesting to find that under both BC the fracture stresses on (110) planes are nearly the same. The values are about 13.4 GPa which is the same magnitude as in the theoretical analysis.

By analysing the results of crystal orientation No 1 and No 2, it can be concluded that cleavage fracture on plane (110) is one of the primary fracture mechanisms in α-Fe. This agrees with results from [1, 4, 10, 11], too.

Crystal orientation No 3 (x:(1 0 1), y:(0 1 0), z:(−1 0 1))

A twinning deformation mechanism is obviously observed in the crystal with crystal orientation No 3 under the stress-free BC from the atomic configurations in figure 5.15(a). With increasing load the twinning deformation on [111] (−1 2−1) extends from the middle part to the whole of the specimen. Analysing the atomic configuration shown in figure 5.15(b), it is found that the crystal has rotated by 90°, i.e. from (101) to (010) and from (010) to (101) after twinning deformation. The large-stress drops and low-stress strains in the curves of figure 5.10 illustrate that twinning is followed by a large and fast strain. The low-stress strain from 6% to 42% in figure 5.10 shows that twinning deformation is of the same magnitude as the given displacement increments.

During twinning deformation, the displacement of atoms has to increase proportionally to their distances to the twinning surface. However, periodic BC do not allow such displacements of atoms. Therefore, under periodic BC no twinning deformation occurs during the whole deformation process. Additionally, no dislocations nucleate in this case. The crystal cleaved on a (010) plane at a strain of 51%. From this it can be derived that a twinning deformation on <111>{-1,2,-1} takes place more easily than the slip of a dislocation in the crystal with orientation No 3.

Crystal orientation No 4 (x:(1 0 0), y:(0 1−1), z:(0 1 1))

For this crystal orientation, two slip bands formed under the stress-free BC. Several dislocations nucleated and slipped along [1 1−1] directions on (−2 1−1) planes. A lot of steps were left on the free surface as shown in figure 5.16. The steps on the stress–strain curve of figure 5.10 are simply related to these dislocation slips. In addition, it can also be seen that the yield stress of the crystal with dislocations becomes smaller than that of a perfect crystal.

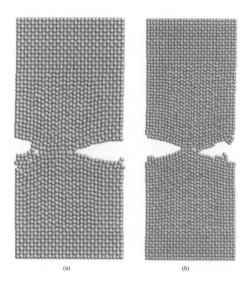

(a)   (b)

Fig 5.14 Atomic configurations for orientation No 2: (a) with periodic boundary conditions, strain = 0.168; (b) with free boundary conditions, strain = 0.168.

Under the periodic BC two slip bands developed although the periodic BC constrained the slip of dislocations (see figure 5.17). Since the dislocation could not slip out of the crystal, however, the slip bands widened and the dislocations piled up at the intersection region of two slip bands. It is worthwhile to study the diagonal atomic rows in figure 5.17(a) which led to the two marked closely neighboured edge dislocations in figure 5.17(b). Microcracks initiated finally at this place. The step at 14% strain in figure 5.9 stemmed from these microcracks. From the atomic configuration one can see that the slip bands are also along the [1 1−1] direction on a (−2 1−1) plane. For this crystal orientation, the presented results indicate that slip of dislocations on <111>{-1,2-1} occurs more easily than twinning deformation.

Analysing the stress components on {-1 2 -1} planes between the crystals with orientations No 3 and No 4 under stress-free BC, it turned out that at the yield point the normal stresses are nearly same in both crystal orientations.

However, the shear stress in crystal orientation No 4 is three times as large as that in crystal orientation No 3. Therefore, we can conclude that when the normal stresses are the same, higher shear stress will activate dislocation slipping on <111>{-1,2-1}. Nevertheless, lower shear stress will activate twinning deformation on <111>{-1,2-1}.

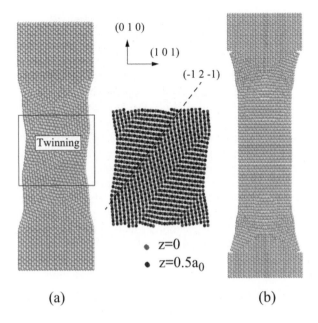

Fig 5.15 Atomic configurations for orientation No 3 with free boundary conditions: (a) strain = 0.136, (b) strain = 0.48.

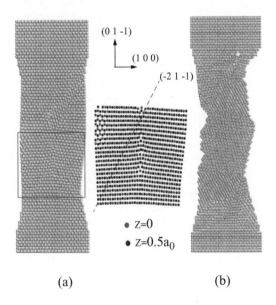

Fig 5.16 Atomic configurations for orientation No 4 with free boundary conditions: (a) strain = 0.152; (b) strain = 0.384.

Crystal orientation No 5 (x:(1−1 0), y:(1 1 1), z:(−1−1 2))

For the case of crystal orientation No 5 the simulation cell consisted of six atomic layers within a period in the z-direction. Under stress-free BC several deformation mechanisms, such as void formation and coalescence, dislocation nucleation and movement etc took place during the deformation process. The curve with saw-tooth shape in figure 5.10 resulted from the interaction of dislocation slips and small voids developing. In addition, it was also found that the dislocations consisted of edge and screw components (see figure 5.18). Thus, it is difficult to separate the main deformation mechanisms. In the case of periodic BC, it can be observed from figure 5.19 that the main damage mechanism was void formation and coalescence. The atoms did not have any displacement in the z-direction, which is different from that in the case of stress-free BC. The reason is that the periodic boundary conditions in the x-direction constrained screw dislocation nucleation. The nanoscale tensile test for orientations 1 and 2 yield the most realistic simulation results. In agreement with [1, 4, 10, 11], the cleavage fracture on plane (110) was confirmed to be one of the primary fracture mechanisms in α-Fe.

The present study demonstrates the success of modelling in reproducing the various essential mechanisms of plasticity and damage on the atomic scale thereby offering the opportunity to observe and understand them in detail.

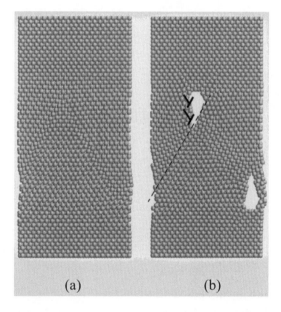

(a)   (b)

Fig 5.17 Atomic configurations for orientation No 4 with periodic boundary conditions: (a) strain = 0.136; (b) strain = 0.146.

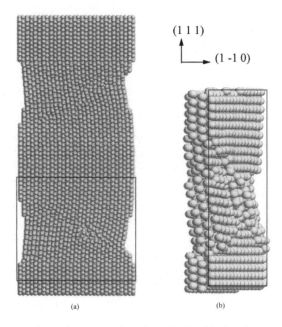

(1 1 1)

(1 -1 0)

(a)   (b)

Fig 5.18 Atomic configurations for orientation No 5 with free boundary conditions at strain = 0.144.

Table 5.7
Summary of the stress–strain relationships and of the atomic configurational results belonging to different crystalline orientations.

Orientation/ boundary conditions	Stress–strain curve, overall shape (not the absolute value)	Fracture strain	Main deformation mechanism
1/free	Realistic onset, but remains plastic until much too high strains	Too high	Expansion and cross-contraction $\langle 111\rangle(110)$ dislocation movement near stress concentration
1/periodic	Realistic		
2/free	Drop after initialization of damage	Realistic	Cleavage fracture on (110) plane
2/periodic	Very similar results for free and periodic boundary conditions	Realistic	Same result for free and periodic boundary conditions
3/free	After damage initialization, resistance against stress very low	Too small	No clear fracture; $\langle 111\rangle\{-1,2,-1\}$ twinning nucleation and growth
3/periodic	Realistic	Too high	Cleavage fracture on (0 1 0) plane

4/free	See below. In addition, saw-tooth shape due to dislocation nucleation and slips	No clear fracture	Slip bands → $\langle 111 \rangle \{-1,2,-1\}$ dislocation nucleation and slipping
4/periodic	Sharp drop after initialization of damage, but crystal still gluestogether up to high strains	No clear fracture	Slip bands → $\langle 111 \rangle \{-1,2,-1\}$ dislocation nucleation and pile up → microcracks
5/free	Saw-tooth shape due to interaction of dislocation slip and voids	Realistic	Dislocation slip → void formation → dislocation slip → void growth → fracture
5/periodic	Sharp drop after initialization of damage	Realistic	Void formation → coalescence → fracture

Conclusions

The following conclusions can be drawn from our molecular statics analyses of uniaxial straining at different boundary conditions.

- for relatively larger models as used in this work, it can be concluded that the model size affects only the threshold value of stress at which the associated deformation mechanism activates or inactivates, but not the deformation mechanism for a given crystal orientation and boundary condition. In addition, it is also found that the stress–strain curves obtained from the molecular statics simulation are well reproducible. Furthermore, the stress–strain curves approach an identical distribution with increasing model sizes. The simulation results of stress–strain curves in the elastic deformation stage are in good agreement with theoretical analyses,
- various deformation evidence such as dislocation movement, dislocation piling up, twinning, formation and coalescence of voids are clearly observed under uniaxial tensileload. The simulation results indicate, that the stress state controls, which deformation mechanism is activated, although a given crystal material has intrinsic deformation mechanisms;
- the results of different displacement increments, i.e $\Delta\varepsilon = 0.008$ and $\Delta\varepsilon = 0.001$, show that displacement increment does not affect elastic deformation and yield mechanism, but does affect follow-up plastic deformation because of the randomness of defect nucleation and movement,
- the essential results of stress–strain relations and of atomic configurational changes are summarized in Table 5.7.

The nanoscale tensile test for orientations 1 and 2 yield the most realistic simulation results. In agreement with [1, 4, 10, 11], the cleavage fracture on plane (110) was confirmed to be one of the primary fracture mechanisms in α-Fe.

The present study demonstrates the success of modelling in reproducing the various essential mechanisms of plasticity and damage on the atomic scale thereby offering the opportunity to observe and understand them in detail.

Acknowledgement

This work was supported by the Bundesministerium für Bildung, Wissenschaft, Forschung und Technologie (BMBF) under grant No 15010129.

References

[1] Mullins M and Dokainish M A (1982), Phil. Mag. A 46, p. 771.
[2] Mullins M (1984), Acta Metall. 12, p. 381.
[3] Doyama M (1995), Nucl. Instrum. Methods Phys. Rev. B 102, p. 107.
[4] Kohlhoff S and Schmauder S (1989) Atomistic Simulation of Materials ed V Vitek and D J Srolovitz (New York: Plenum).
[5] Kohlhoff S, Gumbsch P and Fischmeister H F (1991) Phil. Mag. A 64 851, p. 586 S Y Hu et al.
[6] Cheung K S and Yip S (1990), Phys. Rev. Lett. 65, p. 2804.
[7] Machova A (1992), Mater. Sci. Eng. A 149, p. 153.
[8] Machova A (1996), Mater. Sci. Eng. A 206, p. 279.
[9] Holian L and Ravelo R (1995), Phys. Rev. B 51 11, p. 275.
[10] Cheung K S and Yip S (1994), Modelling Simul. Mater. Sci. Eng. 2, p. 865.
[11] Yanagida N and Watanabe O (1996), JSME Int. J. A 39, p. 321.
[12] Baskes M I, Hoagland R G and Needleman A (1992), Mater. Sci. Eng. A 159, p. 1
[13] Wang N J (1996), Mater. Sci. Eng. A 206, p. 259.
[14] Simonelli G, Pasianot R and Savino E J (1993) Mater. Res. Soc. Symp. Proc. 291, p. 567.
[15] Courtney T H (1990), Mechanical Behaviour of Materials (New York: McGraw-Hill).
[16] Sato A, Nakamura Y and Mori T (1980) Acta Metall. 28, p. 1077.
[17] Stein D F, Low J R Jr and Seybolt A U (1966) Acta Metall. 16, p. 1183.
[18] Stein D F and Low J R Jr (1963) Acta Metall. 11, p. 1253.
[19] Brunner D and Diehl J (1992) Z. Metall. 83, p. 827.

5.3 Atomistic study of void growth in single crystalline copper[3]

Even the purest real material contains a large amount of defects in its crystal structure. The variety and complexity of defects in a real material makes it unpractical to simulate its entire behaviour on the atomic level. However, it is only on this level that the actual mechanisms of such important macroscale processes as plastic deformation, void growth and crack propagation can be seen and understood from a fundamental point of view.

Many researchers have employed static, quasistatic and dynamic techniques to the study of cracks on an atomistic scale in a crystalline material. They have obtained information about the fracture processes and the influence of temperature and crack orientation on this process [1-9]. However, there are no similar studies available for voided crystals. Atomistic stress-strain curves have been plotted for both single crystal and nanocrystal materials [10]. Despite the fact that some of these curves show very good qualitative agreement with experimental results, none of them, to date, have been able to predict reasonable stress levels. Stresses in the range of the theoretical strength of the material are the norm for these type of calculations, as not enough microstructural irregularities can be taken into account in one model.

Typically, materials are modelled on a larger size scale by neglecting local imperfections and the resulting anisotropic nature of the material, replacing it instead with a homogeneous continuous isotropic representation whose properties are defined as an average of the real structure. These properties are easily determined from experiment. The advantage of this type of analysis is that it allows second-phase inclusions, grain boundaries and macrodefects to be considered. On this level, continuum mechanics is used extensively.

Crystal plasticity theory which is also used in this work to compare with atomistic methods is a continuum theory which has the additional feature of taking the first steps to accommodate the actual local anisotropic behaviour resulting from the crystallographic arrangement of the material [11, 12]. This is done by defining certain preferred directions in which the material can deform more easily. These directions are determined from theoretical considerations and are the crystallographic directions along which dislocations can most easily move.

An advantage of atomistic methods over continuum methods is that once a suitable interatomic potential has been chosen to describe the material behaviour and once a set of boundary conditions are applied, the method is completely self-consistent. For example, in an atomistic calculation cracks can grow naturally, whereas in the continuum finite element framework crack growth must be

[3] Reprinted from L. Farrissey, M. Ludwig, P.E. McHugh, S. Schmauder, "An Atomistic Study of Void Growth in Single Crystalline Copper", Computational Materials Science 18, pp. 102-117 (2000) with kind permission from Elsevier

prescribed introducing extra criteria to release nodes or elements at certain stages. In addition, atomistic models have their own intrinsic failure criteria whereas in the continuum models failure is implied when the limit of certain calculated and predefined values are exceeded.

Stress-strain curves and deformation contour plots are calculated for copper (Cu) single crystals using atomistic methods. The influences of crystal orientation and initial void orientation are considered. These results are compared with similar arrangements developed within the continuum crystal plasticity framework. An additional aim of the work is to verify and improve the veracity of the crystal plasticity method. This can be done by using the failure strain as predicted by the atomistic models as a failure criterion for the continuum.

5.3.1 Modelling approach

In atomistic calculations, the material behaviour is almost completely determined by the interaction potential. The potential is an energy function which governs the interaction between the neighbouring atoms. To accommodate this, potentials are parameterised to include many body effects in the pair wise scheme. The potential used in this work is the embedded atom method (EAM) potential and was proposed by Daw and Baskes [1] in the mid-1980s.

The potential is developed within the basis of the quantum mechanical density functional theory. The theory states that the ground state energy and properties of a system are uniquely determined by the electron density. The total electronic energy for an arbitrary arrangement of atoms can be written as a unique function of the total electron density. In metals, the electron density at any point can be approximated with reasonable accuracy as a linear superposition of the individual atoms and so we can write the total energy as

$$E_{tot} = \sum_i F(\rho_i) + \sum_{\substack{i,j \\ i \neq j}} V(r_{ij})$$

(5.16)

where F .qi. is the embedding energy function of an atom i at the position where it has background electron density q. This electron density is calculated by superposition of the electron densities of nearby atoms by

$$\rho_i = \sum_{j \neq i} \phi_j(r_{ij})$$

(5.17)

Here $\Phi_j(r_{ij})$ is the electron density contribution by atom j at the position of the atom i. This is assumed to depend only on the separation distance between the atoms. $V(r_{ij})$ is the repulsive central force between the atoms.

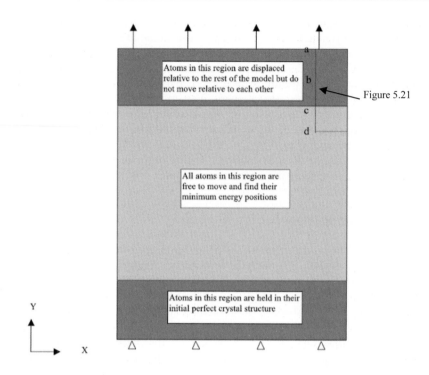

Fig. 5.20 Schematic representation of the model used. Periodic or free boundary conditions can be applied in the x and z directions.

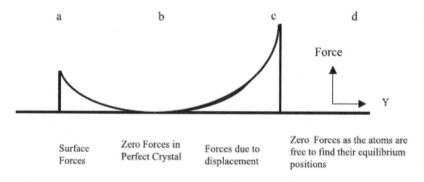

Fig. 5.21 Schematic stress distribution in the cut-out from Fig. 5.20

The functions F and Φ are established empirically from the physical properties of the solid. This is done using the lattice constant, elastic constants, vacancy formation energies and sublimation energies for the pure metal.

The parameters for the potentials used in this work are fitted to the experimental data and made available for use in table form by Voter [13].

The potential energy of the system under consideration is minimised iteratively with respect to the position of all the atoms within the system using the conjugate gradient algorithm. This procedure is implemented using a program initially written by Kohlhoff and coworkers at MPI Stuttgart [3, 14, 15].

The stress-strain curves for a material are calculated by applying uniaxial displacements iteratively and calculating the corresponding stress. The system used to apply displacement steps and calculate the crystal stress for each increment is shown in Fig. 5.20. A block of atoms has to be held in their original perfect crystal structure at both ends of the model in the direction of the applied displacement increments. The purpose of this step is to provide a buffer zone to isolate the stress as a result of the applied displacements from the stress that results from surface effects. A schematic of the forces seen through the thickness of the cut-out in Fig. 5.20 is shown in Fig. 5.21. Surface effects cause forces to penetrate a thickness equal to the cut-o. potential into the material. While the strained atoms at the interface cause resultant stresses to similarly penetrate a distance of one cut-o. potential into the material, these two stress systems are separated by a stress-free buffer zone. The stress can be calculated by summing all the interatomic forces in the isolated section just above the free atoms and dividing this by the relevant area. The strain is calculated by dividing the total applied displacements by the original length of the free atom section. This procedure is carried out for each displacement increment until the material fails giving an overall stress-strain curve. As the Poisson contraction is not accounted for in the fixed block of atoms, these boundary conditions do not replicate pure uniaxial tension.

Crystal plasticity void growth model

The finite element crystal plasticity void growth model used in this study was originally set up by O'Regan et al. [16] and Quinn and McHugh [17]. The material considered is assumed to contain a periodic array of unit cells, each containing a void, as shown in Fig. 5.22. For simplicity and practicality in computation, a two-dimensional (2D) representation was adopted. In certain symmetry conditions it is possible to restrict analysis to a quadrant of one unit cell, outlined in Fig. 5.22(b). The material was assumed to be a single crystal. The 2D assumption meant that a full 3D face centre cubic (fcc) crystal structure with 12 slip systems was not modelled.

Instead idealised 2D slip system geometry was used, shown in Fig. 5.22(d), where three slip systems with equivalent constitutive properties were oriented at angles of 60° to each other. The angle Ψ corresponds to the grain angle of the slip system, while the three angles of β correspond to the orientation of the three slip planes to each other. Simple velocity boundary conditions, shown in Fig. 5.22(b), were set up to allow variation in the overall biaxiality of the strain state developed during deformation.

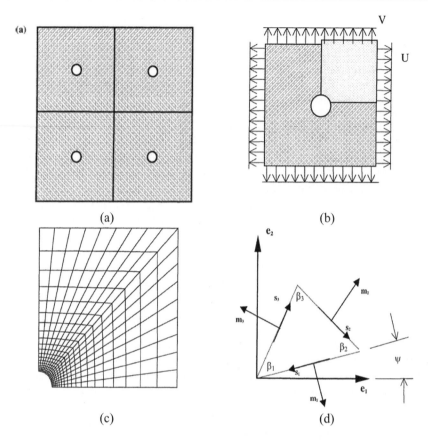

Fig. 5.22 (a) Periodic arrangement of voids, (b) unit cell of material with constrained boundaries, (c) finite element model of the void and (d) slip system arrangement (for details see text).

The biaxiality was quantified as $U = V$, where U is the velocity in the horizontal direction and V is the velocity in the vertical direction. The external boundaries of the quadrant were assumed to remain straight during deformation. These are the correct boundary conditions for the case of a symmetric slip system configuration, i.e., $\Psi = 0$. However, these boundary conditions are approximate for $\Psi \neq 0$. The idealised 2D slip system geometry reflects the redundancy that occurs in 3D when, for an fcc crystal, 12 slip systems are available but only five are required to represent a strain increment.

Here in 2D there are three slip systems but only two are required to represent a strain increment. The crystal plasticity model requires the plastic behaviour of the material to be expressed on a slip system basis, as discussed in [16,17], in terms of the equation under the conditions of each slip system having the same properties and self- and latent hardening of slip systems being equal.

$$g = g_0 \left(1 + \frac{\gamma}{\gamma_0}\right)^n \tag{5.18}$$

This equation provides a relationship between the slip system strain hardness g and the accumulated plastic shear strain (accumulated slip) after yielding. The parameter g_0 is effectively τ_0 the slip system yield stress under shear while γ_0 is the shear strain value at this stress, n is the hardening exponent of the curve. Information about the elastic behaviour of the material is also required in the form of λ and μ, the Lame constants, which can be expressed in terms of the more used elastic modulus (E) and the Poisson's ratio (v) as

$$\lambda = \frac{Ev}{(1+v)(1-2v)} \tag{5.19}$$

$$\mu = \frac{E}{2(1+v)} \tag{5.20}$$

Again this is discussed in detail in [16, 17]. Since during large-scale ductile deformation plastic strains greatly exceed elastic strains, it was considered sufficient to use isotropic elasticity for the crystal. The information for the strain hardening is calculated from a shear stress-shear strain curve calculated at 293 K.

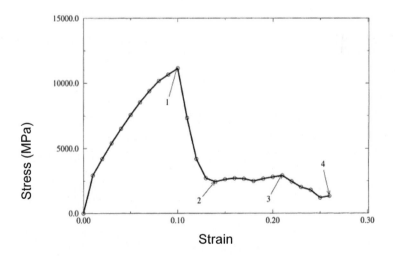

Fig. 5.23 Stress-strain curve for voided crystal no. 1 oriented in the [100] [010] [001] crystal direction.

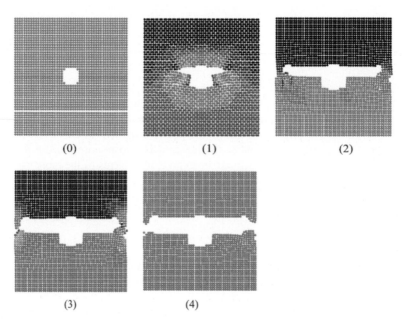

(0)   (1)   (2)

(3)   (4)

Fig. 5.24 Atomistic deformation plots of void case 1: [100] [010] [001] crystal orient-tation.

Table 5.8
Minimum periodic lengths of the crystal directions considered

Crystal direction	Minimum periodic length
$\left[\overline{1}11\right]$	$\sqrt{3}/2a_0$
Error! Objects cannot be created from editing field codes.	$\sqrt{6}/2a_0$
Error! Objects cannot be created from editing field codes.	$\sqrt{2}/2a_0$
Error! Objects cannot be created from editing field codes.	$1/2a_0$

The lattice was assumed to deform in plane strain. The initial volume fraction of the void was assumed to be 0.8%. As mentioned above, the plastic strain hard-ening properties were required on a slip system basis. Experimental data describ-ing the behaviour of a single crystal of Cu were not used. Data for shear tests on polycrystals of Cu were used to calculate the n parameter. Clearer tensile stress-strain curves for polycrystals were found where it was easier to determine the yield stress and yield strain. These data were converted to single crystal shear format by using a Taylor factor (TF) shown as follows:

$$\sigma_{tensile} = TF \times \tau_{resolved\ shear\ stress} \qquad (5.21)$$

The TF is the ratio between the overall tensile stress of a polycrystal and the average resolved shear stress on a slip system. This was converted into equivalent individual slip system data by assuming a TF of 3.03 corresponding to the lattice geometry of the fcc grains [18]. In fitting Eq. (5.18) to this data the material parameters obtained were: g_0 = 69.3 MPa, γ_0 = 1.981 and n = 0.3. Standard elastic material properties for pure copper were used [19].

5.3.2 Results: influence of the crystal orientation of void growth

The first effect studied using the copper system was the influence of the crystal orientation on void growth. To do this, two copper single crystals of approximately the same volume fraction are created. The first crystal is oriented such that the Miller indices representing the crystal directions [100], [010] and [001] line up with the x, y and z axes. The second crystal is aligned such that the [111], [110] and the [112] directions are aligned to the positive x, y and z crystal directions.

In orienting the second crystal away from the simple [100] system of directions, special care has to be taken. If periodic boundary conditions are to be used correctly the simulation cell must contain an integer number of times the periodicity of the crystal in that direction. These distances are shown in Table 5.8.

The deformation plots produced are contour labelled. Darker coloured atoms represent those whose coordinate positions have changed most with the application of that particular load step, while the lighter coloured atoms represent the opposite case of the atoms whose coordinate position has least changed. Atoms between these two positions are graded accordingly.

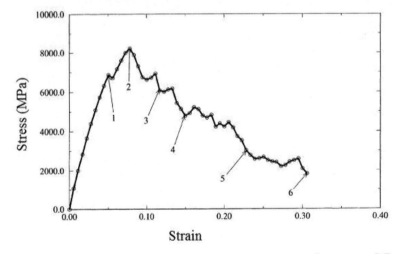

Fig 5.25 Stress-strain curve for voided crystal oriented in the $[\bar{1}11]$, $[110]$,$[\bar{1}\bar{1}2]$ crystal direction.

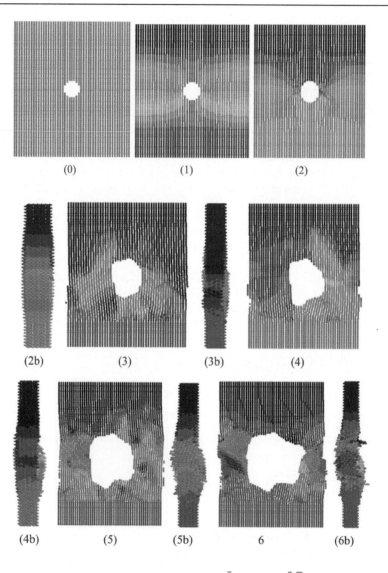

Fig. 5.26 Atomistic deformation plots of void case 2: $[\bar{1}11]$, $[110]$, $[\bar{1}\bar{1}2]$ crystal orientation.

Orientation [100], [010], [001] (orientation 1)

The stress-strain curve for void case 1 is shown in Fig. 5.23. It has an initial, almost linear region where the void grows only slightly as the material bears more and more stress. The reason this region is not completely linear is that we use a nonlinear interaction potential to describe the atomic behaviour over all separation distances. There is a sudden growth in both the shape and size of the void as it reaches the peak stress sustainable or the maximum load. At this load, cracks propagate from the four corners of the void as seen in frame 1 of Fig. 5.24. These cracks appeared along the <010> directions. As the crack further propagates the bottom series of cracks closes while the top two continue to propagate quickly towards the edges. This growth in the length of the crack has a serious weakening effect on the material as can be seen from the steep drop o. in the stress-strain curve after the maximum stress is reached. The crack continues to propagate towards the right-hand edge of the model until it experiences boundary forces due to the effective interaction of the voids from the neighbouring cells. This interaction between the voids results in a change in orientation of the undeformed crystal in the region of the voids, this in turn halts the crack temporarily until further strain increments result in the remaining atoms being pulled cleanly apart.

Crystal orientation [$\bar{1}$11], [110],[$\bar{1}\bar{1}$2] (orientation2)

The stress-strain curve in Fig. 5.25 shows a much more ductile failure of the crystal. The load-carrying capacity is reduced gradually from the initial peak as opposed to the sharp drop of in the stress seen in the other crystal orientation. This increased ductility is due to some extent to the fact that the <111> directions come into play. These are the close packed directions in fcc crystals and so slip is most likely to occur in these directions. Theoretically, the 12 fcc slip systems are aligned to each other at 60° angles. This 60° angle is seen in the model where an originally circular shaped void transforms to more of a hexagonal shape (Fig. 5.26). Frame 3 shows very clearly the development of the top hexagonal corner by the movement of a full dislocation along a slip system aligned at 60° to the loading direction. A further comparison between the two crystal orientations reveals the presence of full partial and screw dislocations in this orientation (seen in the side view plots labelled b). The significance of the appearance of these screw dislocations is discussed in the next section. A comparison of the two orientations shown in Fig. 5.27 shows that variation of the orientation of the crystal effects both the strength and failure mechanism of the material.

Orientation 2 exhibits a much more ductile behaviour than orientation 1. This increased ductility may be due to the additional planes in the thickness direction. The development of dislocations allows new material to be brought to the surface of the void.

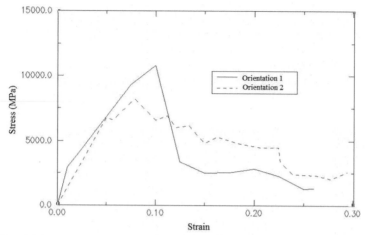

Fig. 5.27 Stress-strain comparison of the two voided crystal orientations.

This can occur in reality but is not allowed in continuum models where, for example, finite element connectivity is predetermined and remains fixed throughout an analysis.The crystal oriented in the [11 1], [1 10],[1l2] direction is modelled with six atoms in the thickness direction. As before, the thickness is an integer number of times the periodicity in this direction. This is necessary to implement periodic boundary conditions.

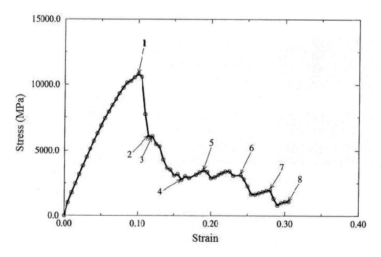

Fig. 5.28 Stress-strain curve for void case 1 modified to include three times the periodicity in the thickness direction.

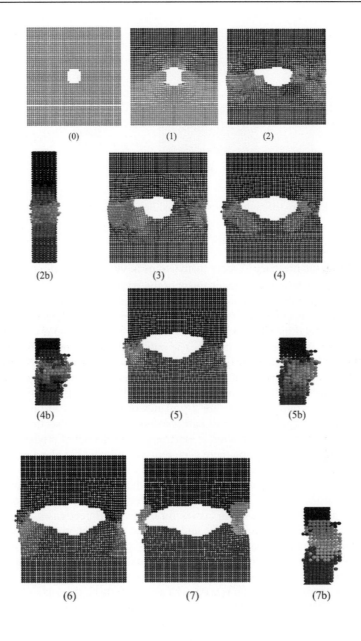

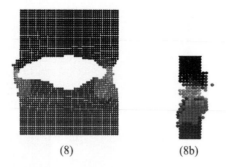

(8)   (8b)

Fig. 5.29 Atomistic deformation plots for modified void case 1: crystal orientation [100] [010] [001].

However, the presence of these many atoms in the thickness direction may have an effect on the deformation mechanism of the crystal.

To investigate this, a second model with crystal orientation in the [100] [010] [001] direction is created with three times the periodicity in the thickness direction. The model dimensions in the x and y directions and the void volume fraction are the same as in the initial model. The only difference between both models is that the new version has six atoms in the thickness direction as opposed to two. The same loading conditions are applied and the resulting stress-strain curve is plotted in Fig. 5.28. The pictures in Fig. 5.29 labelled b are of the z-y plane which is a side view while the others are x-y planes or front views as previously used. The side view images for the most part just concentrate on the section in the vicinity of the void where deformation is experienced. In the initial model with two atoms in the thickness direction there is no movement in the y-z plane while, as seen in Fig. 5.29, there is considerable movement in this plane for the thicker model. This suggests that increasing the thickness of the model results in allowing other deformation mechanisms and in particular screw dislocations to become active. Fig. 5.30 shows the stress-strain curves for the two different cases. The initial region between the undeformed state and the point where the maximum stress level is reached is similar in both cases. There is a slight difference in the slope of the curve to the maximum stress level but the actual magnitude of the peak stress is almost identical. At this stress level, the deformation plots are almost identical. The stress levels tend to drop o. as cracks propagate from the top and bottom corners in the [010] direction. During the initial stages of void growth after the peak stress is reached both the deformation plots and the stress levels remain very similar. The quick void growth and drop in the load-carrying capacity of the model is due to crack propagation. In the thicker model, this crack growth is impeded by the development of screw dislocations (Fig. 5.29). In effect, the crack tip is blunted and more ductile deformation is experienced (Fig. 5.30). The inclusion of the four additional planes of atoms in the model allows for a movement of the internal planes relative to each other.

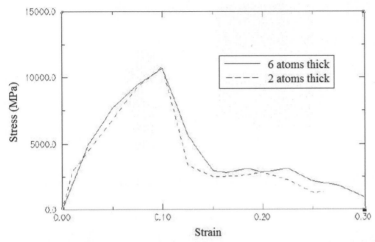

Fig. 5.30 Comparison of different model thicknesses for void case 1: crystal orientation [100] [010] [001].

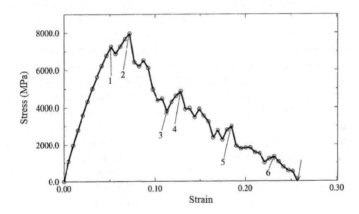

Fig. 5.31 Stress-strain curve of void case 2 elongated in the loading direction: crystal orientation $[\bar{1}1\,1], [1\,10], [\bar{1}\bar{1}2]$.

As the interplanar spacing is a factor in determining the stress level required for the activation of slip, the fact that planes can move relative to each other must affect the deformation process. The extra option of out-of-plane deformation results in a more ductile behaviour as seen in Fig. 5.27.

This is witnessed by the more rounded shape of the void as it grows in Fig. 5.26. The modi fied thicker model is a closer approximation to a 3D model and as such should give a better representation of reality.

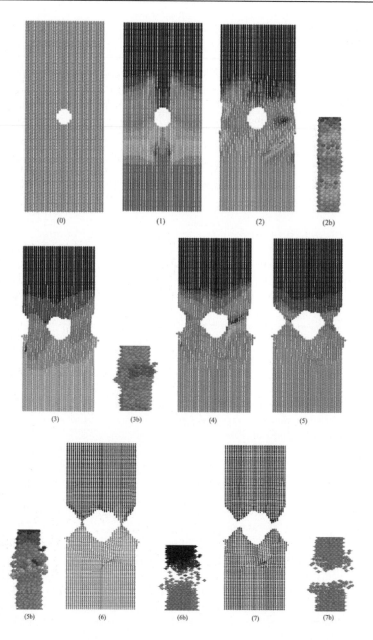

Fig. 5.32 Atomistic deformation plots for the model extended in the y direction: crystal orientation $[\bar{1}11], [1\bar{1}0], [\bar{1}\bar{1}2]$

In the modified model, it was noticed that the deformation had reached the constrained atoms that are used to apply the boundary conditions before the void had grown by any significant amount. That is to say that distorted atoms have reached the effective cell boundary before the void begins to take on a definite shape; as a result they are not free to move as they would in a bigger crystal. The fact that the deformation reaches the boundary is not unrealistic in that there are immovable barriers such as grain boundaries in real crystals. However, this potentially has an influence on the shape of the void as what can be called "back stresses" result from the deformation pattern meeting a solid boundary. The atoms at the surface cannot deform anymore because the surface is held fixed. However, they can sustain higher stress, and therefore, allow higher resultant forces back into the crystal. To see what effect, if any, these back stresses have on the shape of the void a model with larger dimensions in the y direction is used. For the purposes of this investigation, the model oriented in the $[\bar{1}11], [110],[1\bar{1}2]$ crystal direction is used. The resulting stress-strain curve is shown in Fig. 5.31 and the corresponding deformation plots are included in Fig. 5.32. The model dimensions in the x and z directions are similar to those used in the first model but the y dimension is increased significantly. As the volume fraction of both models is different, it does not make sense to plot the two stress-strain curves against each other. However, it is worth looking at the shape of the void as it grows and the comparative trends in the stress-strain curves.

The overall pattern of the stress-strain curves are similar, however, there is a difference in the void shape. In the larger model, where the void is isolated from the fixed surfaces and so the back stresses, the void grows in a more definite hexagonal shape. This happens as the void in this instance is growing only under the influence of the applied displacement and thus deformation along the theoretically more favoured directions are not hindered by back stresses.

Comparison with crystal plasticity

A comparison is made between the purely atomistic models already discussed in this chapter and crystal plasticity finite element models as developed by Quinn and coworkers [16, 17]. In as much as possible the geometry of the models are kept as close to each other as possible. The actual mesh of the model used is shown in Fig. 5.33. Normally for this type of analysis, only one quarter of the unit cell needs to be modelled with symmetry boundary conditions being applied to the two cut surfaces. However, in the case where the slip systems are oriented at an angle other than 0° or 90° to the loading axis the full unit cell cannot be obtained by reflecting a quarter through the x and y axes and the unit cell must be modelled in its entirety.

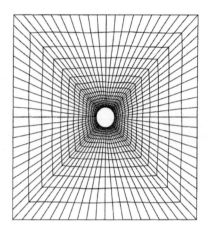

Fig. 5.33 Finite element mesh of crystal plasticity model as pictured in PATRAN.

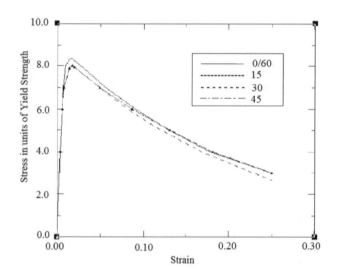

Fig. 5.34 Stress-strain curves for various Ψ angles where the yield strength of the material is 69 MPa.

The mesh is made finer around the void as it is here that the most deformation will occur. The magnitude of the stress in the continuum is expected to be much less than in the atomistic model as the atomistic curve is defined from theoretical considerations for pure single crystals while the continuum code includes parameters fitted to experiments on polycrystals.

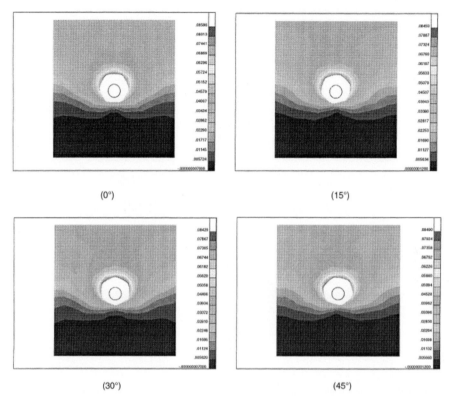

Fig. 5.35 Deformation plots showing accumulated slip at 4% overall strain for 0°, 15°, 30° and 45° orientations.

The crystal orientation is changed by varying the angle w, between the three slip systems, which are oriented at 60° to each other, and the axis perpendicular to the direction of loading as shown in Fig. 5.22(d). The four Ψ angles considered are 0°, 15°, 30° and 45°.

There is a symmetric overlap if further angles are included. The stress-strain curves for the four orientations are shown in Fig. 5.34 where the average true stress, σ_{22}, and true strain, τ_{22}, in the X_2 direction for the unit cell are plotted. As the unit cell is strained, the stress level in each material increases to a maximum value between 2% and 3% strain. On reaching this point, the stress drops very rapidly because matrix strain hardening no longer compensates for the geometric softening due to the growth of the void. The void growth rate increases considerably from this point and eventually reaches a fairly constant value at higher strain. Plots of displacement for the different crystal orientations are shown in Figs. 5.35 and 17 for 4% and 20% overall true strain.

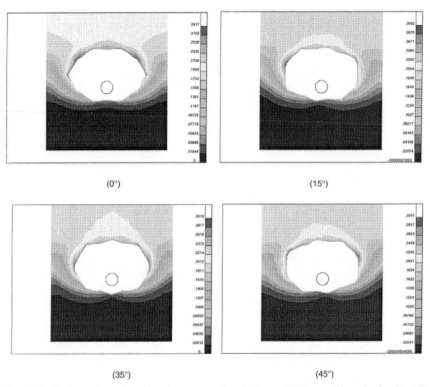

(0°)   (15°)

(35°)   (45°)

Fig. 5.36 Deformation plots showing accumulated slip at 20% overall strain for 0°, 15°, 30° and 45° orientations.

During deformation, the void grows more in the transverse direction than in the loading direction for each lattice arrangement. The void surface tends to develop facets that become aligned with the different slip systems. This results in a non-uniform distribution of plastic strain along the void surface, with peaks occurring at the `corners' between facets. The plastic strain flows in the transverse direction with the result that interaction with neighbouring voids occurs in the transverse direction. There is a significant drop in material strength after the peak stress is reached. This is due to the boundary conditions at the right edge of the unit cell which constrain the side to remain fixed during straining and cause the void to grow faster than would be the case for plane strain tension. The final void volume fraction of each case considered is similar because the boundary conditions are strain controlled.

The material behaviour, as predicted by the crystal plasticity code, proves to be much less sensitive to variations in the crystal orientations than does the atomistic code. This is to be expected as material defined as a continuum cannot be expected to take into account the various irregularities which develop during the course of the analysis and are included from the start. The shape of the voids developed in the crystal code compare very well with the second crystal orientation considered. An almost hexagonal shape is seen in both instances. This is appropriate as the crystal plasticity code works with slip systems aligned at 60° to each other. On an atomic level, the preferred direction of slip on the (111) plane which is active in the second model is along the closed packed directions which are also at 60° to each other. Hexagonal void shapes have been found experimentally [20] in hexagonal close packed (hcp) structures. While the material used in this analysis is fcc, there is a possibility for slip systems to activate at 60° in both cases which suggests these hexagonal shapes also exist in fcc structures.

Conclusions

The results of this work have shown how sensitive crystalline materials really are to the nanostructure. While the atomistic models fail to give an accurate prediction of the global stress-strain response of the material, it is probable that the quantitative information about the shape and size of voids and the direction of crack propagation is very realistic. This type of information cannot be got from the more commonly used phenomenological based constitutive equations. The development of a void from a circular initial shape to an irregular shape is thought to be very realistic. It compares relatively well with the physically based crystal plasticity calculations which contain a certain amount of directional inhomogeneity. Most other constitutive theories are isotropic and so do not have the facility for irregular shapes. In this area atomistic models can be useful in helping to create the correct direction sensitivities for the easier to use phenomenological constitutive theories.

This work can be further used to complement the crystal plasticity code by helping to provide a failure criterion. The crystal code allows voids to grow to volume fractions of over 80%. A real material would have experienced failure long before this volume fraction as the remaining material in the ligament regions between the voids would be unable to sustain the stresses required to carry the applied load. The atomistic code has a built-in failure criterion. Each interatomic bond is continuously updated and when certain stress levels are reached, the bonds are free to break and reform with other atoms if necessary. When a voided crystal is subjected to a critical applied load, the material will break using one or more of a variety of fracture mechanisms. Atomisitic calculations can be run to ascertain the strain levels or void volume fraction to which the crystal code should be used. Likewise, they can tell us what mechanisms of failure occur under what conditions.

Various atomistic deformation mechanisms which are interesting in themselves are clearly seen throughout the work. Dislocation formation movement and piling up, void formation and coalescence and vacancy formation are all seen under uniaxial tensile loading. The simulation results indicate that a number of factors determine which deformation mechanism contributes in what way to the failure of the crystal.

References

[1] Daw M.S., Baskes M.I. (1984), Phys. Rev. B 29, p.12.
[2] Kohlhoff S., Schmauder S. (1989), in: V. Vitek, D.J. Srolovitz (Eds.), Atomistic Simulation of Materials, Plenum, New York.
[3] Kohlhoff S., Gumbsch P., Fischmeister H.F. (1991), Philos. Mag. A 64, p. 851.
[4] Fischmeister H.F., Exner H.E., Poech M.H., Kohlho S.., Gumbsch P., Schmauder S., Sigl, L.S. Spiegler R. (1989), Zeitschrift für Metallkunde, Bd. 80 H. 12, p. 839.
[5] Kitagawa H., Nakatani A. (1991), Mechanical Behaviour of Materials VI, p. 111.
[6] M. Mullins (1984), Acta Metall. Mater. 32, p. 381.
[7] Schmauder S., Ludwig M., Farrissey L., Hu S. (1997), in: Proceedings of Research in Structural Strength and NDE-Problems in Nuclear Engineering, Stuttgart.
[8] Hu S.Y., Ludwig M., Kizler P., Schmauder S. (1998), Modelling Simul. Mater. Sci. Eng. 6, p. 567.
[9] De Hosson J.Th.M. (1996), Computer modelling of dislocations, grain-boundaries and interfaces: its relevance for mechanical properties, in: Workshop Notes, Ameland, Holland.
[10] Spaczer M., Van Swygenhoven H., Caro A. (1998), in: MRS'98 Fall Meeting, Boston.
[11] Asaro R.J., Appl J.(1983), Mech. 50, p. 921.
[12] McHugh P.E., Asaro R.J., Shih C.F. (1993), Acta Metall. Mater. 41, p. 1461.
[13] Voter A.F. (1993), Los Alamos unclassified tech report 93-9001, Los Alamos National Laboratory.
[14] Kohlhoff S., Schmauder S., Gumbsch P. (1990), Bonding, Structure and Mechanical Properties of Metal-Ceramic Interfaces, Pergamon, Oxford, pp. 63-70.
[15] Kohlhoff S. (1988), LARSTRAN - a tool in material sciences, in: Proceedings of the International Finite Element Method Congress, Baden-Baden, Germany.
[16] O'Regan T.L., Quinn D.F., Howe M.A., McHugh P.E. (1997), Comput. Mech. 20, p. 115-121.
[17] Quinn D.F., McHugh P.E., Thesis Ph.D. (1998), National University of Ireland, Galway.
[18] Talyor G.I. (1938), J. Inst. Met. 62, p. 307.
[19] Cahn R.W., Haasen P., Kramer E. (1993), Materials Science and Technology, vol. 1: Structure of Solids, Weinheim.
[20] Crepin J., Breatheau T., Caldemaison D. (1996), Acta Metall. Mater. 44 (12), p. 4927.

5.4 Atomic scale modelling of edge dislocation movement in the α-Fe–Cu system[4]

A detailed understanding of the behaviour of dislocations is essential in determining the mechanical properties of metals and alloys. In recent years a number of molecular dynamic and static simulations [1] were performed on α-Fe single crystals under tensile loading, to account for yield stress, work hardening as well as defect nucleation and growth in this material. Taking into account the presence of precipitates, as for example copper, and the already present dislocations, the flow-stress depends upon the interaction between the dislocation and the obstacle [2, 3]. Ultimately, the strength of the material depends upon the crystallographic structure and dimensions of the precipitates in it [4–6].

While continuum theory can describe the long-range strain fields of cracks and dislocations, atomistic simulations are required to characterize dislocation core structure and dislocation-precipitate interactions. In continuum theory, the dislocations are considered to be smooth flexible strings with a line tension [7]. When large dislocation curvatures are encountered, for instance in the presence of precipitates, the results provided by the line tension approximation can be inaccurate. Here an atomistic simulation model is presented that allows for a smooth movement of two edge dislocations in the absence of applied stresses and thus permits to observe the elastic behaviour of a dislocation line during   the interaction with an obstacle. The use of atomistic modelling also may offer the chance to simulate dislocation phenomena relying on basic atom-scale data, without empirical data on the mesoscopic scale. The case of a spherical coherent Cu precipitate with a body-centred cubic (bcc) lattice identical to that of the α-Fe matrix is considered.

In the following section the computational model is described. In section 3, the critical resolved shear stress of Fe is calculated and a comparison with related values found in literature is provided. Some discussions upon the elastic behaviour of the dislocation line, pinned to the centre or trapped in the Cu precipitate as the size of the obstacle diameter changes follow in section 4.

The computational model

In bcc metals slip occurs in close-packed $\langle 110 \rangle$ directions [8]. The Burgers vector of the perfect slip dislocation is of the type $\frac{1}{2}\langle 111 \rangle$. The motion in the glide plane is that which constitutes the macroscopic phenomenon of slip in crystals [9]. This kind of motion is very easy along the glide direction $\langle 111 \rangle$ in the α-Fe crystal and assures a smooth movement of the dislocations. Starting from this observation and

---

[4] Reprinted from S. Nedelecu, P. Kizler, S. Schmauder, N. Moldovan, "Atomic Scale Modelling of Edge Dislocation Movement in the α-Fe-Cu System", Modelling and Simulation in Materials Science and Engineering 8, pp. 181-191 (2000) with kind permission from Elsevier

from Frank's rule [8] for determining whether or not it is energetically feasible for two dislocations to react and combine to form another one, the present simulation model was constructed. The movement of edge dislocations in the absence of externally applied shear stress will be investigated. Edge dislocations in bcc metals, unless they are locked by impurities, are much more mobile than screw dislocations [10]. As a result, in high-purity specimens, yielding takes place first by motion of edge dislocations at a low stress [10]. Only at a later stage of deformation, which is beyond the scope of the present work, the specimens are exhaustion hardened because multiplication cannot occur without the motion of screw dislocations.

In the present model, the atoms were initially placed on perfect bcc crystal lattice sites of α- Fe. The coordinate axes were chosen parallel to the sides of the simulation cell with the x-axis along $\langle 110 \rangle$ y-axis along $\langle 111 \rangle$ and z-axis along, $\langle \overline{1}12 \rangle$. Two initial straight edge dislocations, ending at free surfaces, with the same slip plane $(1\overline{1}0)$ and opposite Burgers vectors b $= \langle 111 \rangle$ were introduced along the z-direction by removal of two half-planes of atoms.

The initial points where the line of the first and the second dislocation intersected the y-axis were chosen at the origin of the coordinate system and at 115 Å along the y-axis, respectively. In order to preserve the invariance to free translations and rotations, at the left and at the right ends of the sample, see figure 5.37, six atom layers perpendicular to the y-axis were kept fixed at their initial positions. A schematic representation of a section through the sample, showing the initial position of the edge dislocations and the Cu atoms is presented in figure 5.37. In the present molecular dynamics (MD) simulation a number of 82 600 atoms was considered, and the MD program FEAt developed by Kohlhoff and Schmauder [11] was employed. As input to FEAt, a data file with the undisturbed coordinates of the atoms together with the initial displacement field of the two unlike edge dislocations was created. To calculate the initial displacement field of one single dislocation the theory of a moving edge dislocation described by Stroh [12] was used. For an edge dislocation parallel to the z-axis and with the glide plane z–y, the only displacement components are $u_i = (u_x, u_y, 0)$.

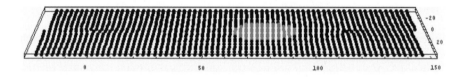

Fig 5.37 Schematic representation of a section through the sample, showing the initial position of the edge dislocations and the Cu atoms (grey).

A general expression for the displacements, according to anisotropic elasticity, may then be written as

$$u_i = \mathrm{Re}\left(\frac{1}{2\pi d}\sum_{\alpha=1}^{3} A_i(p_\alpha)D_\alpha \log(x + p_\alpha y)\right) \qquad (i = x, y) \qquad (5.22)$$

where A_α and D_α are tensors of material properties expressed in the chosen coordinate system, depending on the components of the Burgers vector and the elastic constants. The numerical values used [12] are given in table 5.9.

Table 5.9
Numerical values of the coefficients which define the displacement field (1) of an edge dislocation along the $\langle\overline{1}12\rangle$ direction in the α-Fe crystal.

$A_x (p_1)$	0.101	D_1	-0.685-0.183I
$A_x (p_2)$	-1.414	D_2	0
$A_x (p_3)$	0.101	D_3	0.685-0.183
$A_y (p_1)$	0.778+0.184 I	p_1	-1.378-1.298
$A_y (p_2)$	0	P_2	0
$A_y (p_3)$	-0.778+0.187 I	P_3	1.378-1.298

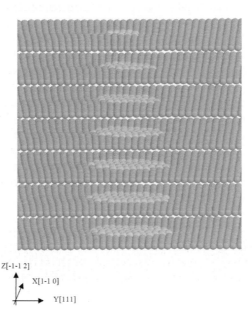

Z[-1-1 2]

X[1-1 0]

Y[111]

Fig 5.38 Detailed structure of one of the dislocation cores during dislocation migration in the vicinity of the obstacle. The Fe atoms are yellow, the Cu atoms are grey. The distance along the z-axis between the upper and the bottom plane is 6 x 1.76 Å = 10.56 Å.

The displacement field of the second dislocation was simply considered as having the form (1), with the proviso that the initial distance between the dislocations is large enough such that the displacement field of it matches well with that of the first dislocation and interactions do not appear. The introduced Cu precipitate, with a bcc structure and with the same lattice parameter for Cu as for α-Fe, is placed close to one of the edge dislocations (figure 5.37). The MD simulation is carried out for two values of the diameter of the Cu precipitate: 13.2 and 30.4 Å. These precipitates consist of 121 and of 1254 Cu atoms, respectively.

The interaction potential for the α-Fe-Cu system taken into consideration was recently constructed [13] using the embedded atom method (EAM). It had been shown previously that EAM potentials allow us to follow metallic systems where fracture, surfaces, impurities and alloying additions (additives) are included [14]. In addition to pair wise interactions, using the EAM method the total energy includes an embedding energy as function of the local atomic density. The actual parameters used by the EAM in the case of iron and copper are described elsewhere [15, 16].

The model set-up containing the two initial straight edge dislocations and the Cu precipitate was then equilibrated for 20 000 time steps equivalent to 40 ps at a temperature of $T_0 = 10$ K.

5.4.1 The movement of an edge dislocation hitting a Cu precipitate

A detailed view of the core of one dislocation, which has already penetrated the obstacle, is presented in figure 5.38 for the regions inside and outside the precipitate. For each slice, the intersection between the dislocation line and the lattice plane can be recognized visually. During the interaction between the edge dislocation and the obstacle, the dislocation line does not remain straight. The dislocations do not move as rigid entities, but via the kink pair mechanism. In order to identify the position of the dislocation lines also by means of an automatized computing algorithm, the maximum of the Burgers vector density distribution [17] was calculated for a cut along a $(1\bar{1}0)$ plane for all 65 x-y layers of the simulation cell and for both precipitate examples. The calculations were performed for a large number of simulational results with equidistant time steps. The most interesting snapshots are presented in figures 5.39 and 5.40. Comparing the results obtained by running the MD simulation of α-Fe for different situations - with and without precipitates - it could be seen that the presence of the obstacles on the glide plane of a moving dislocation reduces the internal shear stress and impedes the movement.

The simplest case of a simulation model containing two unlike edge dislocations and without the Cu precipitate was considered first. In this case, the initial straight lines of the dislocations pre-served their shapes until the end of the relaxation process, when the two dislocations were no longer distinguishable. The movement of the dislocations was taking place under no external shear stress. This result is in good agreement with Frank's rule describing several dislocations that might 'associate' to form a single dislocation.

The difference in elastic moduli between copper and iron can account for the observed influence of the copper precipitate on the movement of the edge dislocation. From this point of view the copper precipitate has an attractive influence on the edge dislocation.

Case 1. Smaller precipitate, diameter 13.2 Å, figures 5.39(a)-3.39(i):

Starting from the initial position, the movement of the dislocation line takes place such that it is curved toward the precipitate, see figure 5.39(b) in comparison to figure 5.39(a). Further on, the edge dislocation passes through the precipitate and after passing, a backward bowing can be recognized, see figure 5.39(h), indicating the persisting attractive force between the precipitate and the dislocation line. Altogether, the movement of the dislocation takes place almost without any impediment, see also the discussion of the Peierls stress in the previous section and the appendix.

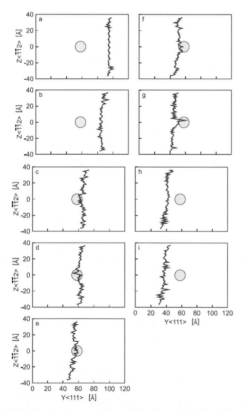

Fig 5.39 Projections on the glide plane (1–10) of the nearest atoms above and below the glide plane that constitute the edge dislocation lines. The Cu precipitates (diameter 13.2 Å) are represented by the circle.

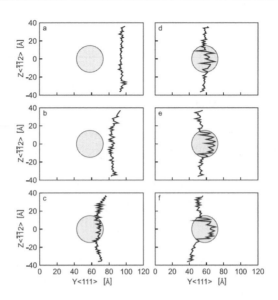

Fig 5.40 Projections on the glide plane (1−10) of the nearest atoms above and below the glide plane, that constitute the edge dislocation lines. The Cu precipitates (diameter 30.4 Å) are represented by the circle.

Case 2. Larger precipitate, diameter 30.4 Å, figures 5.40(a)-5.40(f):

In the case of the 30.4 Å diameter Cu precipitate a very elastical behaviour of the dislocation line hitting the obstacle could be observed. Passing does not happen and the dislocation line is pinned by the precipitate, with the free ends oscillating. The dislocation is not able to cut the obstacle. It can only pass through the precipitate completely as soon as an external shear stress is applied to increase the strain beyond the Peierls stress. During such calculations (figures not included), the angle between the dislocation arms did not become sharper than that in figure 5.40(f).

The Peierls stress of edge dislocations in the presence of the Cu precipitate

The difference in the shear moduli μ_{Fe} = 86 GPa and μ_{Cu} = 54.6 GPa [7] of the α-Fe matrix and of the Cu particle, respectively, gives rise to an elastic interaction force between particle and dislocations. In the following this force will be calculated for the two cases of Cu precipitate diameters and the resulting Peierls stress—the minimum stress required to move a dislocation by one lattice plane distance. The interaction energy E_μ between the Cu precipitate and the edge dislocations can be obtained using the following formula [18]:

$$E_\mu = \sum_{i=1}^{N} (E_{Cu}^i - E_{Fe}^i) \qquad (5.23)$$

where N is the total number of the Cu atoms for a given diameter of the precipitate, E_{Cu}^i is the energy of the ith Cu atom and E_{Fe}^i is the energy of the ith Fe atom, when only Fe atoms are present.

The values E_{Cu}^i and E_{Fe}^i of the ith atom energy were calculated using the EAM potential and running two MD simulations for the same geometry of the model, firstly containing a Cu precipitate and secondly for a simulation cell consisting only of Fe atoms, without Cu atoms. The numbering rule of the atoms was kept the same in both cases. Considering the position y of the edge dislocation as the point where the line crosses the plane $z = 0$, one can represent the interaction energy $E\mu$ as function of the distance y along the $\langle 111 \rangle$ axis, see figures 5.41(a) and 5.41(c).

The dotted lines denote the boundaries of the Cu precipitate and the symbols and the connecting lines denote the position of the dislocation core along the y-axis. Several of the points correspond to the images of figures 5.39 and 5.40, respectively. For the precipitate with a diameter of 13.2 Å, an abrupt change of the energy per Cu atom can be observed after the dislocation has reached the precipitate. The same kind of change takes place later when the dislocation exits the precipitate again.

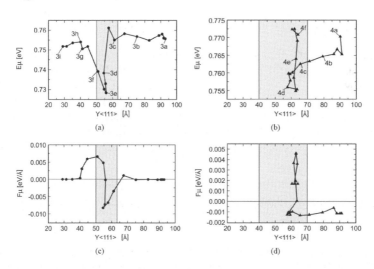

Fig 5.41 Interaction energy per Cu atom $E_\mu=N$ as function of the distance y along the <111> axis of the simulation cell (a and c), and the corresponding atom-dislocation interaction force F_μ (b and d). The pair plots a and b, c and d correspond to a 13.2 Å and 30.4 Å diameter Cu precipitate, respectively. The shaded area marks the location of the precipitate. Movement of the dislocation is from right to left. The energy plots contain also the figure numbers of the corresponding figure numbers in figures 5.39 or 5.40.

Deriving $E\mu$ with respect to y yields the particle—dislocation interaction force $F\mu$ [18], see figures 5.41(b) and 5.41(d):

$$F_\mu = -\frac{\partial E_\mu}{\partial y} \tag{5.24}$$

The maximum interaction force that the whole 13.2 Å diameter Cu precipitate exerts on the dislocation is attained when the edge dislocation is exiting the precipitate. The maximum value equals the forces as displayed in figure 5.41(b), multiplied with the total number of 121 Cu atoms of the precipitate and amounts to about 0.80 eV Å^{-1}. However, the maximum negative value of the interaction force when the edge dislocation is first in contact with the precipitate is not attained at the particle-matrix interface. This value is reached at the time when the edge dislocation is stopped for a short moment inside the precipitate, before completely shearing the particle and moving further.

For the 30.4 Å diameter Cu precipitate, the maximum positive value of the interaction force was calculated as the force per atom (see figure 5.41(d)) times the total number of 1254 Cu atoms in the precipitate and amounts to 5.75 eV Å^{-1}. At this size of the obstacle, the edge dislocation cannot pass through and its motion will be blocked. Additional external force has to be applied to continue the dislocation movement. The energy calculation was currently not possible within the frame of the calculational arrangment presently used. The upper bound of the Peierls stress τ_P was calculated using the basic equation [18] of strengthening by shearable particles:

$$b\tau_p = \frac{F_0}{L_C} \tag{5.25}$$

where b is the Burgers vector, τ_P is the calculated Peierls stress, F_0 is the maximum value of the particle-dislocation interaction force, derived as explained above, see (3), and L_c is the minimum length of the dislocation line, 75.2 Å for the considered simulation cell.

The calculated Peierls stress τ_P required to move the dislocation through the Cu precipitate in the case of the 13.2 Å diameter Cu precipitate is 0.0084 times the iron shear modulus, i.e. τ_P = 0.72 GPa. A value of 0.059 times the iron shear modulus (τ_P = 5.07 GPa) results in the case of the 30.4 Å diameter Cu precipitate. The core structure and Peierls barrier for an edge dislocation lying in the $\{110\}$ plane with the Burgers vector along $\langle 111 \rangle$ in bcc iron without Cu was also investigated by Chang and Graham [19] using an anharmonic potential. The calculated Peierls barrier was about 0.03 eV and the Peierls stress for dislocation motion at absolute zero temperature was 5.36 x 10^9 dyn cm^{-2} or 0.0066 times the shear modulus of Fe (0.567 GPa). In the present results of τ_P for the smaller precipitate compare closely with those of Chang and Graham, taking into account the presence of the Cu atoms in our model, and also with experimental values for pure Fe at very low temperatures [20], which range between 0.34 and 0.42 GPa. In other words, for the case of the small precipitate, the Peierls Stress is quite close to the case of pure Fe. The present calculations rely on no other physical assumptions than on the interatomic potentials, which base themselves on basic elastic constants. This suggests that the strengthening of the iron-rich iron-copper system derives from the modulus mismatch between particle and matrix and from no other strengthening mechanism such as, for example, lattice constant mismatch strengthening.

5.4.2 Derivation of dispersion strengthening from modelling

The bent dislocation with the backlash inside the precipitate is to be regarded as a part of a whole dislocation line, typically bowing between two obstacles, such as it is known from TEM images [21]. Considering a dislocation bowing between a pair of such obstacles, together with the assumption of a conventional constant line tension approximation, the angle between the two dislocation line branches on either side of the precipitate, together with the distance between the obstacles, is the key parameter to calculate the increase in matrix strength due to precipitation strengthening [2, 22]:

$$\tau = \frac{Gb}{L}\left(\cos\frac{\Phi}{2}\right)^{3/2} \qquad (5.26)$$

where τ is the matrix strengthening by precipitates, G is the shear modulus of the matrix, b is the burgers vector in the matrix, L is the obstacle spacing in the slip plane and Φ is the critical angle between the arms of the dislocation at which the obstacle is cut.

A detailed study of the dislocation line (see, e.g. figure 5.40(f)) permits one to determine the angle between the arms of the dislocation line on the glide plane. This smallest achievable angle between the two dislocation branches corresponding to the maximum of strengthening amounts to 140° in the case of the precipitate with a diameter of 30.4 Å. This value agrees very well with the critical angle as calculated from the mesoscopic continuum theory formalism by Russell and Brown, using as input the precipitate radius together with several empirically determined parameters (formula (2) and figure 5.38 in [2]), see appendix. For the smaller precipitate, the calculation along the Russell-Brown formalism yields a critical angle of $\Phi = 171°$, which means negligible strengthening, in agreement with the simulation results of figure 5.39.

In a recent study on the relationship between structural information about Cu precipitates in a steel as derived from TEM images and macroscopical mechanical data of the same steel, the approach of Russell-Brown has been proven to be reliable [5, 6].

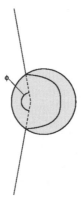

Fig 5.42 Scheme of a dislocation cutting a precipitate to explain the definition of the critical angle.

The present investigations have shown that these calculations can also base on results from nanosimulation instead of such from the mesoscopic calculations (Russell-Brown). In contrast to the Russell-Brown formula (formula (2) in [2]), which uses empirically determined mesoscopic material parameters [2], in the case of nanosimulation the atomistic calculations are based solely on physical interatomic potentials.

Conclusions

A MD simulation was performed to understand the detailed mechanism of the complex interaction between a moving edge dislocation and differently sized Cu precipitates in the α-Fe crystal. The model set-up contained two edge dislocations, sufficient by itself to permit an attractive movement under no external stress, and main attention was paid to the interaction of one of them with the Cu precipitate. Based on Frank's rule, the constructed model revealed the elastic behaviour of the edge dislocations and its strong dependence on the size of the obstacle. For a 32 Å diameter of the precipitate, the pinning process of the dislocation centre and also the trapping of the dislocation line in contrast to an obstacle diameter of 13 Å could be made evident. The calculated Peierls stress from the present MD simulation compares closely with other published values.

The precipitates acting as obstacles to dislocation movement induce bowing of the dislocation lines. The present calculations enabled us to derive the critical angles of the dislocation lines at the Fe-Cu interface, which are in perfect agreement to that obtained from mesoscopic dislocation theoretical calculations. These angles provide a direct connection to the numerical values of the increase in strength of such a model crystal. This means, that the present modelling of dislocation movement through a precipitate provides, for the first time in materials science, a way to simulate the precipitation hardening from basic principles (atomic properties) to a macroscopically relevant material's property.

Appendix

The critical angle between two edge dislocation branches can be calculated following the mesoscopic continuum theory formalism by Russell and Brown [22]. The strength of an alloy in the overaged condition can be calculated following the methods of Brown and Ham [2, 22]. The stress, at which a dislocation can move through an array of obstacles is identified with the yield stress τ and is a function of the obstacle spacing L in the slip plane and the critical angle Φ at which a dislocation can cut an obstacle. The shear stress is given by

$$\tau = \frac{Gb}{L}\left(\cos\frac{\Phi}{2}\right)^{3/2} \tag{5.27}$$

where G is the shear modulus and b is the Burgers vector of the dislocation. Russell and Brown derived the shear stress from the relationship between the energies of the dislocation per unit length inside (E_1) and outside (E_2) the precipitate as

$$\tau = \frac{Gb}{L}\left(1 - \frac{E_1^2}{E_2^2}\right)^{3/4} \tag{5.28}$$

Therefore,

$$\left(\cos\frac{\Phi}{2}\right)^{3/2} = \left(1 - \frac{E_1^2}{E_2^2}\right)^{3/4} \tag{5.29}$$

where the energies of the dislocation length inside ($E1$) and outside ($E2$) the precipitate depend on the precipitate radius as

$$\frac{E_1}{E_2} = \frac{E_1^\infty \log(r/r_0)}{E_2^\infty \log(R/r_0)} + \frac{\log(R/r)}{\log(R/r_0)} \tag{5.30}$$

where E_1^∞ and E_2^∞ refer to the energies per unit length of a dislocation in infinite media (Fe or Cu, respectively), r is the precipitate radius and R and $r0$ are the outer and inner cut-off radii used to calculate the energy of the dislocation. Russell and Brown verified the validity of the following values for the Fe-Cu system:

$E_1^\infty / E_2^\infty = 0.6$, $r_0 = 2.5b$ with b = Burgers vector of the dislocation = 2.48 Å, and, finally $R = 10^3 r_0$. For precipitate radii of 16 Å and 6.5 Å, the formalism results in critical angles of $\Phi = 140°$ and $171°$, respectively.

References

[1] Hu S Y, Ludwig M, Kizler P and Schmauder S (1998), Modelling Simul. Mater. Sci. Eng. 6, p. 567 and further references therein.

[2] Russell K C and Brown C M (1972), Acta Metall. 20, p. 969.

[3] Pretorius T, Rönnpagel D and Nembach E (1998), Proc. 19th Risø Int. Symp. on Materials Science: Modelling of Structure and Mechanics of Materials from Microscale to Product ed J V Christensen et al (Roskilde, Denmark: Risø National Laboratory), p. 443.

[4] Kelly A and Nicholson (1971), Strengthening Methods in Crystals (Amsterdam: Elsevier).

[5] Uhlmann D, Kizler P and Schmauder S (1998), Proc. 19th Risø Int. Symp. on Materials Science: Modelling of Structure and Mechanics of Materials from Microscale to Product ed J V Christensen et al (Roskilde, Denmark: Risø National Laboratory) p. 529.

[6] Kizler P, Uhlmann D and Schmauder S (2000) Nucl. Eng. Design 196, p. 175.

[7] Hirth J P and Lothe J (1982), Theory of Dislocations (New York: Wiley).

[8] Hull D and Bacon D J (1984), Introduction to Dislocations (Oxford: Pergamon).

[9] Suzuki T, Takeuchi S and Yoshinaga H (1985), Dislocation Dynamics and Plasticity (Berlin: Springer).

[10] Suzuki T, Takeuchi S and Yoshinaga H (1985) Dislocation Dynamics and Plasticity (Berlin: Springer) ch 6.3, p. 1.

[11] Kohlhoff S and Schmauder S (1989), Atomistic Simulation of Materials ed V Vitek and D J Srolovitz (New York: Plenum) Kohlhoff S, Gumbsch P and Fischmeister H F 1991 Phil. Mag. A 64, p. 851.

[12] Stroh A N (1962), J. Math. Phys. 41, p. 77.

[13] Ludwig M, Farkas D, Pedraza D and Schmauder S (1998), Modelling Simul. Mater Sci. Eng. 6, p. 19.

[14] Daw M S and Baskes M I (1984) Phys. Rev. B 29, p. 6443.

[15] Voter A F and Chen S P (1987) Mater. Res. Soc. Symp. Proc. 82, p. 175.

[16] Voter A F Los Almos Unclassified Technical Report #LA-UR-93-3901

[17] Schroll R, Vitek V and Gumbsch P (1998) Acta Mater. 46, p. 917.

[18] Nembach E (1997) Particle Strengthening of Metals and Alloys (New York: Wiley) ch 5, p. 2.

[19] Chang R and Graham L J (1966) Phys. Status Solidi 18, p. 99.

[20] Brunner D and Diehl J (1992) Z. Metallkd. 83, p. 828.

[21] For example, Othen P J, Jenkins M L and Smith G D W (1994) Phil. Mag. 70, p. 1.

[22] Brown L. M. and Ham R K (1971) Strengthening Methods in Crystals ed A Kelly and R B Nicholson (Amsterdam: Elsevier) p. 12.

5.5 Molecular dynamics study on low temperature brittleness in tungsten[5]

The brittleness and ductility of materials have been major subjects of materials science. Research into brittle and ductile characteristics has advanced greatly in recent years. One source of this progress is the Rice- Thomson formulation (Rice and Thomson, 1974) of a dislocation emission, with its later improvement (Rice, 1992). The formulation, based on the competition between dislocation emission from a crack tip and cleavage, has successfully explained the intrinsic ductility of most fcc metals and cleavability in most bcc metals. However, a thermal activation process had not been considered, and the formulation was not sufficient to explain the brittle to ductile transition (BDT).

Fracture toughness in most bcc metals is influenced by temperature (Ha et al., 1994; Gumbsch et al.,1998). The toughness of metals increases with temperature, and the materials never cleave above a critical temperature. Even semiconductors (lohn, 1975) and ionic crystals (Narita et al., 1989), which are concluded to be intrinsically brittle materials in the Rice-Thomson formulation, show BDT characteristics. What mechanisms cause the temperature dependency of toughness and the brittle to ductile transition, is still a question that has not been answered satisfactorily until now. Some groups (Zhou and Thomson, 1991; Rice and Beltz,1994; Xu and Argon,1995) insist that dislocation emission is the controlling factar, whereas others (Hirsch and Roberts, 1991; Maeda, 1992) insist on dislocation mobility. Several remarkable models have been proposed in the discussion of dislocation emissions and dislocation mobility. Zhou and Thomson (1991) proposed a dislocation emission model from the ledge of a crack front, which enables dislocations to be emitted at much lower extern al loading than in the Rice- Thomson formulation. The dislocation emission from the ledge of a crack front provides a good explanation for the river patterns on the fracture surfaces (George and Michot, 1993), as well as the observation of ten or fewer dislocations per slip plane (Michot et al., 1994). Hirsch (1991) proposed a computer simulation method far the generation and motion of the dislocations from crack tips, where the dynamics of emitted dislocations were taken into account.

In this study molecular dynamics (MD) has been applied to investigate the process of brittle fracture and the temperature dependency of fracture toughness. Molecular dynamics is an effective tool for the analysis of a crack. The technique enables us to analyse directly the events occurring on an atomic scale, such as dislocation emissions and cleavage in the crack tip region. However, modem computers are only capable of treating nano-scale material specimens - in the order of

[5] Reprinted from Y. Furuya, H. Noguchi and S. Schmauder, "Molecular Dynamics Study on Low Temperature Brittleness in Tungsten Single Crystals", International Journal of Fracture 107, pp. 139-158 (2001) with kind permission from Kluwer/Springer

10^6 atoms. This problem is fatal in the simulation of a crack because periodic boundary conditions can not be assumed in all directions. In such case it is necessary to combine molecular dynamics with continuum mechanics. Molecular dynamics should be applied only to the crack tip region and continuum mechanics should then be applied to the surrounding region.

In early research into finding a method of combining molecular dynamics with a continuum, a major area of investigation had been how to synchronise the deformation of a continuum region with that of a molecular dynamics region. Several groups (Mullins and Dokanish, 1982; Mullins, 1982; Kohlhoff and Schmauder, 1988; Kohlhoff et al.,1991) proposed a method to combine molecular dynamics with a finite element method (FEM), where nodes of finite elements were synchronised with atoms of MD in a boundary region. Others (Sinclair et al., 1978; Hirth et al., 1974) proposed a method of correcting the boundary conditions using Green's function. However, the problem with these methods is that the emitted dislocations from a crack tip can not pass smoothly through the boundary between the molecular dynamics and the continuum regions. Yang et al. (1994) firstly proposed a method where emitted dislocations could pass through the boundary. In their method the continuum region was divided into two regions. The outer region was calculated with a finite element method, and the inner region was calculated with an elastic continuum where the movement of dislocations was analysed dynamically. Yang's method, however, has a limitation in the number of emitted dislocations, and in the validity of the method, which was not examined satisfactorily.

The authors developed a new method (Noguchi and Furuya, 1997; Furuya and Noguchi, 1998) in which molecular dynamics was combined directly with linear elastic theory, that is micromechanics (Eshelby et al., 1951; Chou, 1967; Mura, 1968; Lekhnitski, 1968). A thorough examination of the validity of the method was undertaken. In the new method the dislocations emitted in the molecular dynamics region can pass through the boundary of the two regions smoothly, and are distributed at the equilibrium positions in the micromechanics region according to the elastic solution. That is to say that the dynamics of dislocations was not under consideration. The simulation was then presumed to be quasi-static. The limitation of dislocation emissions was removed by moving the molecular dynamics region with the crack propagation.

Crack tip opening displacements calculated in the simulation with the method showed good agreement with an analytical solution derived by Rice (1974). The combined model is limited to two-dimensional and quasi-static simulations. It means that the ledge of a crack front and the effect of strain rate can not be taken into account. However, the limitation does not mean that the temperature dependency of dislocation mobility is neglected, because friction forces acting on each dislocation, which reflect dislocation mobility, depend on temperature. The difference between a dynamic simulation and a quasi-static simulation is merely whether the distribution of dislocations is dynamic or in equilibrium.

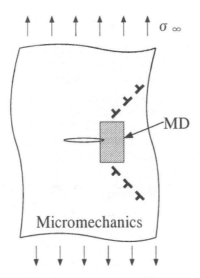

Fig 5.43 Combined molecular dynamics and micromechanics model.

In this section, brittle fracture processes at low temperature are simulated with the combined model of molecular dynamics and micromechanics in tungsten single crystals. The mechanisms of brittle fracture toughness are investigated.

5.5.1 A combined model of molecular dynamics with micromechanics

The principle of the combined model

The combined model of molecular dynamics with micromechanics is shown in Figure 5.43. An infinite plate exhibiting a crack and dislocations is subjected to uniform tension applied at infinity. The deformation of the hatched region in Figure 5.43 is analysed with molecular dynamics, and that of the surrounding region is analysed with micromechanics. A periodic boundary condition is applied to the molecular dynamics region in the direction of plate thickness, and the micromechanics region is analysed as a plane strain problem in two dimensions.

A model in the molecular dynamics region is shown in Figure 5.44. A free atom is defined as an atom that moves according to the molecular dynamics algorithm with thermal oscillations. A boundary atom is defined as an atom that moves without thermal oscillation according to the displacement calculated with micromechanics, that is the boundary atom layer is apart of the micromechanics region. As shown in Figure 5.44, the crack in the molecular dynamics region is expressed by removing two layers of atoms. A quasistatic simulation, the detail of which is explained elsewhere (Furuya and Noguchi, 1998), is presumed in this model. Temperature of the molecular dynamics region is kept constant using a velocity scaling technique.

Remarkable points in this model are:

- The boundary condition to combine two regions is flexible and both displacement and stress fields are continuous at the boundary.
- Emitted dislocations in the molecular dynamics region pass through the boundary smoothly.

The molecular dynamics region moves with the crack propagation. The details of these three points are explained in the following sections.

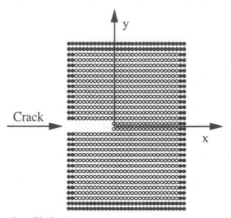

○: Free atom with thermal oscillation
●: Boundary atom without thermal oscillation

Fig 5.44 Molecular dynamics model.

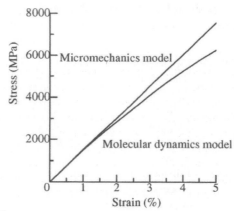

Fig 5.45 Stress-Strain curves.

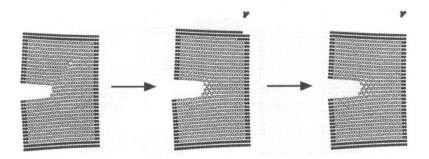

Fig 5.46 Transformation from an atomistic dislocation to an elastic dislocation.

Flexible Boundary Conditions using Body Forces

Figure 5.45 shows stress-strain curves of smooth specimens in the process of elastic deformation. A stress-strain curve calculated with molecular dynamics is shown together with a linear line used in micromechanics. As shown in Figure 5.45, the deformation of the molecular dynamics region is intrinsically non-linear. In turn a rigid boundary condition, where only a displacement field is continuous at the boundary is not satisfactory, because a stress field is discontinuous in this case.

In our combined model the boundary condition, which is basically rigid, is corrected with body forces (Eshelby, 1957). Body farces distributed at the boundary influence both the stress field and the displacement field. This procedure is similar to Flex-II (Sinclair et al., 1978). In Flex-II the balance of farces acting on each atom is considered at the boundary and Green's functions are used far correction. In our model the balance of stress at the boundary is considered and the body forces are used for correction.

5.5.2 Transformation from an atomistic dislocation to an elastic dislocation

Dislocations are distributed at equilibrium positions in quasi-static simulations, that is the dynamics of dislocation movements are not considered in the combined model. In the case when the equilibrium position of an emitted dislocation is in the micromechanics region, the dislocation must move across the boundary of the two regions. A method far moving an emitted dislocation from the molecular dynamics region to the micromechanics region is illustrated in Figure 5.46. A displacement field caused by slip is applied to the boundary atom's layer. After that, the molecular dynamics region is smoothed. The re-smoothing procedure contributes to avoidance of difficulties arising from hard distortion of the molecular dynamics region after several slips have occurred. The dislocation from the molecular dynamics region is transformed into an elastic dislocation and distributed at the equilibrium position.

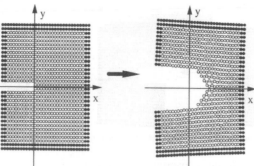

Fig 5.47 Simulation result of crack propagation.

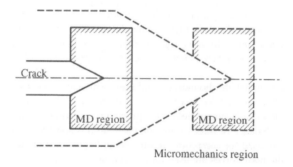

Fig 5.48 Movement of the molecular dynamics region with crack propagation.

The dislocations become edge type. In the case when the equilibrium position of the emitted dislocation is in the molecular dynamics region, the above procedure should not be applied. Equilibrium positions of the dislocations are derived from a balance between driving forces and friction farces acting on each dislocation. The driving forces include loading stress, interaction with other dislocations and an attractive force from the free surface of a crack. The friction force should be Peierls stress in the case of a perfect crystal. However, critical resolved shear stresses (CRSS), obtained in experiments, and may be better for simulations in comparison with experimental results. This is because practical crystals used in experiments include defects such as pre-exiting dislocations and inclusions.

Movement of a Molecular Dynamic Region with Crack Propagation

Figure 5.47 shows the simulated result of the crack propagation process with emission of dislocations. The crack tip moves with crack propagation so quickly reaches the boundary between the molecular dynamics region and the micromechanics region. The crack propagation simulation must be stopped when the crack tip reaches the boundary.

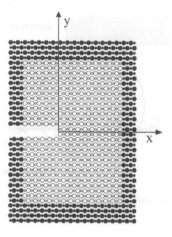

○ : First layer of free atoms
∘ : Second layer of free atoms
● : First layer of boundary atoms
• : Second layer of free atoms

Fig 5.49 Molecular dynamics model for tungsten single crystals.

This means that the length of crack propagation that can be simulated and the number of dislocations that can be emitted, depends on the size of the molecular dynamics region. This limitation is fatal because of the limited capacity of computers.

In order to remove the limitation, the molecular dynamics region moves with crack propagation. The basic idea is illustrated in Figure 5.46. The crack tip could be kept in the molecular dynamics region with this procedure. Therefore, the simulation is no longer limited by the size of a molecular dynamics region

Table 5.10

Calculation conditions and material properties

Temperature	77 K
Pre-crack length	2 mm
Young`s modulus (plane strain)	445.7 GPa
Poisson`s ratio (plane strain)	0.390
Shear modulus	160.6 GPa
CRSS (Bucki et al., 1979)	450 MPa

5.5.3 Simulation of a brittle fracture process in tungsten single crystals

Calculation conditions and additional procedures for the Simulation of tungsten single crystals

Tungsten single crystals are appropriate specimens for brittle fracture simulations because of the brittleness in spite of being single crystals. An N-body potential derived by Finnis and Sinclair (1984) was used in the simulations. Figure 5.49 shows a molecular dynamics model for tungsten single crystals. Two layers of atoms were accumulated in the direction of plate thickness and a periodic boundary condition was applied. In this model the crack face was in the (110) plane and the crack front direction was (110). Temperature, pre-crack length, bulk moduli and CRSS for the simulations are shown in Table 5.10. The CRSS value, obtained in simulations in which the crystallographic orientations corresponded to those of Riedle's experiments, (121) cleavage was expected to occur.

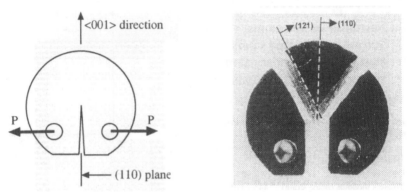

(a) Geometry and orientation of the round compact tension specimen.

(b) Broken specimen

Fig 5.50 Broken specimen of a fracture toughness test at 77 Kin a tungsten single crystal (Riedle et al., 1994).

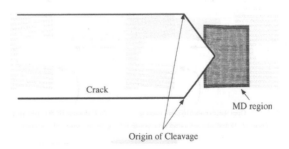

Fig 5.51 Origin of cleavage in ca se of (121) cleavage.

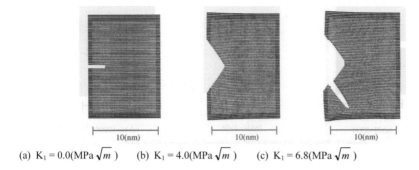

(a) $K_1 = 0.0(MPa \sqrt{m})$ (b) $K_1 = 4.0(MPa \sqrt{m})$ (c) $K_1 = 6.8(MPa \sqrt{m})$

Fig 5.52 Result (Furuya and Noguchi, 1999) of a simulation for cleavage in case the crack tip is blunted. In this case friction forces acting on each dislocation were 1000 MPa.

To achieve (121) cleavage the combined model required an additional procedure. Figure 5.51 shows crack tip shape and the position of the molecular dynamics region in the crack tip after the crack has been opened. In the case of (121) cleavage, the origins of the cleavage are not expected to be the center of the crack tip but the edges of the crack tip as indicated in Figure 5.51. The problem then is that origins exist outside of the molecular dynamics region. In the combined model, cleavage occurrence is dependent on the molecular dynamics calculation. In turn simulations would contain errors if the origins were not in the molecular dynamics region. Figure 5.52 shows the result of one such simulation run without any additional correcting procedure (Furuya and Noguchi, 1999). In this case cleavage occurred from a boundary between the molecular dynamics region and the micromechanics region. The result is obviously wrong because the boundary between the two regions does not exist in real materials. The additional procedure for correcting this problem is illustrated in Figure 5.53. Dislocation emissions cause crack opening and crack tip blunting. The more the dislocations are emitted, the more the crack tip is blunted. This crack tip blunting leads to the crack tip edges escaping from the molecular dynamics region.

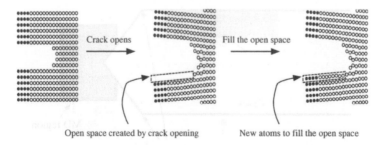

Fig 5.53 Method to prevent a crack from opening by filling the open space with new atoms.

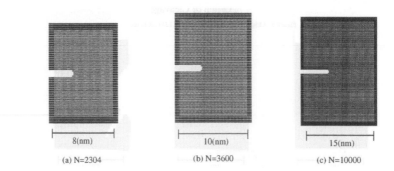

| 8(nm) | 10(nm) | 15(nm) |
| (a) N=2304 | (b) N=3600 | (c) N=10000 |

Fig 5.54 Three sizes of molecular dynamics regions *(N:* number of free atoms).

The point of the additional procedure is to keep the crack closed. In the procedure shown in Figure 5.53, the open space created by a dislocation emission is filled with new atoms and relaxation calculations are performed.This procedure prevents the crack from opening and keeps the crack tip edges in the molecular dynamics region. This is an approximation and is accompanied by the problem that now the influence of crack tip blunting is removed. However, the procedure is hardly expected to influence the intrinsic crack behavior, and the radius of a crack tip with blunting would remain so small compared to the crack length that its effects are negligible in comparison to the brittleness of tungsten single crystals.

Simulation results and size dependency of the molecular dynamics region on the results

Three sizes of molecular dynamics regions (see Figure 5.54) were used in simulations to investigate size dependency. The numbers of free atoms in each were 2304, 3600 and 10000. Figure 5.55 shows the result of a simulation containing 3600 free atoms. In the simulation, cleavage occurred not from the boundary between the molecular dynamics region and the micromechanics region but from the edge of the crack tip along a (121) plane.

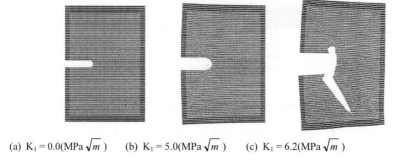

(a) $K_1 = 0.0(MPa \sqrt{m})$ (b) $K_1 = 5.0(MPa \sqrt{m})$ (c) $K_1 = 6.2(MPa \sqrt{m})$

Fig 5.55 Result of a simulation with a N = 3600 model (N: number of free atoms).

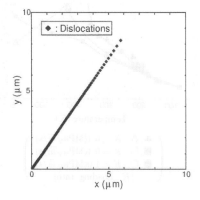

Fig 5.56 Distribution of dislocations at the loading when the material was fractured at 77 K. Only upper side is shown and the origin means a crack tip.

The (121) cleavage corresponds to experimental results (see Figure 5.50). An interesting feature in this result was the presence of backward twins from the edges of the crack tip at $K_1 = 5.0$ MPa $\sqrt{m}$. Backward twins have also been observed in other simulations (Kohlhoff and Schmauder 1988) for α-iron.In this simulation, 155 dislocations were emitted before cleavage occurred although the crack tip in Figure 5.55 was not blunted due to the additional procedure applied as explained in the previous section. The distribution of dislocations is shown in Figure 5.56 where only the upper half of the model $(y \geq 0)$ is plotted, both halves of the model being considered in the simulation.

Values of fracture toughness evaluated for each size of the molecular dynamics regions are displayed in Table 5.11. The differences in fracture toughness were quite small. It was concluded that the simulation result was independent of the size of the molecular dynamics region.

Table 5.11
Fracture toughnesses obtained in simulations (N: number of free atoms)

Type of model	Fracture toughness – K_{1c}
N = 2304 model	6.3 (MPa $\sqrt{m}$)
N = 3600 model	6.2 (MPa $\sqrt{m}$)
N = 10 000 model	6.2 (MPa $\sqrt{m}$)

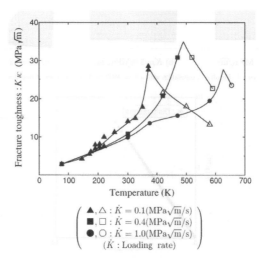

Fig 5.57 Experimental results (Gumbsch et al., 1998) of fracture toughnesses in tungsten sigle crystals. Solid marks show fracture toughnesses and open marks show stress intensities at failure in ductile manner.

Investigation of brittle fracture processes and temperature dependency of fracture toughness at low temperature

Simulation results at low temperature

Figure 5.57 shows experimental results (Gumbsch et al., 1998) for fracture toughness in tungsten single crystals. The brittle to ductile transition temperature and the fracture toughness at high temperature are both influenced by strain rate. However, fracture toughness at low temperature (77-225 K) is not influenced by strain rate. In the present study strain rates can not be taken into account because of the quasi-static simulations. At high temperatures, near the transition temperature, there are expected to be too many dislocations emitted to simulate, i.e. the more dislocations that are emitted, the greater the number of calculations to be performed with the combined model.

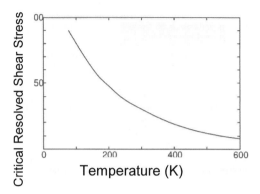

Fig 5.58 Temperature dependency of CRSS obtained in experiments (Bucki et al., 1979).

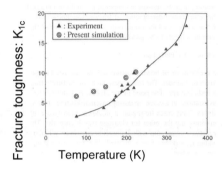

Fig 5.59   Comparison of fracture toughnesses between simulations and experiments (Gumbsch et al., 1998). Loading rates in experiments: $K_1 = 0.1$ MPa $\sqrt{m}$ .s^{-1}.

This is because the transformation from an atomic dislocation to an elastic dislocation requires relaxation calculations of molecular dynamics with 10000 steps or more. The experimental results and the limitations of the simulations mean that low temperatures (77-225 K) are appropriate conditions for the simulations.

The model with 2304 free atoms was used under these conditions. The temperature dependency of critical resolved shear stress (CRSS) (Bucki et al., 1979) is shown in Figure 5.58. Brittle fracture processes were simulated in the low temperature regime (77-225 K). Figure 5.59 shows fracture toughness values obtained from the simulations together with previous experimental results (Gumbsch et al., 1998). The number of dislocations emitted and the plastic zone lengths at failure are displayed in Table 5.12. In the simulations, fracture toughness showed clear dependency on temperature and the tendency of the fracture toughness to increase showed good correlation to experimental data. The values of fracture toughness, however, varied from experimental results. Also, it was still unknown whether the steep increase in fracture toughness at high temperature, near the transition temperature, could be obtained in the simulations.

Table 5.12
Number of emitted dislocations and lengths of plasic zones at the loading when the materials were fractured in simulations

Temperature (K)	Dislocation	Plastic zones (μm)
77	155	10.0
120	298	25.4
150	452	46.3
200	907	120.9
225	1266	208.5

It was good that the temperature dependency of fracture toughness was obtained at low temperature. However, the accuracy of the simulations and performance of the simulations at high temperature still remain subjects to be addressed.

A brittle fracture process

In this section the mechanism of brittle fracture will be discussed, based on the simulation results using the combined model. The main phenomena of the brittle fracture process are dis10cation emissions and cleavage. The point of the process is to determine whether the material cleaves or emits dislocations at a crack tip during external loading. Two simple models, based on local stress analyses in the crack tip region at an atomic scale, were introduced, one model for dislocation emissions and the other for cleavage (see Figure 5.60). The driving force τ_{local} is a resolved shear stress causing slip with dis10cation emission. When the driving force τ_{local} reaches a critical value τ_c, a dis10cation is emitted. The driving force σ_{local} is a normal stress that causes cleavage, i.e. when the driving force σ_{local} reaches a critical value σ_e then cleavage occurs.

The driving forces τ_{local} and σ_{local}, which were calculated elastically through continuum mechanics, consist of a loading K-field and the shielding forces of dislocations. The information about the K-field and the dislocations had already been obtained from simulations with the combined model. The problem with these models was where to calculate τ_{local} and σ_{local}, because the crack tip was a singular point, i.e. τ_{local} and σ_{local} depended not only on the loading but also on the calculation points. In the present study, the calculation points were one or two atoms from the crack tip. The details of calculation points are shown in Figure 5.61. σ_{local} was calculated on a (121) plane because cleavage occurred along (121) planes in this case. The radius of the crack tip was assumed to be two atoms space $(\rho = \sqrt{2}a)$. Although the models have the problem of where driving forces should be calculated from, they are useful for understanding the brittle fracture process.

Figure 5.62 shows the driving forces τ_{local} and σ_{local} at 77 K obtained from the calculations. Young's modulus E', displayed in Table 5.10, is for a plane strain problem.

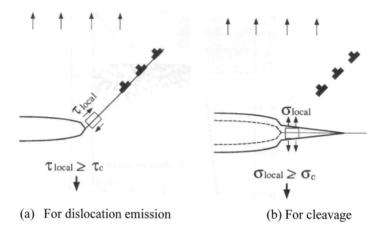

(a) For dislocation emission   (b) For cleavage

Fig 5.60 Simple models for dislocation emissions and cleavage considering driving forces in an atomistic scale. τ_c and σ_e are critical values for dislocation emissions and cleavage, respectively .

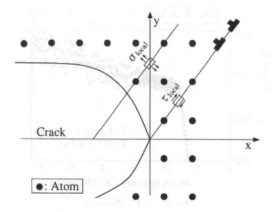

Figure 5.61 Continuum model for σ_{local} and σ_{local} calculations, compared with atom positions. A radius of the crack tip was $\rho = \sqrt{2}a$. Calculation points of τ_{local} and σ_{local} were $(\frac{1}{2}a, \frac{\sqrt{2}}{2}a)$ and $(0, \sqrt{2}a)$, respectively (a: lattice constant).

The relationship between E' and E. is $E' = E/(1-v^2)$. The driving force τ_{local} for dislocation emissions, which increased linearly with K-field in elastic deformation, was saturated after the first dislocation emission.

The saturation is a result of the shielding force of the emitted dislocation. Dislocations begin to be emitted when τ_{local} exceeds a critical value τ_c ($\tau_{local} \geq \tau_c$). The driving force τ_{local} is, however, immediately relaxed because of the shielding forces of emitted dislocations.

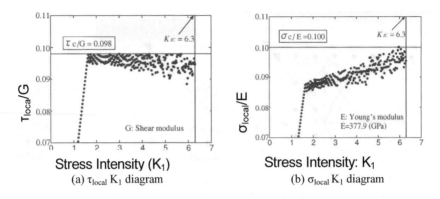

(a) τ_{local} K_1 diagram   (b) σ_{local} K_1 diagram

Fig 5.62 Calculation results of τ_{local} and σ_{local}

Table 5.13
Temperature dependencies of τ_c and σ_c

Temperature (K)	τ_c/G	σ_c/E
77	0.098	0.100
120	0.096	0.099
150	0.096	0.098
200	0.097	0.100
225	0.096	0.099

In turn dislocations continue to be emitted while τloeal exceeds τ_c ($\tau_{local} \geq \tau_c$), and the emissions are stopped when τ_{local} is relaxed below τ_c ($\tau_{local} < \tau_c$). The driving force σ_{local} for cleavage, which is also relaxed by the shielding forces of dislocations, gradually increased with K-field even after the first dislocation emission. In the simulation at $K_1 = 6.3$ MPa ~, the driving force σ_{local} reached a critical value σ_e and cleavage occurred. In summary, the brittle fracture process of the simulation would be as follows. Whilst loading K_1 from zero to K_{1C}, the driving force τ_{local} for dislocation emissions reaches τ_c. After that τ_{local} is saturated by the shielding forces of emitted dislocations, while the driving force σ_{local} for cleavage continually increases with K_1. In turn cleavage, which leads to fatal fracture, occurs when σ_{local} reaches σ_c.

It is interesting to note that from the results in this section that the critical values τ_c and σ_c were quite close to the theoretical shear strength *(G/2n)* and the theoretical tensile strength (E/10) of the materials. This implies that the strength in the atomic scale region around the origin of fracture might be close to the theoretical strength of the material even if the macroscopic strength is much lower than the theoretical strength. 4.3. Temperature dependency of fracture toughness

Some groups insist that dislocation emission is the controlling factor in brittle to ductile transition and others insist on dislocation mobility (see Introduction). In the present study, dislocation emission was determined from the critical value τ_c, and the dislocation mobility from friction forces acting on each dislocation

(CRSS). The critical values τ_c and $(\sigma_e$ calculated from simulation results at each temperature are displayed in Table 5.13. Both values show no dependency on temperature. In comparison, the friction forces (CRSS) do vary with temperature (see Figure 5.58). These calculation results mean that the temperature dependency of fracture toughness in the simulations is due to the temperature dependency of dislocation mobility. Temperature dependency of fracture toughness is a characteristic in the low temperature region of brittle to ductile transition. The results of the present study, therefore, support the latter insistence of dislocation mobility. The mechanism is as follows. Increasing temperature causes an increase of dislocation mobility, which causes differences in dislocation distribution, and hence leads to differences in the shielding forces of dislocations. The difference in shielding forces causes the slope of the $(\sigma_{local}$ to decrease. In turn the fracture limit (K_{Ic}), which is the load at which σ_{local} reaches $(\sigma_e$, increases with temperature.

Discussion

As mentioned above, the accuracy of simulations and performing simulations at high temperatures remain problems still unsolved by this study. The key points to solving these problems are an extension to three-dimensional simulations and an increase in calculation speed. Two dimensional simulations lead not only to a decline of accuracy but also to principal limitations in the simulations. With a pre-crack on a (100) plane, which is a primary cleavage plane, the material fractured in a perfect brittle manner with the combined model, i.e. the material cleaved before it emitted dislocations. In turn it was observed that fracture toughness was independent of temperature. The perfect brittle fracture might be caused by an absence of ledge sites (Zhou and Thomson, 1991) on the crack front because of the two-dimensional nature of the model. The extension to three-dimensional simulations, involving ledges of a crack front, might be absolutely necessary both to improve the accuracy and to extend the applicability of the simulations. Increasing the speed of the calculations is necessary both to extend to three-dimensional simulations and to simulate at high temperatures. The speed increase may require the improvement not only of computers but also of software.

In this study the molecular dynamics simulations brought much benefit to solving the problem of brittle fracture and the temperature dependency of fracture toughness. This might be a good demonstration of the usefulness and applicability of simulations in research into material strength. The present study is successful in showing the range of possibilities of molecular dynamics simulations, while the simulations still have several deficiencies.

Conclusion

The combined model of molecular dynamics with micromechanics was applied to simulations of brittle fracture processes in tungsten single crystals at low temperatures. The pre-cracks were introduced on (110) planes and cleavage was observed along (121) planes in the simulations. The cleavage along (121) planes had previously been observed in experiments (Riedle et al., 1994). In the simulations the material was twinned backwards from edges of the crack tip. The backward twins have also been observed in other simulations (Kohlhoff and Schmauder, 1988). Three sizes of molecular dynamics regions were tested at 77 K, and the results of the simulations were found to be independent of the sizes used. Brittle fracture processes were simulated at the temperatures between 77 K and 225 K. The fracture toughness values obtained in these simulations showed clear temperature dependency, but did not show good agreement with those from experiments. The main problem with the simulation was the limitation in the number of calculations that could realistically be performed which affected both accuracy and limited the simulations to low temperatures only. A brittle fracture process at 77 K was discussed by considering the driving forces for dislocation emissions and cleavage in an atomic scale region of a crack tip. It was found that the driving force for dislocation emissions was saturated after the first dislocation emission, whereas the driving force for cleavage gradually increased with loading K-field. When the driving force for cleavage reached a critical value, the material cleaved. The critical values of driving forces for both dislocation emissions and for cleavage, which were quite close to the theoretical strengths of the material, were not influenced by temperature. This means that the temperature dependency of fracture toughness is not caused by a temperature dependency of dislocation emissions but by that of dislocation mobility.

References

Bucki, M., Novak, Y., Savitsky, Y.M., Burkhanov, G.S. and Kirillova, Y.M. (1979), Work-hardening in tungsten single crystals between 77 and 1343K. Strength of Metals and Alloys. Proceedings of the 5th international Conference, xxx+760, pp. 145–150.

Chou, Y.T. (1967), Dislocation pile-ups against a locked dislocation of different Burgers vector. Journal of Applied Physics 38, pp. 2080–2085.

Cordwell, J.E. and Rull, D. (1972), Observation of {110} cleavage in (110) axis tungsten single crystals. Philosophical Magazine 26(1), pp. 215–224.

Eshelby, J.D. (1957), The determination of the elastic field of an ellipsoidal inclusion, and related problems. Proceedings, Royal Society London A241, pp. 376–396. MD study on low temperature brittleness 157

Eshelby, J.D., Frank, F.C. and Nabarro, F.R.N. (1951), The equilibrium of linear arrays of dislocations, Philosophical Magazine 42, pp. 351–364.

Finnis, M.W. and Sinclair, J.E. (1984), A simple empirical N-body potential for transition metals. Philosophical Magazine ASO, pp. 45–55.

Furuya, Y. and Noguchi, H. (1998), A combined method of molecular dynamics with micromechanics improved by moving the molecular dynamics region successively in the simulation of elastic-plastic crack propagation. International Journal of Fracture 94, pp. 17–31.

Furuya, Y. and Noguchi, H. (1999), Simulation of crack propagation with (molecular dynamics + micromechanics) model. Proceedings of 9th CIMTEC 18, pp. 57–64.

George A. and Michot, G. (1993), Dislocation loops at crack tips: nucleation and growth – an experimental study in silicon. Materials Science and Engineering A164, pp. 118–134.

Gumbsch, P., Riedle, J., Hartmaier, A. and Fischmeister, H. (1998), Controlling factors for the brittle-to-ductile transition in tungsten single crystals. Science 282, pp. 1293–1295.

Ha, K.F., Yang, C. and Bao, J.S. (1994), Effect of dislocation density on the ductile-brittle transition in bulk Fe-3%Si single crystals. Scripta Metallurgica et Materialia 30(8), 1065-1070.

Hirsch, P.B. and Roberts, S.G. (1991), The brittle-to-ductile transition in silicon. Philosophical Magazine A64, pp. 55-80.

Hirth, J.P., Hoagland, K.G. and Gehlen, P.C. (1974), The interaction between line force arrays and planar cracks. International Journal of Solids and Structures 10(9), pp. 977-984.

John, C.St. (1975), The brittle-to-ductile transition in precleaved silicon single crystals. Philosophical Magazine 32, pp. 1193-1212.

Kohlhoff, S., Gumbsch, P. and Fichmeister, H.F. (1991), Crack propagation in b.c.c. crystals studied with a combined finite-element and atomistic model. Philosophical Magazine 64(4), pp. 851-878.

Kohlhoff, S. and Schmauder, S. (1988), A new method for coupled elastic-atomistic modeling. (Edited by . Vitek and DJ. Srolovitz), Large Atomistic Simulation of Materials - Beyond Pair Potentials - Plenum Press, New York, pp. 411-418.

Lekhnitski, S.G. (1968), Anisotropie Plates Gordon and Breach, New York.

Maeda, K. (1992), Effect of crack blunting on dislocation-mobility-controlled brittle-ductile transition. Scripta Metallurgica et Materialia 27, pp. 805-809.

Michot, G., Loyola de Oliveria, M.A. and George, A. (1994), Dislocation loops at crack tips: control and analysis of sources in silicon. Materials Science and Engineering Al76, pp. 99-109.

Mullins, M. and Dokanish, F. (1982), Simulation of the (001) Plane Crack in a-Iron Employing a New Boundary Scheme. Philosophical Magazine 46(5), pp. 771-787.

Mullins, M. (1982), Molecular dynamics simulation of propagating cracks. Scrip ta Metallurgica 16, pp. 663-666.

Mura, T. (1968), The continuum theory of dislocations. Advances in Materials Research (edited by H. Herman H.), Interscience Publ., New York, pp. 1-107.

Narita, N., Higashida, K., Torii, T. and Miyaki, S. (1989), Crack-tip shielding by dislocations and fracture toughness in NaCl crystals. Materials Transactions, JIM 30, pp. 895-907.

Noguchi, H. and Furuya, Y. (1997), A method of seamlessly combining a crack tip molecular dynamics enclave with a linear elastic outer domain in simulating elastic-plastic crack advance. International Journal of Fracture 87, pp. 309-329.

Rice, J.K. (1974), Limitations to the small scale yielding approximation for crack tip plasticity. Journal of the Mechanics and Physics of Solids 22, pp. 17-26.

Rice, J.R. (1992), Dislocation nucleation from a crack tip: an analysis based on the Peierls concept. Journal of the Mechanics and Physics of Solids 40, pp. 239-271.

Rice, J.R. and Beltz, G.E. (1994), The activation energy for dislocation nucleation at a crack tip. Journal of the Mechanics and Physics of Solids 42, pp. 333-360.

Rice, J.K. and Thomson, R. (1974), Ductile versus brittle behavior of crystals. Philosophical Magazine 29, pp. 73-97.

Riedle, J., Gumbsch, P. and Fischmeister, H.F. (1994), Fracture studies of tungsten single crystals. Materials Letters 20, pp. 311-317.

Sinclair, J.E., Gehlen, P.C., Hoagland, K.G. and Hirth, J.P. (1978), Flexible boundary conditions and nonlinear geometric effects in atomic dislocation modeling. Journal of Applied Physics 49(7), pp. 3890-3897.

Xu, G. and Argon, A.S. (1995), Nucleation of dislocations from crack tips under mixed modes of loading: implications for brittle against ductile behavior of crystals. Philosophical Magazine A72, pp. 415-451.

Yang, W., Tan, H. and Guo, T. (1994), Evolution of crack tip process zones. Modeling and Simulation in Materials Science and Engineering 2, pp.767-782.

Zhou, S.J. and Thomson, R. (1991), Dislocation emission at ledges on cracks. Journal of Materials Research 6, pp. 639-653.

5.6 Simulation of the formation of Cu-precipitates in steels[6]

Cu-alloyed steels are applied to large extent as pipe materials. These steels are typically alloyed with 0.6–1.5 wt.% Cu in order to improve the yield stress at higher temperatures. In German power plants, the service temperature typically lies below 300 °C, in pressure vessels below 340 °C and in conventional power plants it ranges up to 450 °C. After long time in service, strengthening with toughness decrease (embrittlement) was found. This undesired change in material properties is due to Cu-precipitation at service temperatures above 300 °C [1]. At MPA Stuttgart, simulations have been performed in order to correlate the amount and size of Cu-precipitates which have formed under service conditions and the hardening [3]. The aim of the present work is a deeper understanding of the mechanical properties of Cu-alloyed steels by an atomistic simulation of the formation and growth of precipitates.

Cu-precipitates

At temperatures below 910 °C, pure iron posesses a body centered cubic (bcc) crystal structure while pure copper posesses a face centered cubic (fcc) crystal structure. The crystal structure of Cu precipitates is known to be dependent on their size:

- Small precipitates with radii smaller than about 2 nm are coherent and possess the bcc structure of iron [6,8].
- Precipitates with radii between 2 and 9 nm possess a twinned 9R structure [6].
- Precipitates with radii larger than about 9 nm are present in a 3R structure, a distorted fcc structure which continually changes to the fcc crystal structure of pure copper for increasing precipitate radii [6].

Small angle neutron scattering (SANS) was applied on several specimen of material 15 NiCu- MoNb 5 in order to analyse these precipitates in a defined initial state as well as in a thermally aged state [14]. In the initial state the SANS measurements show Cu-precipitates with a maximum in the radii distribution at R = 2.2–2.5 nm, while some copper is still dissolved in the matrix. In the thermally aged state (e.g. 57 000 h, 350 °C) the radii distribution of the initial state remains nearly constant, but additionally particles with a maximum in the radii distribution at R = 1.3–1.7 nm have precipitated, while nearly no copper is dissolved in the

---

[6] Reprinted from S. Schmauder, P. Binkele, "Atomistic Computer Simulation of the Formation of Cu-Precipitates in Steels", Computational Materials Science 24, pp. 42-53 (2002) with kind permission from Elsevier

matrix. Due to the large number of newly formed small precipitates in the thermally aged case, the mean distance between the precipitates is drastically reduced. In summary, thermal ageing due to a secondary Cu precipitation can be understood: precipitates represent obstacles for moving dislocations, and therefore, dislocation movement is strongly impeded by the precipitates. The material is hardened as there are more Cu precipitates present and the mean distance between two neighbouring precipitates is reduced in the aged state.

5.6.1 Monte Carlo simulations

Model

In the present simulations, a fixed bcc crystal lattice is used with periodic boundary conditions and a fixed size of $L = 64$ and 128 lattice constants, respectively. Thus, $N = 2L^3$ lattice sites are available. The lattice is occupied with N_A atoms of type A, N_B atoms of type B and one vacancy ($N_V = 1$). For the total number of lattice sites the following equation holds: $N = N_A + N_B + N_V$. The chemical binding between atoms has been described by first and second neighbour pair interactions, $\varepsilon^{(i)}_{AA}$, $\varepsilon^{(1)}_{BB}$, $\varepsilon^{(1)}_{(AB)}$ with $i \in \{1,2\}$ denoting the ith neighbours. The binding between atoms and the vacancy has been described by first neighbour interactions $\varepsilon^{(1)}_{AV}$, and $\varepsilon^{(1)}_{RV}$. The movement of the atoms occurs by change of the vacancy with a nearest neighbour atom. Such a change is thermally activated and the transition rates $\Gamma_{A,V}$ for an A-atom and $\Gamma_{B,V}$ for a B-atom are given by

$$\Gamma_{A,V} = v_A \exp\left(-\frac{\Delta E_{A,V}}{kT}\right) \tag{5.31}$$

$$\Gamma_{B,V} = v_B \exp\left(-\frac{\Delta E_{A,V}}{kT}\right) \tag{5.32}$$

Here v_A and v_B represent attempt frequencies for an A-atom and B-atom respectively. The activation energy ΔE_{XV}, $X \in \{A, B\}$ is the energy difference between the stable position and the saddle point position of a diffusing atom A or B, which is situated next to the vacancy (Fig. 5.63). The activation energies depend on the local atom configuration, and a simple model is applied to calculate them, which is presented in the following for atom type A. The activation energies depend on the saddle point energy $E_{sp,A}$ and the interatomic binding energies of the first and second neighbours of the A-atom, and on the interaction energies of the first neighbours of the vacancy.

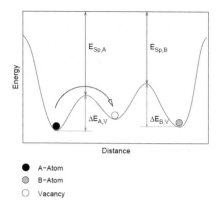

Fig. 5.63 Schematic representation of activation energies and saddle point energies.

$$\Delta E_{A,V} = E_{Sp,A} - n_{AA}^{(1)}\varepsilon_{AA}^{(1)} - n_{AB}^{(1)}\varepsilon_{AB}^{(1)} - n_{AA}^{(2)}\varepsilon_{AA}^{(2)} - n_{AB}^{(2)}\varepsilon_{AB}^{(2)} - n_{AV}^{(1)}\varepsilon_{AV}^{(1)} - n_{BV}^{(1)}\varepsilon_{BV}^{(1)} \quad (5.33)$$

Here $n^{(i)}{}_{AA}(n^{(i)}{}_{AB})$ is the number of AA-bonds (AB-bonds) at the ith neighbours of the A-atom ($i \in \{1,2\}$), and $n^{(i)}{}_{Av}(n^{(i)}{}_{Av})$ is the number of AV- 'bonds' (BV-'bonds') at the first neighbours of the vacancy.

As there are $z_1 = 8$ first neighbour lattice sites, and $z_2 = 6$ second neighbour lattice sites on a bcc lattice, the following equations hold:

$$n_{AA}^{(1)} + n_{AB}^{(1)} = z_1 - 1 \quad (5.34)$$

$$n_{BB}^{(2)} + n_{AB}^{(2)} = z_2 \quad (5.35)$$

$$n_{AV}^{(1)} + n_{BV}^{(1)} = z_1 \quad (5.36)$$

For the B-atoms an analogues consideration provides the following equations:

$$\Delta E_{B,V} = E_{Sp,B} - n_{AB}^{(1)}\varepsilon_{AB}^{(1)} - n_{BB}^{(1)}\varepsilon_{BB}^{(1)} - n_{AB}^{(2)}\varepsilon_{AB}^{(2)} - n_{BB}^{(2)}\varepsilon_{BB}^{(2)} - n_{AV}^{(1)}\varepsilon_{AV}^{(1)} - n_{BV}^{(1)}\varepsilon_{BV}^{(1)} \quad (5.37)$$

And

$$n_{BB}^{(1)} + n_{AB}^{(1)} = z_1 - 1 \quad (5.38)$$

$$n_{AA}^{(2)} + n_{AB}^{(2)} = z_2 \quad (5.39)$$

$$n_{AV}^{(1)} + n_{BV}^{(1)} = z_1 \tag{5.40}$$

The energies $\varepsilon_{AA}^{(1)}$ and $\varepsilon_{BB}^{(1)}$, $i \in \{1,2\}$ were estimated from the cohesive energies of the pure metals, using the assumptions $\varepsilon_{AA}^{(2)} = \varepsilon_{AA}^{(1)}/2$ and $\varepsilon_{BB}^{(2)} = \varepsilon_{BB}^{(1)}/2$.

$$E_{coh,A} = \frac{z_1}{2}\varepsilon_{AA}^{(1)} + \frac{z_2}{2}\varepsilon_{AA}^{(2)} \tag{5.41}$$

$$E_{coh,B} = \frac{z_1}{2}\varepsilon_{BB}^{(1)} + \frac{z_2}{2}\varepsilon_{BB}^{(2)} \tag{5.42}$$

The energies $\varepsilon_{AB}^{(i)}, i \in \{1,2\}$ are related to the mixing energy ω_{AB} which is defined as

$$\omega_{AB} = \frac{z_1}{2}\left(\varepsilon_{AA}^{(1)} + \varepsilon_{BB}^{(1)} - 2\varepsilon_{AB}^{(1)}\right) + \frac{z_2}{2}\left(\varepsilon_{AA}^{(2)} + \varepsilon_{BB}^{(2)} - 2\varepsilon_{AB}^{(2)}\right) \tag{5.43}$$

For mixing energies $\omega_{AB} < 0$ the system has a tendency to form precipitates, for $\omega_{AB} = 0$ A- and B-atoms are ideally solvable, and for $\omega_{AB} > 0$ a tendency to form superstructures exists [12]. The energies $\varepsilon_{AV}^{(1)}$ and $\varepsilon_{BV}^{(1)}$ are related to the vacancy formation energies as

$$E_{V,for,A} = z_1\varepsilon_{AV}^{(1)} + E_{coh,A} \tag{5.44}$$

$$E_{V,for,B} = z_1\varepsilon_{BV}^{(1)} + E_{coh,B} \tag{5.45}$$

With the knowledge of $E_{coh,A}$, $E_{coh,B}$, ω_{AB}, $E_{V,for,A}$, $E_{V,for,B}$ the energies $\varepsilon_{AA}^{(1)}$, $\varepsilon_{BB}^{(1)}$, $\varepsilon_{AB}^{(1)}$, $i \in \{1,2\}$ and $\varepsilon_{AV}^{(1)}$, $\varepsilon_{BV}^{(1)}$ can be calculated.

An approach to the Fe–Cu system

In order to simulate the system Fe–Cu, material data are required. The kinetic parameters were adjusted to diffusion data, assuming an Arrhenius law.

$$D = D_0 \exp\left(\frac{-Q}{kT}\right) \qquad \text{with } Q = E_{V,for} + E_{V,mig} \tag{5.46}$$

The attempt frequencies were estimated using the Debye frequencies v_D in the pure metals: $v_A = v_B = v_D$. The saddle point energies determine the vacancy migration energies in the pure metals and the following estimations were used where $E_{V,mig,A}$ and $E_{V,mig,B}$ are the vacancy migration energies of atom types A and B, and c_A and c_B are the concentrations of A and B.

Table 5.14
Material data

Cohesive energy Fe	$E_{coh, Fe} = -4.28$ eV	[2]
Cohesive energy Cu (in Fe)	$E_{coh,Cu} = -4.28$ eV	Sym. Model
Lattice constant	$a = 0.287$ nm	[2]
Mixing energy	$\omega = -0.515$ eV	Adjust to Cu sol. limit
Vacancy formation energy Fe	$E_{V, for, Fe} = 1.60$ eV	[4,5]
Vacancy formation energy Cu	$E_{V, for,Cu} = E_{V, for, Fe} = 1.60$ eV	Assumption
Vacancy migration energy Fe	$E_{V, mig, Fe} = 0.90$ eV (770-884°C)	Calc. with [5]
Vacancy migration energy Cu (in Fe)	$E_{V, mig, Cu} = 0.90$ eV (300-450°C)	Calc. with [10]
Diffusion constant Fe	$D_0^{Fe} = 2.01 \times 10^{-4}$ m²s⁻¹	[5,11]
Diffusion constant Cu	$D_0^{Cu} = (2.16 \pm 0.9) \times 10^{-4}$ m²s⁻¹	Calc. with [10]
Debye frequency Fe	$\nu_D = 8.70 \times 10^{12}$ s⁻¹	Calc. with [2]

Table 5.15
Simulation parameters

$\varepsilon_{AA}^{(1)} = -0.778$ eV	$\varepsilon_{AA}^{(2)} = -0.389$ eV
$\varepsilon_{BB}^{(1)} = -0.778$ eV	$\varepsilon_{BB}^{(2)} = -0.389$ eV
$\varepsilon_{AB}^{(1)} = -0.731$ eV	$\varepsilon_{AB}^{(2)} = -0.366$ eV
$\varepsilon_{AV}^{(1)} = -0.335$ eV	$\varepsilon_{BV}^{(1)} = -0.335$ eV
$E_{Sp,A} = -9.557$ eV	$E_{Sp,B}$ -9.098 eV
$\nu_D = 8.70 \times 10^{12}$ s⁻¹	$\nu_B = 8.70 \times 10^{12}$ s⁻¹

In the present simulations, a symmetrical model was used, i.e. $\varepsilon_{AA}^{(i)} = \varepsilon_{BB}^{(i)}$ ($i = 1$; 2). The applied material data are listed in Table 5.14, and the simulation parameters in Table 5.15.

$$E_{Sp,A} = E_{V,mig,A} + z_1\varepsilon_{(AV)}^{(1)} + c_A\left((z_1 - 1)\varepsilon_{AA}^{(1)} + z_2\varepsilon_{(AA)}^{(2)}\right) + c_B\left((z_1 - 1)\varepsilon_{AB}^{(1)} + z_2\varepsilon_{AB}^{(2)}\right) \quad (5.47)$$

$$E_{Sp,B} = E_{V,mig,B} + z_1\varepsilon_{(BV)}^{(1)} + c_A\left((z_1 - 1)\varepsilon_{AB}^{(1)} + z_2\varepsilon_{(AB)}^{(2)}\right) + c_B\left((z_1 - 1)\varepsilon_{BB}^{(1)} + z_2\varepsilon_{BB}^{(2)}\right) \quad (5.48)$$

Residence-time-algorithm

In the simulations a rejection free residencetime - algorithm is applied, which shows significant calculation time advantages in comparison to a Metropolis algorithm [12]. Eight nearest neighbours are surrounding the vacancy. A jump rate is now calculated for each of these eight jump candidates depending on their first and second neighbours. This provides eight independent high jump frequencies Γ_1, Γ_2, ... Γ_8. One of these eight possibilities is now selected on the basis of its probability by a random number and the site change is then performed, see Fig. 5.64. The number of performed jumps during the simulations is in the order 10^{12}.

In order to define a time scale, the averaged residence time was used which is given by

$$ t_{MC} = \left(\sum_{i=1}^{8} \Gamma_i \right)^{-1} \tag{5.49} $$

However, the different vacancy concentrations in the simulation ($c_{V,sim} = 1/(2 \cdot L^3)$) and in reality (we used $c_{V,real} = 280 \cdot \exp(E_{V,for}/(kT)))$ have still to be correlated. In order to obtain the time t, the Monte Carlo time t_{MC} is multiplied by a time adjusting factor as follows:

$$ t = \left(\frac{c_{v,sim}}{c_{V,real}} \right) t_{MC} \tag{5.50} $$

The thus calculated time scale is sensitively dependent on the used energies and attempt frequencies. Therefore, the calculated time periods for the simulation results have to be considered with some care. In any case, the thus calculated pseudo time is directly proportional to the real time.

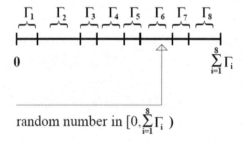

Fig. 5.64 A random number directs to one of the eight possible vacancy jumps as signified by its jump rate.

The random numbers have to fulfill the following requirements:

- They shall be random numbers.
- They shall be reproducable in order to reproduce simulation results.
- They shall be long periodic in order to avoid interdependencies in the simulation.

Some pre-installed random number generators on computers do not fulfill these requirements. Some good random number generators are described in [7]. Therefore, an appropriate one is installed which also guarantees a nearly independence of the used computer type.

5.6.2 Simulation results: formation and growth of precipitates at different temperatures

Simulation at T = 600 °C

First a study is presented which demonstrates how the precipitates form and grow. The used base centered cubic lattice with periodic boundary conditions posseses a side length of L=64 lattice constants which means in reality a side length of $64 \cdot 0.287 = 18.4$ nm. It consists of $N = 2 \cdot L^3 = 524.288$ lattice sites which are occupied by 99% A-atoms ($N_A = 519.044$), 1% B-atoms ($N_B = 519.044$) and one vacancy ($N_V = 1$). At the beginning, the B-atoms are randomly distributed on the crystal lattice. At the beginning of the simulation, all B-atoms (1%) are dissolved in the matrix, (point A in Fig. 5.66). At the end of the simulation (Fig. 5.65) only 0.2 at.% of B remains dissolved while 0.8 at.% have formed precipitates according to the Fe–Cu phase diagram (point B in Fig. 5.66, equilibrium state).

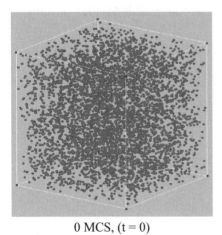
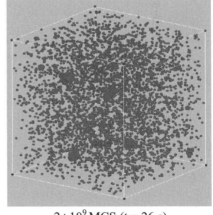

0 MCS, (t = 0) $2 \cdot 10^9$ MCS (t = 26 s)

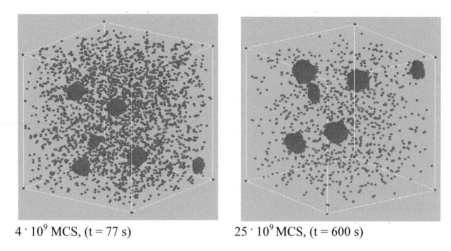

4 · 10⁹ MCS, (t = 77 s)   25 · 10⁹ MCS, (t = 600 s)

Fig. 5.65 Simulation of nucleation and growth of precipitates; B-atoms are shown only for visibility reasons (T = 600 °C).

Now a function is defined which gives a measure for the deviation from the equilibrium state:

$$f(t) = \frac{c_B^m(t=0) - c_B^m(t)}{c_B^m(t=0) - c_B^m(\infty)} \tag{5.51}$$

Here, $c_B^m(t)$ is the B-concentration in the matrix at time t. For f(t) = 1 there are as many B-atoms dissolved in the matrix as can be expected from the Fe–Cu phase diagram.

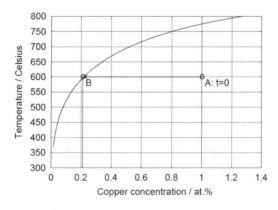

Fig. 5.66 The Fe–Cu phase diagram (calculated with Thermo- Calc). At a temperature of 600 °C the solubility limit of Cu in Fe is 0.21 at.%.

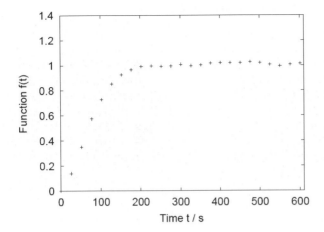

Fig. 5.67 The function f (t) converges against 1 (for details see text).

Values of f(t) < 1 mean, that too many B-atoms are dissolved in the simulation, values of f(t) > 1 mean, that not enough B-atoms are dissolved in the matrix. The values of f(t) have been calculated for different times during the simulation and are shown in Fig. 5.67. It can be seen that f(t) converges against 1 for t → ∞. This means that this simulation is in good agreement with the Fe–Cu phase diagram, calculated with Thermo-Calc [13], [15].

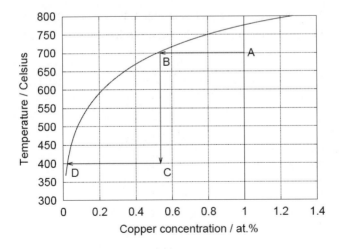

Fig. 5.68 The Fe–Cu phase diagram (calculated with Thermo- Calc). At a temperature of 700 °C the solubility limit of Cu in Fe is 0.536 at.%, and at a temperature of 400 °C the solubility limit of Cu in Fe is 0.023 at.%.

Simulation for two different temperatures

In this simulation a crystal lattice with periodic boundary conditions and a side length of $L = 128$ lattice constants has been used which corresponds to an absolute value of 36.2 nm. The model volume posesses $N = 2 \cdot L^3 = 4.194.304$ lattice sites and is occupied with 99% A-atoms ($N_A = 4.152.360$), 1% B-atoms ($N_B = 41.943$) and one vacancy ($N_V = 1$). In comparison to the previous calculation, the model volume is increased by a factor of 8 and an initial distribution with a few very small precipitates was used. In the first part of the simulation the temperature amounted to 700 °C and in the second part the simulation was continued with a temperature of 400 °C, according to the Fe–Cu phase diagram (Fig. 5.68).

Part 1, simulation at T = 700 °C

At 700 °C precipitates of radii between 1.1 and 1.7 nm formed within seconds and 0.5% B-atoms remain dissolved at that time (Fig. 5.69). In the present context a 'dissolved' B-atom means that no other B-atom is present among the eight first with the dissolved atoms when the temperature is decreased at once to 400 °C. In order to follow the precipitation behaviour of these atoms they have been given a brighter colour.

Part 2, simulation at T = 400 °C

At 400 °C a part of the dissolved atoms produces many small precipitates, another part increases the precipitates which have formed at 700 °C (Fig. 5.70). After 15×10^{10} Monte Carlo steps (MCS) only a few B-atoms are still in solution and nearly all B-atoms have precipitated in accordance to the Fe–Cu phase diagram. In a further step, the size distributions of the precipitates which have formed after 1×10^{10}, 15×10^{10}, 35×10^{10}, 75×10^{10} MCS have been calculated.

Considered geometrically, the simulated precipitates are polyhedrons. Now, a method has to be developed to define the radii of the simulated precipitates. We calculated the radius of the sphere having the same volume as the precipitate. An elementary cell of a base centered cubic lattice includes two atoms and posesses the volume $V_{ec} = a^3$. As the number of atoms N in each precipitate is known, the following equation holds for large values of N ($N > 20$).

$$V_{Sphere} = \frac{4}{3} \pi R^3 = \frac{N}{2} a^3 \tag{5.52}$$

This equation provides the precipitate radius R which was used to calculate the size distributions:

$$R = \left(\frac{3a^3}{8\pi} N \right)^{1/3} \qquad\qquad (5.53)$$

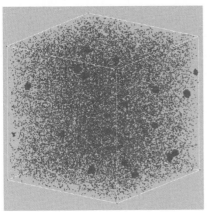

2 · 10^{10} MCS, (t = 1 s), T = 700°C   7 · 10^{10} MCS, (t = 10 s), T = 400° C

Fig. 5.69 After 10 s precipitates with radii between 1.1 and 1.7 nm have been formed; approximately 0.5% B-atoms are still dissolved in the matrix (brighter colour).

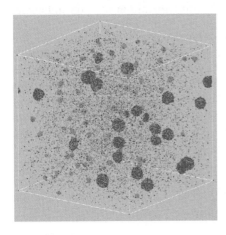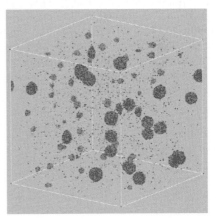

1 · 10^{10} MCS, (t$_1$ = 140 h), T = 400°C   15 · 10^{10} MCS, (t$_2$ = 1933h), T = 400°C

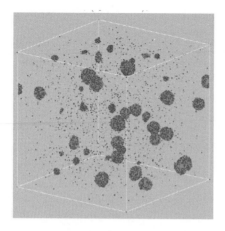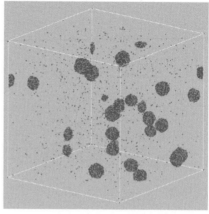

$35 \cdot 10^{10}$ MCS, ($t_3 = 4504$h), T = 400°C $75 \cdot 10^{10}$ MCS, ($t_4 = 9680$h), T = 400°C

Fig. 5.70 The simulation is continued at T = 400 °C. A part of the still dissolved atoms produce small precipitates, another part increases the precipitate which have been formed before at T = 400 °C. In the long run the number of small precipitates decreases, and the average particle size increases.

After 7 x 10^{10} MCS at a temperature of 700 °C nineteen precipitates with radii between 4.5 and 6.5 lattice constants ($1.1 < R < 1.7$ nm) were formed. The mean radius of the precipitates in this configuration was calculated to be 1.43 nm. After 1 x 10^{10} MCS at a temperature of 400 °C, 74 new small precipitates with radii between 1 and 3 lattice constants ($0.3 < R < 0.9$ nm) have formed.

Due to the large number of small precipitatxes the mean radius of the precipitates decreased to 0.79 nm. In the course of the simulation the radii of small and large precipitates grow while the number of small precipitates is reduced drastically and the mean radius of the precipitates increases (Fig. 5.71).

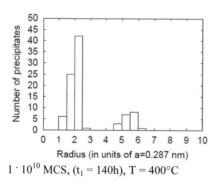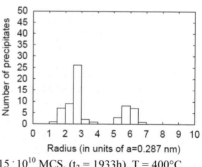

$1 \cdot 10^{10}$ MCS, ($t_1 = 140$h), T = 400°C $15 \cdot 10^{10}$ MCS, ($t_2 = 1933$h), T = 400°C

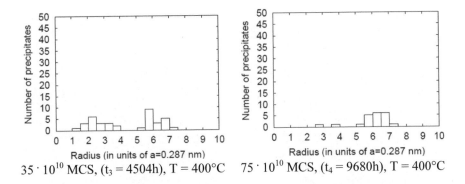

$35 \cdot 10^{10}$ MCS, ($t_3 = 4504$h), T $= 400°$C $75 \cdot 10^{10}$ MCS, ($t_4 = 9680$h), T $= 400°$C

Fig. 5.71 Size distribution of the precipitates after 1 1010 MCS, 15 1010 MCS, 35 1010 MCS and 75 1010 MCS.

At the end of the simulation after 75×10^{10} MCS only two small precipitates are still existent, 72 small precipitates vanished and increased the 19 large precipitates, which now have radii between 5 and 7.5 lattice constants ($1.4 < R < 2.2$ nm). The mean radius of the precipitates increased to 1.72 nm. Fig. 5.72 shows the decrease of the number of precipitates and Fig. 5.73 the increase of the average precipitate radius during the simulation.

In order to calculate the precipitation strengthening of Fe by Cu particles a theory of Russel and Brown [9] is applied to describe the increase in yield stress $\Delta\sigma = 2.5 \Delta\tau$ (as assumed for Fe; Russel and Brown [9]) due to cutting of precipitates, which are softer than the embedding matrix

$$\Delta\tau = \frac{Gb}{D}\left(1 - \frac{E_1^2}{E_2^2}\right)^{3/4} \tag{5.54}$$

with

$$\frac{E_1}{E_2} = \frac{E_1^\infty \log(r_{ppt}/r_0)}{E_2^\infty \log(r_c/r_0)} + \frac{\log(r_c/r_{ppt})}{\log(r_c/r_0)} \tag{5.55}$$

where G is the shear modulus of the matrix; b, the Burgers vector of the dislocation; D, the distance between the precipitates E_1^∞ and E_2^∞, the energies per unit length of a dislocation in infinite media (Fe or Cu respectively); r_{ppt}, the precipitate radius; r_c and r_0 the outer and inner cut-off radii, respectively, used to calculate the energy of the dislocation. Russel and Brown verified the validity of the following values for the Fe–Cu system. $E_1^\infty / E_2^\infty \approx 0.6$, b $= 0.248$ nm, $r_0 = 2.5$b, G $= 83$ GPa, and finally $r_c = 1000 r_0$.

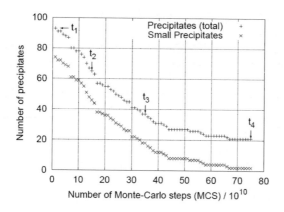

Fig. 5.72 Decrease of the number of precipitates during the simulation at T = 400 °C.

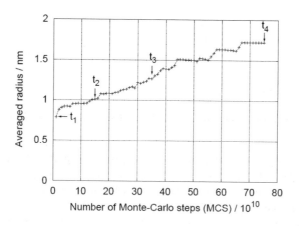

Fig. 5.73 Increase of the average precipitate radius during the simulation at T = 400 °C

    The formula points out the impact of the distance D. The dislocations are poorly impeded in the case of largely distant obstacles, and strongly impeded in the case of obstacles with a short distance. In the idealised case with homogeneously distributed particles, D depends on r_{ppt} as $D = \sqrt{\pi} r_{ppt} f^{-1/2}$, where f is the atomic concentration of Cu. Using this $D(r_{ppt})$ law, Dr can be directly calculated as a function of the mean particle radii. For the present Cu-concentrations (0.5% and 1.0%) these functions peak for r_{ppt} around 1.3 nm, see Fig. 5.74. In real steels as well as in the simulation, neither the radii nor the distances between the precipitates are identical, and the above linking D with r_{ppt} does not hold. Nevertheless, in a simple approach, averaged particle radii and the corresponding distances according to the above $D(r_{ppt})$ law were used. A detailed treatment with respect to particle radii distributions and varying distances will be future work.

In part 1 of the simulation, after 7×10^{10} MCS at a temperature of 700 °C nineteen precipitates with a mean radius of 1.43 nm were formed. The concentration of the atoms which have precipitated is 0.5%, while the remaining (other) 0.5% are still solved in the matrix. This configuration corresponds to the time t_0 in Fig. 5.74.

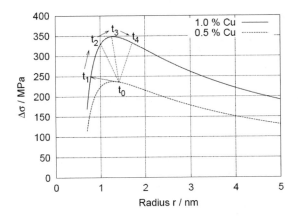

Fig. 5.74 Increase of yield stress according to Russel–Brown for f = 0.5% Cu and 1.0% Cu.

In part 2 of the simulation, after 1×10^{10} MCS at a temperature of 400 °C a total number of 93 precipitates (19 large ones and 74 small ones) is formed with a mean radius of 0.79 nm, and the concentration of precipitated atoms is close to 1.0%. This corresponds to the time t_1 in Fig. 5.74. After 15×10^{10} MCS a total number of 65 precipitates (19 large ones and 46 small ones) exist and the mean radius in this configuration is 1.01 nm, corresponding to the time t_2 in Fig. 5.74. After 35×10^{10} MCS a total number of 37 precipitates (19 large ones and 18 small ones) is formed with a mean radius of 1.32 nm, corresponding to the time t_3 in Fig. 5.74. At the end of the simulation after 75×10^{10} MCS a total number of 21 precipitates (19 large ones and 2 small ones) and a mean radius of 1.72 nm is found, corresponding to the time t_4 in Fig. 5.74.

Using the results of the simulation and the theory of Russel and Brown, strengthening can be calculated as a function of time and is plotted in Fig. 5.75. At the beginning of the simulation at a temperature of 400 °C the strengthening is weak, despite the existence of 74 small precipitates, but the mean radius of 0.79 nm is too small to contribute to strengthening (at time t_1 in Figs. 5.74 and 5.75). After 15×10^{10} MCS a strong strengthening contribution of $\Delta\sigma = 94.4$ MPa due to an average particle radius of 1.01 nm is found (at time t_2 in Fig. 5.67.).

The strengthening reaches a maximum of $\Delta\sigma = 112.7$ MPa after 35×10^{10} MCS (at time t_3 in Fig. 5.75) and then decreases to

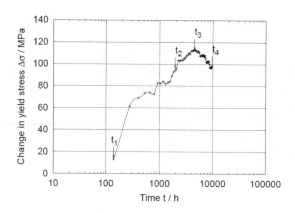

Fig. 5.75 Change in yield stress during simulation at T = 400 °C using the theory of Russel–Brown.

$\Delta\sigma = 97.1$ MPa after 75 x 10^{10} MCS (at time t_4 in Fig. 5.75_1). Such simulation results are in close agreement to recent experimental observations [14].

The thus calculated values for $\Delta\sigma$ have to be reduced by the change in the contribution from solution hardening. The contribution from solution hardening vanishes during ageing when the remaining dissolved copper atoms precipitate to form particles. In the initial state of the material when 0.5 at.% Cu are still dissolved in the matrix, solution hardening amounts to approximately 10 MPa [3], leaving a total increase in the maximum yield stress of 102 MPa for the aged state.

Summary

Atomistic computer simulations of the formation of precipitates can contribute to a deeper understanding of the mechanical behaviour of Cu-alloyed steels. The underlying Monte Carlo method was presented and a binary system with components A and B was considered. Starting with a random distribution of atoms, the formation and growth of precipitates was simulated at a constant temperature of 600 °C. In a second simulation, an initial temperature of 700 °C was lowered to 400 °C. At 700 °C precipitates with radii between 1.1 and 1.7 nm were formed within seconds. At 400 °C a part of the still dissolved atoms formed smaller precipitates while other atoms increased the size of the larger precipitates. At longer simulation times a significant decrease of the number of small precipitates and an increase of the averaged precipitate radius was found. The Russel–Brown theory was applied on the simulation results in order to calculate the increase of the yield stress in the thermally aged state. The change in yield stress as a function of annealing time was calculated.

References

[1] Altpeter I., Dobmann G., Katerbau K.-H., Schick M., Binkele P., Kizler P., Schmauder S. (2001): Copper precipitates in 15 NiCuMoNb 5 (WB36) steel: material properties and microstructure, atomistic simulation, and micromagnetic NDE techniques, Nucl. Eng. Des. 206, pp. 337–350.

[2] Kittel C. (1976), Introduction to Solid State Physics, Wiley, New York.

[3] Kizler P., Uhlmann D., Schmauder S. (2000): Linking nanoscale and macroscale: calculation of the change in crack growth resistance of steels with different states of Cu precipitation using a modification of stress-strain curves owing to dislocation theory, Nucl. Eng. Des. 196, pp. 175–183.

[4] Landolt-Börnstein (1991), Numerical Data and Functional Relationships in Science and Technology, in: H. Ullmaier (Ed.), Atomic Defects in Metals, vol. 25, Springer-Verlag, Heidelberg.

[5] Landolt-Börnstein (1991): Numerical Data and Functional Relationships in Science and Technology, in: H. Mehrer (Ed.), Diffusion in Solids, Metals and Alloys, vol. 26, Springer- Verlag, Heidelberg.

[6] Othens P.J., Jenkins M.L., Smith G.D.W. (1994): High-resolution electron microscopy studies of the structure of Cu precipitates in a-Fe, Philos. Mag. A 70, pp. 1–24.

[7] Press W.H., Teukolsky S.A. , Vetterling W.T., Flannery B.P. (1992): Numerical Recipes in Fortran, second ed., Cambridge University Press, Cambridge.

[8] Pizzini S., Roberts K.J., Phythian W.J., English C.A., Greaves G.N. (1990): A fluorescence EXAFS study of the structure of copper-rich precipitates in Fe–Cu and Fe–Cu–Ni alloys, Philos. Mag. Lett. 61, pp. 223–229.

[9] Russel K.C., Brown L.M. (1972): A dispersion strengthening model based on different elastic moduli applied to the Fe– Cu system, Acta Met. 20, pp. 969–974.

[10] Schick M., Wiedemann J., Willer D. (1997), Untersuchungen zur sicherheitstechnischen Bewertung von geschweissten Komponenten aus Werkstoff 15 NiCuMoNb 5 (WB36) im Hinblick auf die Zähigkeitsabnahme unter Betriebsbeanspruchung, Technischer Bericht, MPA Stuttgart.

[11] Smithell C.J. (1983): in: E.A. Brandes (Ed.), Smithells Metals Reference Book, sixth ed., Butterworths, London.

[12] Soisson F., Barbu A., Martin G. (1996), Monte-Carlo simulations of copper precipitates in dilute iron-copper alloys during thermal ageing and under electron irradiation, Acta Mater. 44, pp. 3789–3800.

[13] Thermo-Calc AB, Stockholm Technology Park, Björnnesväagen 21, SE-11347 Stockholm, Sweden, http://www.thermocalc.se.

[14] Willer D., Zies G., Kuppler D., Föhl J., Katerbau K.-H. (2001), Service-Induced Changes of the Properties of Copper- Containing Ferritic Pressure-Vessel and Piping Steels, Final Report, MPA Stuttgart.

[15] S. Ziegler, private communication, 02/2000.

5.7 Atomistic simulation of the pinning of edge dislocations[7]

Ni-base superalloys are important materials for technological applications because of their high strength even at elevated temperatures and their high resistance to creep deformation [1]. The strength of these materials is mainly due to precipitation hardening. For small volume fractions of the precipitate phase (γ') in the matrix phase (γ), the precipitates possess a spherical shape. For higher volume fractions, the precipitates have the shape of cubes or plates.

In order to model precipitation hardening at the atomic level, molecular dynamics (MD) simulations of the dislocation–precipitate interaction are a suitable method [2–4]. For the case of cubic Ni_3Al precipitates in Ni, the behaviour of dislocations has been studied by MD simulations in [5]. In this article, we study the precipitation hardening in a binary Ni-base Ni_3Al alloy due to coherent spherical Ni_3Al precipitates. Precipitation hardening has been extensively studied using continuum mechanics methods (see [6] and references therein).

While Ni has an fcc structure, Ni_3Al possesses an Ll_2 structure. Accordingly, order hardening due to the formation of an antiphase boundary (APB) in a precipitate during its cutting by a dislocation plays an important role. A second dislocation moving in the same glide plane restores the Ll_2 structure. Therefore, the two dislocations are not equivalent. In order to study these processes, we consider the cutting of γ' precipitates by superdislocations. More specifically, a perfect dislocation in the γ'-phase with Burgers vector $a_0[110]$ dissociates into two partial dislocations with Burgers vector $a_0/2[110]$ that are connected by an APB ribbon. We denote these dislocations by D1 and D2. The dislocations D1 and D2 constitute perfect dislocations in the γ-phase. They further dissociate into Shockley partial dislocations with Burgers vectors $a_0/6[211]$ and $a_0/6[12\bar{1}]$ and a stacking fault (SF) ribbon. Accordingly, we have to consider the interaction of four Shockley partial dislocations (denoted by P1–P4) with a precipitate and to determine the corresponding critical resolved shear stresses (CRSSs). Since the system sizes we employ in our simulations are too small to consider the interaction of the dislocations D1 and D2, we study the interaction between each dislocation and the precipitates separately. For the CRSSs we adopt the following definitions: τ_{c1} and τ_{c3} shall be the CRSSs for the penetration of P1 and P3 into a precipitate while τ_{c2} and τ_{c4} shall be the CRSSs for the detachment of P2 and P4, or, equivalently, for the detachment of D1 and D2 from the precipitate.

[7] Reprinted from C. Kohler, P. Kizler, S. Schmauder, "Atomistic simulation of the pinning of edge dislocations in Ni by Ni_3Al precipitates", Mat. Sci. and Engng. A400-401, pp. 481-484 (2005) with kind permission from Elsevier

5.7.1 Molecular dynamics simulations for the analysis of the interaction of dislocations and precipitates

We use classical molecular dynamics simulations with EAM potentials [7] in order to model the interaction of dislocations and Ni_3Al precipitates in Ni crystals. The EAM potential functions are taken from reference [8]. The lattice constants, cohesive energies and shear moduli for Ni and Ni_3Al following from these potentials are given in table 5.16. The lattice mismatch is $\Delta a_0 = 1.5\%$. Note that the shear moduli of Ni and Ni_3Al are nearly equal. Consequently, modulus mismatch hardening will not be relevant.

The starting configuration used in the simulations with the leading dislocations D1 is shown schematically in Fig. 5.76. Within a rectangular block of Ni, an edge dislocation with Burgers vector $b = a_0/2\,[\bar{1}10]$ is created by removing two $(1\bar{1}0)$ half planes and deforming the rest of the crystal in such a way that the fcc structure is restored outside the dislocation core. A precipitate is created by substituting Ni atoms of the fcc structure by Al atoms within a spherical region of radius r to generate the $L1_2$ structure of Ni_3Al. We have considered precipitates with radii $r = 1.25$, 2.5 and 3.75 nm. Within the (111) glide plane, periodic boundary conditions are applied meaning that infinitely long dislocation lines interact with chains of precipitates with distances L equal to the box length L_z in z direction. Perpendicular to the glide plane, the atoms within two boundary regions (shown shaded in Fig. 5.76) are constrained to move in the x–z plane.

For the simulations with the trailing dislocations D2, precipitates sheared by an amount equal to the magnitude of the Burgers vector b (and thus possessing an APB) are used in the starting configuration. The box lengths L_x and L_y are fixed to $L_x = 19.78$ nm and $L_y = 9.75$ nm. For L_z we have chosen the values $L_z = 14.65$, 19.82, 24.99 and 29.73 nm. The number of atoms in the simulation box ranges accordingly from 260.000 to 527.000. The simulations are performed at temperature $T = 0K$ using a relaxation algorithm in which the velocities of the atoms are set equal to zero, whenever the system evolves uphill on the potential energy surface.

Table 5.16
Lattice constant a_0, cohesive energy E_c and shear modulus G for Ni and Ni_3Al

	Ni	Ni_3Al
a_0 (Å)	3.518	3.571
E_c (eV)	4.427	4.559
G (GPa)	73.47	77.32

All values are obtained from the EAM potentials at $T = 0$ K

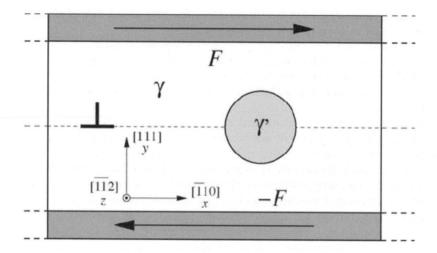

Fig. 5.76 Schematical illustration of the simulation box projection along the z axis is shown.

The CRSSs are determined in the following way. Initially, small external shear stresses are applied such that the dislocations attain stable positions near or inside the precipitate. The shear stresses are then increased in steps until the partial dislocations P1 or P3 enter the precipitate (in the case of the determination of τ_{c1} and τ_{c3}) or the partial dislocations P2 or P4 detach from the precipitate (in the case of τ_{c2} and τ_{c4}). For the stable positions of the dislocations, the systems have been relaxed until the mean force on the atoms is smaller than 5×10^{-7} eV/Å (quasi-static simulation). The final stress increment leading to the penetration or the detachment of the partials is at most 5MPa, which determines the accuracy of the CRSS. The locations of the partial dislocations and the SF ribbon are detected using a common-neighbour analysis [9].

5.7.2 Determination of critical resolved shear stresses

Dislocation D1

For the dislocation D1, the corresponding CRSSs τ_{c1} and τ_{c2} are plotted in Fig. 5.77 as a function of the precipitate distance L. In all cases, we obtain non-zero values for the CRSS τ_{c1}, which means that an external shear stress is necessary to move P1 into the precipitate. Simulations with external shear stresses smaller than τ_{c1} have shown that there is a repulsion of the dislocations D1 by the precipitates. Accordingly, there are stable positions of D1 outside the precipitate. Fig. 5.78(a) shows atoms near the glide plane of the system with $r = 2.5$ nm and $L = 30$ nm at an external shear stress $\tau = 25$MPa slightly below the CRSS $\tau c1$. That the partial dislocations P1 do not enter the precipitate for shear stresses smaller than τ_{c1} can be explained by a higher SF energy in the γ' phase compared to the one in the γ phase. When the external shear stress is increased, P_1 moves into the precipitate. A new stable position is attained when the partial dislocation P2 touches the precipitate (see Fig. 5.79(b and c) for precipitates of radius $r = 2.5$ and 1.25 nm, respectively).

This can be explained by the large APB energy that has to be supplied by moving P2 through the precipitate. As can be seen in Fig. 5.77, the value of τ_{c2} is for all precipitate radii and distances higher than the corresponding value of τ_{c1}, which means that the partial dislocation P2 determines the CRSS of dislocation D1. τ_{c1} and τ_{c2} decrease with increasing precipitate distance L and increase with increasing precipitate radius r. The dependence of τ_{c2} on L can be described to a good approximation by $\tau_{c2} \sim 1/L$.

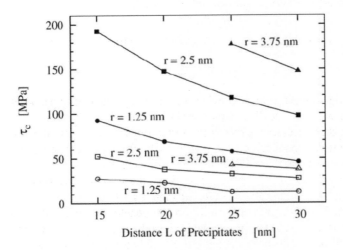

Fig. 5.77 Critical resolved shear stresses of the leading dislocations D1 for precipitates of different radii as a function of the precipitate distance L. The open symbols correspond to τ_{c1} and the full ones to τ_{c2}.

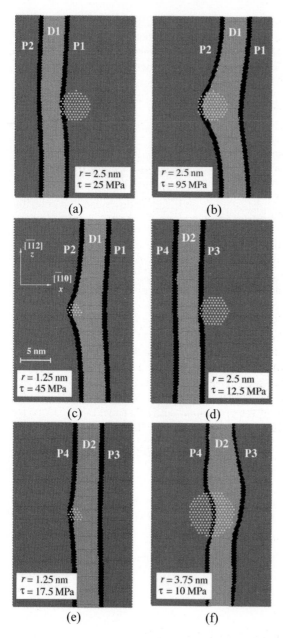

Fig. 5.78 Atoms near the glide plane for systems with precipitate distance $L = 30$ nm. The color scheme is as follows: Al atoms (white), stacking fault (light gray), Ni atoms (dark gray) and partial dislocation (black).

Dislocation D2

The CRSSs for the dislocation D2 are plotted in Fig. 5.79. For all considered radii and distances of precipitates, we obtain non-zero values for the CRSS τ_{c3}. There is, however, no clear dependence of τ_{c3} on the precipitate radius r and distance L, τ_{c3} ranges from 5 to 20 MPa. In contrast to the case of the dislocations D1, the dislocations D2 are attracted by the sheared precipitates. The location of D2 for the system with r = 2.5 nm and L = 30 nm and external shear stress τ = 12.5 MPa is visualized in Fig. 5.78(d). For external shear stresses $\tau > \tau_{c3}$, there is a difference in the behaviour of the precipitates of radius r = 1.25 nm and the precipitates with radii r > 1.25 nm. For r = 1.25 nm, there exist stable positions of the trailing partial dislocation P4 at the left phase boundary of the precipitate (see Fig. 5.78(e)). Correspondingly, there exists a CRSS $\tau_{c4} > \tau_{c3}$ for the detachment of D2 from the precipitate. The CRSS τ_{c4} decreases with increasing precipitate distance. For the larger precipitates of radius r = 2.5 and 3.75 nm, the situation is different: no additional shear stress is needed to move P4 into the precipitate and to detach the dislocation D2. Fig. 5.78(f) shows the case of the precipitate with radius r = 3.75 nm and distance L = 30 nm for an external shear stress τ = 10 MPa slightly larger than τ_{c3} = 7.5MPa. This figure does not show a stable configuration of the dislocation D2 but a snapshot of the moving dislocation. The bowing out of P4 within the precipitate indicates that there is a force that drives D2 out of the precipitate. The difference in the behaviour of the precipitates with radius r = 1.25 nm and the ones with larger radius can be explained in the following way. When P3 enters the precipitate, it destroys the APB with an ensuing energy gain and additional force on the dislocation D2. For the small precipitates, the dissociation width of the dislocation (which is about 4 nm) is larger than the diameter of the precipitate. In this case, P3 has already moved out of the precipitate when P4 touches the left phase boundary. Then no extra force is acting on P3 and P4 can be pinned.

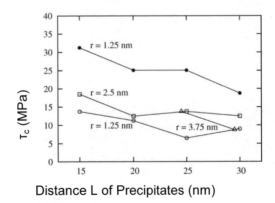

Fig. 5.79 Critical resolved shear stresses of the trailing dislocations D2 for precipitates of different radii as a function of the precipitate distance L. The open symbols correspond to τ_{c3} and the full ones to τ_{c4}.

For the precipitates that are larger than the dissociation width of D2, the extra force on P3 pulls also P4 through the precipitate. In summary, for precipitates smaller than the dissociation width, the CRSS of the dislocation D2 is determined by P4 while for larger precipitates, the CRSS is determined by P3

Conclusions

In this article, we have presented results of simulations of the dislocation–precipitate interaction in Ni with small spherical Ni_3Al precipitates. The CRSSs for all partial dislocations of a superdislocation have been determined for different radii and distances of the precipitates. It has been found that for precipitates that are smaller than the dissociation width of the dislocations, there exists a CRSS of the trailing dislocation which is a fraction of about 0.4 of the CRSS of the leading dislocation while for larger precipitates, the CRSS of the trailing dislocation is negligible.

The simulations have been performed at temperature T = 0K. This has the advantage that the mechanisms of the dislocation–precipitate interaction can be investigated in detail. In particular, the CRSS can be determined for each partial dislocation. At elevated temperatures, it can be expected that due to the activation energy from the temperature motion of the atoms only the CRSS τ_{c2} will be relevant.

References

[1] Ross E.W, Sims C.T, (1987), In: Sims C.T., Stoloff N.S., Hagel W.C. (Eds.), Superalloys II, Wiley, New York.
[2] Nedelcu S., Kizler P., Schmauder S., Moldovan N. (2000), Modell. Simul. Mater. Sci. Eng. 8, p. 181.
[3] Osetsky Y.N., Bacon D.J., Mohles V. (2003), Philos. Mag. 83, p. 3623.
[4] Kohler C., Kizler P., Schmauder S. (2005), Modell. Simul. Mater. Sci. Eng. 13, p. 35.
[5] Yashiro K., Naito M., Tomita Y. (2002), Int. J. Mech. Sci. 44, p. 1845.
[6] Nembach E. (1997), Particle Strengthening of Metals and Alloys, Wiley, New York and Chichester.
[7] Daw M.S., Baskes M.I. (1984), Phys. Rev. B 29, p. 6443.
[8] Voter A.F., Chen S.P. (1987), Mat. Res. Soc. Symp. Proc. 82, p. 175.
[9] Clarke A.S., Jonsson H. (1993), Phys. Rev. E47, p. 3975.

Index